PROGRESS IN BRAIN RESEARCH

VOLUME 196

OPTOGENETICS: TOOLS FOR CONTROLLING AND MONITORING NEURONAL ACTIVITY

SERIES EDITORS

PROGRESS IN BRAIN RESEARCH

VOLUME 196

OPTOGENETICS: TOOLS FOR CONTROLLING AND MONITORING NEURONAL ACTIVITY

EDITED BY

THOMAS KNÖPFEL

Laboratory for Neuronal Circuit Dynamics
RIKEN Brain Science Institute
Hirosawa, Wako City, Saitama, Japan

EDWARD S. BOYDEN

Departments of Biological Engineering and Brain and Cognitive Sciences
MIT, Cambridge, MA, USA

ELSEVIER

AMSTERDAM – BOSTON – HEIDELBERG – LONDON – NEW YORK – OXFORD
PARIS – SAN DIEGO – SAN FRANCISCO – SINGAPORE – SYDNEY – TOKYO

Elsevier
Radarweg 29, PO Box 211, 1000 AE Amsterdam, The Netherlands
Linacre House, Jordan Hill, Oxford OX2 8DP, UK
360 Park Avenue South, New York, NY 10010-1710

First edition 2012

Library of Congress Cataloging-in-Publication Data
A catalog record for this book is available from the Library of Congress

British Library Cataloguing in Publication Data
A catalogue record for this book is available from the British Library

ISBN: 978-0-444-59426-6
ISSN: 0079-6123

Printed and bound by CPI Group (UK) Ltd, Croydon, CR0 4YY

Transferred to Digital Print 2012

List of Contributors

W. Akemann, RIKEN Brain Science Institute, Hirosawa, Wako City, Saitama, Japan

J. Akerboom, Howard Hughes Medical Institute, Janelia Farm Research Campus, Ashburn, VA, USA

F. Anselmi, Neurophysiology and New Microscopies Laboratory, Wavefront Engineering Microscopy Group, CNRS UMR 8154, INSERM U603, Paris Descartes University, Paris Cedex, France

E.S. Boyden, MIT Media Lab; Departments of Biological Engineering, and Brain and Cognitive Sciences, MIT, Cambridge, MA, USA

B.Y. Chow, Department of Bioengineering, University of Pennsylvania, Philadelphia, PA, USA

G.P. Dugué, Champalimaud Neuroscience Programme, Instituto Gulbenkian de Ciência, Oeiras, Portugal

V. Emiliani, Neurophysiology and New Microscopies Laboratory, Wavefront Engineering Microscopy Group, CNRS UMR 8154, INSERM U603, Paris Descartes University, Paris Cedex, France

X. Han, Department of Biomedical Engineering, Boston University, Boston, MA, USA

Y.P. Hung, Department of Neurobiology, Harvard Medical School, Boston, MA, USA

T. Knöpfel, RIKEN Brain Science Institute, Hirosawa, Wako City, Saitama, Japan

J.Y. Lin, Department of Pharmacology, University of California at San Diego, La Jolla, CA, USA

L.L. Looger, Howard Hughes Medical Institute, Janelia Farm Research Campus, Ashburn, VA, USA

L. Madisen, Allen Institute for Brain Science, Seattle, WA, USA

H. Mutoh, RIKEN Brain Science Institute, Hirosawa, Wako City, Saitama, Japan

D. Oron, Department of Physics of Complex Systems, Weizmann Institute of Science, Rehovot, Israel

E. Papagiakoumou, Neurophysiology and New Microscopies Laboratory, Wavefront Engineering Microscopy Group, CNRS UMR 8154, INSERM U603, Paris Descartes University, Paris Cedex, France

A. Perron, RIKEN Brain Science Institute, Hirosawa, Wako City, Saitama, Japan

J. Rossier, Laboratoire de Neurobiologie et Diversité Cellulaire, Centre National de la Recherche Scientifique, Unité Mixte de Recherche 7637, Ecole Supérieure de Physique et de Chimie Industrielles, Paris, France

E.R. Schreiter, Howard Hughes Medical Institute, Janelia Farm Research Campus, Ashburn, VA, USA

E. Scott, School of Biomedical Sciences, The University of Queensland, St. Lucia, Queensland, Australia

J. Simmich, School of Biomedical Sciences, The University of Queensland, St. Lucia, Queensland, Australia

E. Staykov, School of Biomedical Sciences, The University of Queensland, St. Lucia, Queensland, Australia

M. Tantama, Department of Neurobiology, Harvard Medical School, Boston, MA, USA

L. Tian, Howard Hughes Medical Institute, Janelia Farm Research Campus, Ashburn, VA, USA

C.L. Tucker, Department of Pharmacology, University of Colorado School of Medicine, Aurora, CO, USA

A. Urban, Laboratoire de Neurobiologie et Diversité Cellulaire, Centre National de la Recherche Scientifique, Unité Mixte de Recherche 7637, Ecole Supérieure de Physique et de Chimie Industrielles, Paris, France

G. Yellen, Department of Neurobiology, Harvard Medical School, Boston, MA, USA

H. Zeng, Allen Institute for Brain Science, Seattle, WA, USA

Preface

Tools for observing and controlling specific molecular or physiological pathways in intact cells and tissues are opening up new frontiers in the understanding and engineering of complex biological systems and even pointing the way toward novel kinds of therapy and prosthetic for treating human disease. One of the most popular strategies is to utilize what has come to be known as an "optogenetic" strategy—namely, to create a genetically encoded optical reporter of biological activity, or a light-driven actuator of biological signaling, and then to express the gene that encodes for this molecule in a specific cell or set of cells, often within an intact tissue or organism. Then, the investigator uses light to perform the readout of the biological system state by monitoring cells expressing optical reporters, say using a microscope, or to alter the biological system via illuminating it so that cells expressing the light-driven actuator are selectively altered.

The use of genetically encoded reagents insures ease of use, as well as the ability to target specific cells, even within intact tissues or organisms, thanks to the wealth of transgenic and gene-delivery mechanisms available for use in a variety of organisms utilized in modern biological science and preclinical medicine. The use of light insures high temporal fidelity of the observation or of the perturbation, and also supports spatial targeting of the observation or perturbation to defined cells or targets in the brain or body. In this way, observational or causal information, precisely obtained, about how a given molecularly defined pathway plays a role in cell-, tissue-, and organism-level operation can be obtained.

Over the past few years, the number of groups making, and utilizing, such optogenetic tools, has exploded. Several of the participants in this volume were involved with a Minisymposium at the 2010 Society for Neuroscience Meeting in San Diego, California, chaired by Thomas Knöpfel and Ed Boyden, titled "Towards the Second Generation of Optogenetic Tools." The symposium was standing room only throughout, a testament to the power such tools are having on neuroscience, a field in which there is a great desire to observe and see what is happening in a diversity of cells which are in turn assembled into very complex three-dimensional circuit formations within the brain. After the symposium was over, we decided that perhaps a book that was basically an extension of being the proceedings of such an event could be of great interest.

This book focuses on some recent inventions in the space of optogenetic tools, accompanied by critical evaluations of how they differ from past innovations in this space. The first half of this volume is made up of chapters that are organized around the kinds of tools being invented. Chapter 1, by Dugué and colleagues, gives an overview of the space of optogenetics, followed by Chapters 2 and 3 by Lin and by Boyden and colleagues, respectively, which deals with molecules that can be expressed in neurons to make them sensitive to being activated or silenced by pulses of light. We then switch directions, for two chapters, focusing on tools for reading out neural activity, using molecules that change fluorescence in response to changes in voltage (Chapter 4 by Knöpfel and colleagues) or calcium (Chapter 5 by Looger and colleagues). We then continue with a chapter on how protein multimerization can be controlled with light, using light-driven proteins from plants that, when illuminated, bind to one another (Chapter 6 by Tucker), thus enabling optical control of gene expression and other signaling functions.

The second half of this volume explores how tools can be used, in a variety of systems. Chapter 7, by Emiliani and colleagues, focuses on strategies for the use of two-photon microscopy to activate optogenetic reagents. In Chapter 8, Scott and colleagues explore the use of optogenetics in zebrafish;

in Chapter 9, Urban and Rossier explore the use of optogenetics in the mammalian cerebral cortex; and in Chapter 10, Zeng and Madisen explore the use of optogenetics in the context of transgenic mice. In Chapter 11, Han explains how to utilize optogenetics in the context of the living nonhuman primate brain, and in Chapter 12, Yellen and colleagues utilize optogenetics to study metabolism.

Our goal is that this volume not only presents current details about optogenetic tools but also the underlying logic of how they work and how they can be applied, as well as practical information about their use.

Thomas Knöpfel
Edward S. Boyden

Contents

T. Knöpfel and E. Boyden (Eds.)
Progress in Brain Research, Vol. 196
ISSN: 0079-6123

CHAPTER 1

A comprehensive concept of optogenetics

Guillaume P. Dugué[†,*], Walther Akemann[‡] and Thomas Knöpfel[‡]

[†] Champalimaud Neuroscience Programme, Instituto Gulbenkian de Ciência, Oeiras, Portugal
[‡] RIKEN Brain Science Institute, Hirosawa, Wako City, Saitama, Japan

Abstract: Fundamental questions that neuroscientists have previously approached with classical biochemical and electrophysiological techniques can now be addressed using optogenetics. The term optogenetics reflects the key program of this emerging field, namely, combining optical and genetic techniques. With the already impressively successful application of light-driven actuator proteins such as microbial opsins to interact with intact neural circuits, optogenetics rose to a key technology over the past few years. While spearheaded by tools to control membrane voltage, the more general concept of optogenetics includes the use of a variety of genetically encoded probes for physiological parameters ranging from membrane voltage and calcium concentration to metabolism. Here, we provide a comprehensive overview of the state of the art in this rapidly growing discipline and attempt to sketch some of its future prospects and challenges.

Keywords: optogenetics; optical imaging; optical control; introduction; fluorescent proteins; opsins; database; wiki.

A historical perspective on optogenetics

The term "optogenetics" was coined a few years after neurons had first been engineered to express opsins and other light-driven actuator proteins, and photoevoked firing had been obtained in cell cultures (Banghart et al., 2004; Boyden et al., 2005; Li et al., 2005; Zemelman et al., 2002, 2003) and behaving flies (Lima and Miesenböck, 2005). The term initially served as a common denomination for approaches combining "genetic targeting of specific neurons or proteins with optical technology for imaging or control of the targets within intact, living neural circuits" (Deisseroth et al., 2006). A later definition of optogenetics as "the branch of biotechnology which combines genetic engineering with optics to observe and control the function of genetically targeted groups of cells with light, often in the intact animal" (Miesenböck, 2009) continued to gather under the same name two complementary approaches with intertwined histories: one consists in monitoring neuronal activity using

*Corresponding author.
Tel.: +351 210480116; Fax: +351 210480298
E-mail: guillaume.dugue@neuro.fchampalimaud.org

DOI: 10.1016/B978-0-444-59426-6.00001-X

genetically encoded fluorescent reporters (sensors), while the other aims at controlling neuronal activity using genetically addressable light-activated tools (actuators). But with the lightning success of actuators in and outside neuroscience, a more restrictive definition of optogenetics started to gain increasing acceptance. In the *Nature Methods* issue of January 2011 featuring optogenetics as Method of the Year 2010, optogenetics was introduced as "the combination of genetic and optical methods to achieve gain or loss of function of well-defined events in specific cells of living tissue" (Deisseroth, 2011). In this section, we trace back the conceptual roots and history of the field and try to paint a comprehensive and balanced picture of what optogenetics encompasses today.

Early ideas

Scientific intuitions and representations of natural phenomena are often formed and conveyed in a visual form. The fact that neuronal activity relies on primarily invisible electrochemical phenomena makes its representation particularly uneasy. Yet an appealing popular depiction of brain function is an intricate mesh of neuronal processes traveled by evanescent bursts of light symbolizing electrical activity. But as this representation was being adopted by popular media for the sake of simplifying the communication of scientific contents, the idea of optically visualizing neuronal activity was evolving from a mere visionary fantasy to an existing technology. Charles S. Sherrington was probably the first to inspire this notion in an oft-quoted passage from his book *Man on His Nature* in which he imagined neuronal activity as points of light (Sherrington, 1940, pp. 176–178). Sherrington used this metaphor to describe the different stages of a sleep-to-wake transition, upon which the brain gradually becomes "an enchanted loom where millions of flashing shuttles weave a dissolving pattern." Beyond Sherrington's vision, it progressively became clear that light would be an interesting tool not only to interrogate but also to manipulate neuronal activity, an idea which was publicized by Francis H. Crick. In a 1979 article entitled "Thinking about the brain," Crick pointed out the need for "a method by which all neurons of just one type could be inactivated, leaving the others more or less unaltered" (Crick, 1979). Some 20 years later in a paper reviewing the current and future benefits of molecular biology in neuroscience, Crick explicitly envisioned that light might be used to control and monitor the activity of genetically defined neuronal populations (Crick, 1999): "One of the next requirements is to be able to turn the firing of one or more types of neuron on and off in the alert animal in a rapid manner. The ideal signal would be light, probably at an infrared wavelength to allow the light to penetrate far enough. This seems rather far-fetched but it is conceivable that molecular biologists could engineer a particular cell type to be sensitive to light in this way. [...] Most modern theories of brain action stress the firing (in one way or another) of not single neurons but groups of neurons. [...] One way-out suggestion is to engineer these neurons so that when one of them fires it would emit a flash of light of a particular wavelength. The experimenter could then follow the firing of that group of neurons alone." Quite remarkably, Crick's ideas would essentially come true within the next decade.

From organic to genetically encoded reporters of neuronal activity

The chemical approaches to ions and voltage sensing

The first ideas on how to implement an optical measure of neuronal activity emerged in the late 1960s from the study of changes in light scattering, birefringence, and fluorescence associated with action potentials (Cohen et al., 1968). Larger optical signals were obtained in the 1970s by introducing voltage-sensitive organic molecules into neuronal membranes. This opened the field of voltage-sensitive dye imaging which is still intensively explored today (Peterka et al., 2011). Another step toward the realization of Sherrington's vision was taken during the next decade by Roger Y. Tsien who synthesized organic molecules which change their fluorescence with variations in the concentration of intracellular calcium and could therefore be used as reporters of neuronal activity (Göbel and Helmchen, 2007). These calcium-sensitive dyes have opened an avenue for noninvasive imaging of neuronal activity, a field which has exploded during the 1990s with modern imaging techniques such as two-photon microscopy (Denk et al., 1990; Göbel and Helmchen, 2007). Besides voltage and calcium sensing, a variety of other fluorescent indicators were developed to detect variations in sodium, chloride, zinc, and pH but have been much less intensely used in neurobiological research (Johnson and Spence, 2010).

The genetic approaches to sensing chemicals and voltage

Despite tremendously helpful, organic dyes carried intrinsic limitations which restricted the scope of their applicability. First, these dyes usually stain all cell types indiscriminately and therefore do not offer cell-type-specific activity readout. Second, they have to be added externally and useful staining typically lasts for less than a day, prohibiting chronic experiments such as the study of lasting neuronal plasticity. Third, the dyes themselves or the conditions to deliver them into neurons can present some toxicity. These reasons led to orienting efforts toward substituting these dyes for proteinaceous fluorescent indicators which would enable chronic staining of genetically defined neuronal populations. This new generation of fluorescent reporters was almost exclusively engineered based on the green fluorescent protein (GFP) cloned from the jellyfish *Aequorea victoria* (Prasher et al., 1992). It is interesting to note that the work on *A. victoria*'s bioluminescence also led to the cloning of the calcium-sensitive luminescent protein aequorin almost 10 years before (Inouye et al., 1985; Prasher et al., 1985). Quite remarkably, aequorin had provided the first report on the use of an optical protein calcium sensor even earlier, when Ridgway and Ashley (1967) optically recorded calcium transients after microinjecting it into single muscle fibers of the barnacle.

To generate fully genetically encoded fluorescent reporters, the classical approaches consisted in fusing one or more fluorescent proteins (FPs) with various protein moieties offering sensitivity to signals such as transmembrane potential, ions (calcium, pH, chloride, or zinc), neurotransmitters (glutamate), or second messengers molecules like cyclic nucleotides (Chudakov et al., 2010; Tian and Looger, 2008). For the sensing process to be converted into a measurable change in fluorescence output of the FPs, these probes usually rely on two possible design strategies (Fig. 1). In the first one, the photophysical properties of a single FP are modulated by conformational changes imposed by the sensor. In most cases, two portions of the FP are interchanged and reconnected by short spacers (circularly permuted) so that its fluorescence becomes more sensitive to small structural rearrangements at its extremities. In the second one, conformational changes are used to modify the distance or orientation of two FP variants with spectral properties allowing Förster resonance energy transfer (FRET). While the work on certain reporters is still in its early phase, others like voltage-sensitive fluorescent proteins (VSFPs) and genetically encoded calcium indicators (GECIs) have already gone through multiple improvement steps and show promising results. These tools are reviewed in detail in Chapters 4 and 5, respectively. An overview of available optical reporters for probing the activity of genetically defined neurons is given in Fig. 1. A general overview of optogenetic reporters is provided in Chapter 12.

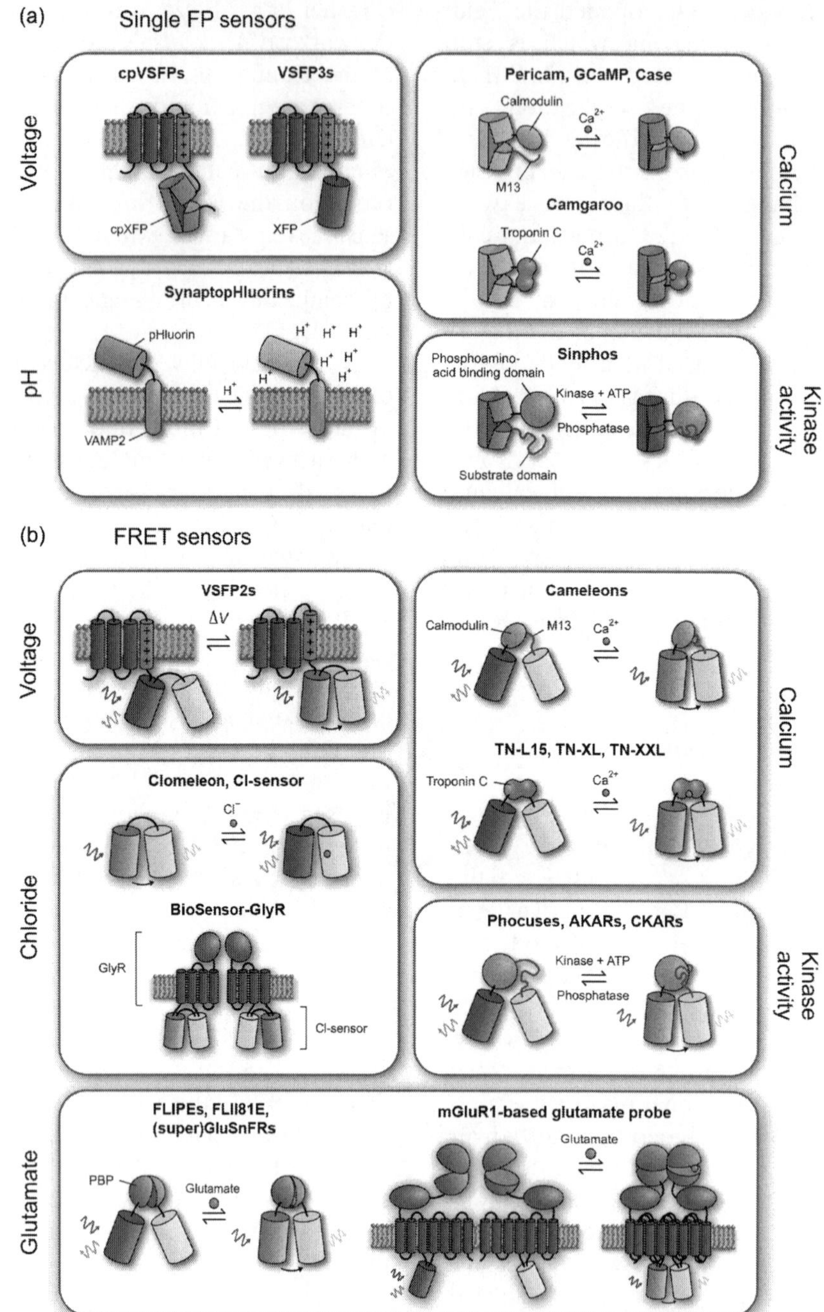

Fig. 1. Optogenetic tools for monitoring cellular signals. (a) Sensors based on single fluorescent proteins (FPs). These sensors usually incorporate intact GFP variants (XFPs) or their circularly permuted version (cpXFPs). Single FP-based voltage-sensitive fluorescent proteins (VSFPs, voltage probes) are derived from a combination of a membrane-integrated voltage sensor domain (gray and purple transmembrane domains) and cpXFPs (cpVSFPs) or XFPs (VSFP3s). Single FP calcium indicators include

Tools for controlling genetically defined neurons

Optogenetic approaches employing microbial rhodopsins for exciting and inhibiting neurons are covered in Chapters 2 and 3. We are providing here a historical overview of how modern strategies for controlling neuronal activity with light emerged over the past 10 years.

Photochemical approaches

The idea of using light-sensitive molecular tools for the optical control of neuronal activity (actuators) had been latently present in the literature well before Crick articulated their theoretical utility. In particular, photochemistry had already provided insights into how to convert a ligand from an inert state into a high-affinity form, a field which would become popular in neuroscience with the use of caged neurotransmitters (Nerbonne, 1996). Using this technique, synthetic photoconvertible ligands can be used to optically modulate neuronal activity through the activation of specific receptor proteins (Fig. 2a). To restrict the action of the ligand to genetically designated neurons, the receptor itself has to be targeted to these neurons (Zemelman et al., 2003). This method was used by Lima and Miesenböck (2005) to elicit specific behaviors in fruit flies using light as a trigger, providing the first example of an optically "remote-controlled" animal.

The photoactivation process can be made more efficient by linking the ligand to the protein through a covalent bond and obtaining a "photoswitched tethered ligand" (PTL, Fig. 2a), a technique used successfully to control nicotinic receptors (Bartels et al., 1971), ionotropic glutamate receptors (Volgraf et al., 2006), potassium channels (Banghart et al., 2004; Chambers et al., 2006; Fortin et al., 2011), and recently a chimeric potassium-selective glutamate receptor called HyLighter (Janovjak et al., 2010). One major drawback of photochemical approaches is the necessity of either delivering the ligand or conjugating the PTL to the target protein, which limits their use to easily accessible preparations like cultured neurons, brain slices, or small organisms such as fruit flies (Lima and Miesenböck, 2005) or zebrafish larvae (Janovjak et al., 2010). Photochemical approaches to control neuronal firing have been reviewed by Gorostiza and Isacoff (2007, 2008) and Miesenböck (2011).

Genetic approaches based on animal opsins

Fully genetically encoded light-gated actuators which do not require the addition of an exogenous cofactor appeared as a more viable solution for controlling neuronal activity *in vivo*. Not surprisingly, the hunt for candidates first concentrated on the phototransduction machineries underlying animal vision. The light-sensitive elements in these systems are membrane-embedded photopigments

scaffolds based on GFP and the calmodulin-M13 complex (such as Pericam, GCaMPs, and Case) and troponin-based scaffolds (Camgaroo). SynaptopHluorins are indicators of vesicle release and recycling, consisting of a pH-sensitive form of GFP (pHluorin) fused to the luminal side of a vesicle-associated membrane protein (VAMP). Sinphos are detectors of protein phosphorylation (kinase activity) made of a fusion between a cpXFP, a phosphorylable substrate peptide, and a phosphoamino acid binding domain. (b) FRET sensors. These sensors are traditionally based on a FRET pair of FPs such as CFP and YFP. VSFP2s are FRET-based voltage sensors. FRET calcium indicators include cameleons (based on the calmodulin-M13 complex) and the TN sensor family (based on troponin). Chloride sensors (Clomeleon and Cl-sensors) take advantage of the fact that chloride can efficiently quench YFP fluorescence, thus reducing the FRET signal of a CFP-YFP pair. In BioSensor-GlyR, Cl-sensor proteins are grafted to the subunits of a glycine receptor (GlyR) in order to sense chloride ions flowing through the receptor. FRET kinase activity sensors (Phocuses and XKARs) have been developed using the same rationale as for Sinphos (see a). Finally, glutamate can be detected using FRET sensors based on bacterial periplasmic binding proteins (PBPs) or on a metabotropic glutamate receptor (mGluR1). Membranes are represented with the cytoplasmic side toward the bottom. (For interpretation of the references to color in this figure legend, the reader is referred to the Web version of this chapter.)

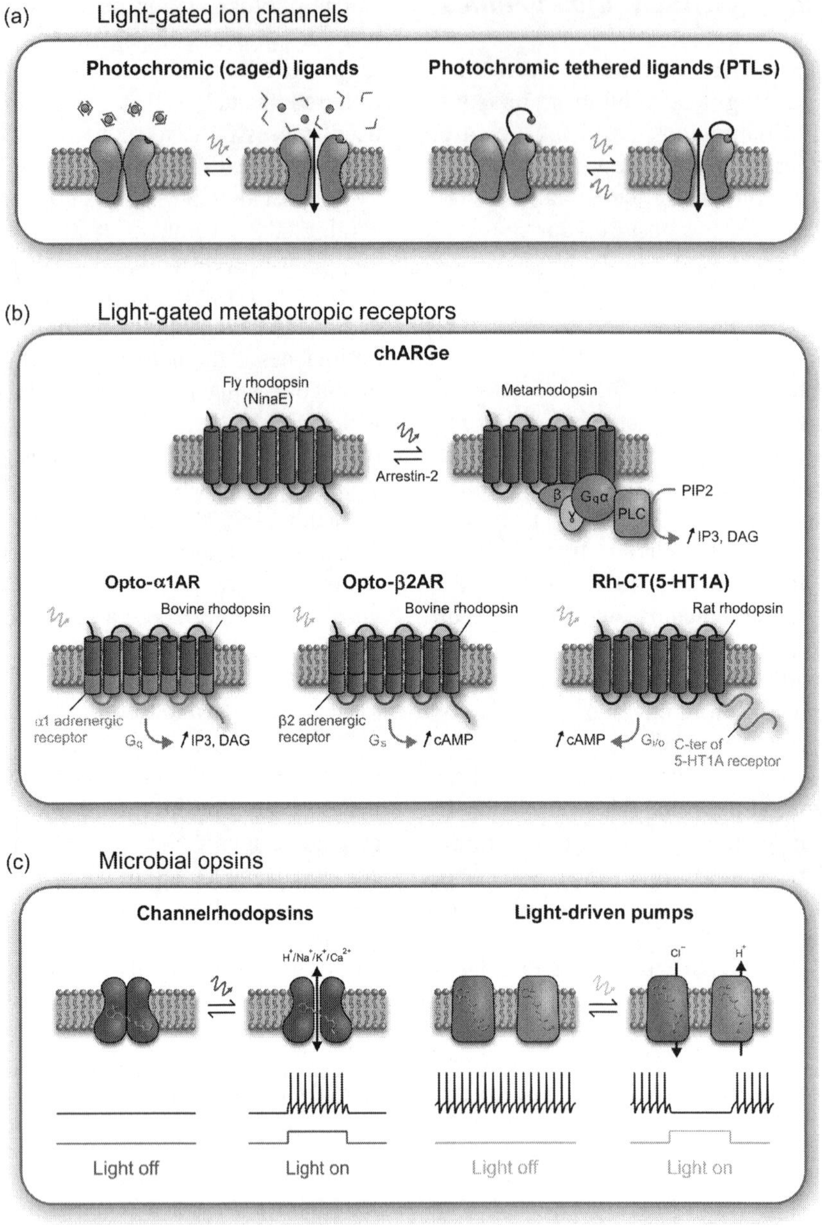

Fig. 2. Optogenetic tools for controlling neuronal activity. (a) Artificial gating of ion channels by light can be accomplished using photochromically caged ligands or tethered ligands with a photochromic activation switch. (b) Endogenous conductances can be modulated through light-dependent activation of intracellular second messenger cascades using animal opsins (type II opsins). This was achieved in mammalian neurons by reconstituting a minimal fly phototransduction machinery through the heterologous expression of three proteins (NinaE, arrestin-2, and Gqα, a system called chARGe) or by expressing single vertebrate rhodopsin genes (not shown). Alternatively, the intracellular loops or C-terminal domain of vertebrate rhodopsins can be exchanged with the intracellular domains of specific metabotropic receptors to gain optical control over specific signaling cascades (Opto-α1AR,

called rhodopsins, each rhodopsin molecule consisting of a protein called opsin (belonging to the family of G-protein-coupled receptors or GPCRs) covalently bound to a chromophore (a vitamin A-related compound called retinal or one of its derivatives). Upon illumination, the bound retinal molecule undergoes isomerization, which induces conformational changes in the opsin backbone and activates a G-protein signaling pathway.

This path was pioneered by Har Gobind Khorana in the late 80s, who observed light-dependent ionic currents in Xenopus oocytes transfected with a bovine rhodopsin gene (Khorana et al., 1988). The next significant step was taken in the early 2000s by the team of Gero Miesenböck who managed to reconstitute a minimal fly phototransduction cascade in mammalian neurons by coexpressing NinaE, a blue-sensitive rhodopsin and two of its natural partners: the αq G-protein subunit and arrestin-2, a protein required for deactivation of rhodopsin. Upon illumination, the excited rhodopsin activates an endogenous phospholipase C through the action of the G-protein, which, in turn, activates nonspecific cation channels through the production of second messengers (Fig. 2b). The system called "chARGe" was used to optically elicit action potential firing in cultured hippocampal neurons (Zemelman et al., 2002) but was fastidious to implement and carried intrinsic limitations like slow and variable activation and deactivation kinetics (a few hundred milliseconds to several tens of seconds). Following a similar rationale, subsequent studies showed that heterologous expression of single mammalian opsins in neurons was enough to modulate endogenous conductances through specific G-protein cascades, but with comparable slow kinetics (Gutierrez et al., 2011; Li et al., 2005; Masseck et al., 2011; Melyan et al., 2005).

The slow kinetics observed in these approaches is inherent to the metabotropic nature of vertebrate and invertebrate opsin signaling, which challenges their relevance as strategies for temporally precise control of neuronal firing. However, in a different perspective, animal opsins were used successfully to gain optical control over specific intracellular transduction pathways. Building up on the work of Kim et al. (2005), the team of Karl Deisseroth engineered chimeric receptors by replacing the intracellular loops of the bovine rhodopsin with those of specific adrenergic receptors (Airan et al., 2009), taking advantage of common structure–function relationships among GPCRs. Using these tools, they were able to optically activate the intracellular pathways normally recruited by these receptors (the cAMP and IP3 pathways; Fig. 2b). Following a similar approach, the team of Stefan Herlitze produced a light-activated receptor which recruits the signaling cascade of a specific serotonin receptor (Oh et al., 2010). These emerging tools might be gathered under the name "opto-XRs" proposed by Airan et al. (2009), where X specifies the particular pathways which is being optically hijacked (e.g., opto-α1AR for α1 adrenergic receptors).

The revolution of microbial opsins

Ideal light-gated actuators would be single proteins rather than effector proteins activated by multicomponent signaling cascades. Unexpectedly, such tools were to be found in branches of biology which could have hardly been more distant from neuroscience: the study of phototropism

---

Opto-β2AR, and Rh-CT(5-HT1A)). (c) Naturally occurring microbial opsins (type I opsins) and their synthetic variants provide a large panel of single-protein actuators which can be used to control neuronal firing with millisecond precision. Channelrhodopsins display an intrinsic light-gated cationic conductance and can be used to depolarize neurons. Microbial light-driven pumps can produce hyperpolarizing currents by translocating chloride ions into the cell (halorhodopsins) or protons outside (bacteriorhodopsins, proteorhodopsins, and archeorhodopsins). Membranes are represented with the cytoplasmic side toward the bottom. (For color version of this figure, the reader is referred to the Web version of this chapter.)

in unicellular algae and of light-driven ion transport in halophilic archaebacteria.

Indeed, animals are not the only realm that possesses retinal opsins. Unicellular photosynthetic organisms like green algae or euglenids also express rhodopsins in a photoreceptive organelle called the "eyespot apparatus" used to initiate phototropic reactions (swimming toward or away from light). In the green algae *Chlamydomonas*, the eyespot contains atypical rhodopsins displaying intrinsic light-gated ion conductance, called channelrhodopsins (Nagel et al., 2005). In 2002–2003, two channelrhodopsins were cloned from the species *Chlamydomonas reinhardtii*. The first one (ChR1) is selectively permeable to protons (Nagel et al., 2002), while the second one (ChR2) is also permeable to other cations (Nagel et al., 2003) and can thus mediate depolarizing currents irrespective of the extracellular pH. Due to their channel-like structure, these proteins provided extremely rapid responses to light when tested in Xenopus oocytes, with photocurrents occurring within tens of microseconds upon illumination with blue light (450–500nm). A couple years later, the teams of Karl Deisseroth and Stefan Herlitze expressed ChR2 in cultured hippocampal neurons and showed that ChR2-mediated photocurrents were rapid and large enough to fire these cells with millisecond precision (Boyden et al., 2005; Li et al., 2005). New channelrhodopsins cloned from two other species of green algae were recently used successfully in mammalian neurons: VChR1 from *Volvox carteri* (Zhang et al., 2008) and MChR1 from *Mesostigma viride* (Govorunova et al., 2011). Initially, there was considerable doubt whether this approach would be successful, as acknowledged in retrospect by Deisseroth himself (Deisseroth, 2010), because no one could tell whether the protein would fold and integrate correctly into the cytoplasmic membrane of mammalian cells and if endogenous retinal would be available at sufficient quantities. But subsequent experiments showed that within reasonable expression levels, ChR2 and its variants could be used safely to control the activity of genetically defined neuronal populations in animal models ranging from flies to monkeys (Fenno et al., 2011). In only 5 years, channelrhodopsins emerged as a technical revolution at almost all levels of neurobiological research.

But channelrhodopsins were not the only players in this revolution. Other microbial rhodopsins behaving as light-driven ion pumps were long known to exist in halophilic archaebacteria (Mukohata et al., 1999) and were discovered recently in bacteria (Beja et al., 2000) and some eukaryotes (Waschuk et al., 2005). Proteins like bacteriorhodopsins, proteorhodopsins, and archaerhodopsins extrude protons from the cytoplasm, building up a proton gradient used for the production of ATP. Others like halorhodopsins are used by certain halobacteria to maintain their osmotic balance by transporting chloride into their cytoplasm (Muller and Oren, 2003). Both types thus generate a hyperpolarizing photocurrent which can be used to silence neuronal activity. Interestingly, these pumps operate at different peak sensitivity wavelengths compared to ChR2, opening the possibility of coexpressing them with ChR2 to achieve bidirectional control of the same cell. Zhang et al. (2007) provided the proof of principle that this is indeed possible by coexpressing ChR2 and the halorhodopsin from *Natronomonas pharaonis* (NpHR) in acute brain slices and in *Caenorhabditis elegans*. But translating these light-driven pumps into usable tools was not as straightforward as for ChR2. A common problem with these proteins was impaired subcellular localization which decreased their tolerability when expressed at high levels. In particular, the proteins would accumulate at successive steps along the secretary pathway to the cell surface. A number of candidates isolated from various species were tested and modified by adding a series of trafficking signal peptides to improve their membrane localization (Gradinaru et al., 2010). Two families of pumps emerged as promising light-activated silencers: the series of proteins derived from NpHR (the latest being eNpHR3.0 described in Gradinaru et al., 2010) and the archaerhodopsins Arch from

Halorubrum sodomense (Chow et al., 2010) and ArchT from *Halorubrum genus* (Han et al., 2011). The diversity and biophysical properties of microbial opsins and their use in neuroscience were reviewed extensively (Boyden, 2011; Fenno et al., 2011; Hegemann and Moglich, 2011; Lin, 2011; Yizhar et al., 2011a,b,c). State-of-the-art methodologies to deploy these tools in mammalian cells were reviewed in detailed by Chow et al. (2011). An overview of available tools for controlling genetically defined neurons is given in Fig. 2.

Emerging optogenetic approaches outside neuroscience

The origins of optogenetics are deeply rooted in neuroscience. As described above, the first efforts to engineer genetically encoded optical sensors were aiming at monitoring neuronal activity. Similarly, the first light-actuated control systems were designed to modulate neuronal firing. But today's developments in the field are addressing a much broader scope of unmet needs in the study of biological systems. Recent progress in bioengineering has provided a new panel of optogenetic readout and control strategies to study a variety of molecular and cellular processes (Miesenböck, 2011).

An expanding toolkit for sensing and monitoring cellular activities

Optical reporters have long been used outside neuroscience to monitor various biomolecules and enzymatic activities (Souslova and Chudakov, 2007). The range of substances and processes which can be optically tracked is quickly expanding. A new family of indicators has been designed from bacterial periplasmic binding proteins (PBPs) to sense metabolites such as carbohydrates and amino acids. These indicators use the Venus flytrap-like conformational change of PBPs upon binding to their substrate in order to generate a FRET signal (Deuschle et al., 2005). Other indicators were engineered to detect second messenger molecules like H_2O_2 (Markvicheva et al., 2011), enzymatic activities (kinase, protease, GTPase), and several cellular processes (cell cycle, actin dynamics). These recent developments are reviewed extensively in Lalonde et al. (2005), Okumoto et al. (2008), Frommer et al. (2009), and Okumoto (2010). Chapter 12 of this issue reviews the wide range of factors for which optogenetic reporters are now available.

An emerging repertoire of light-gated effectors to control cell physiology

A new repertoire of light-activated tools for manipulating identified biochemical events is emerging. These new tools include rhodopsin-based chimeric GPCRs like the opto-XRs which can trigger specific intracellular signaling cascades upon illumination (Airan et al., 2009; Kim et al., 2005; Oh et al., 2010). Following a similar logic, Ye et al. (2011) recently managed to functionally link the signal transduction pathway of a vertebrate rhodopsin to a specific gene transcription control mechanism in order to achieve light-induced transgene expression. Other research lines are exploiting and improving non-membrane-associated photoreceptor protein domains to build photoswitchable cytoplasmic effectors (Losi and Gärtner, 2011; Moglich and Moffat, 2010; Strickland et al., 2010). Such domains can be found in numerous species of bacteria, protists, fungi, and plants where they serve a great variety of functions. Recently, naturally occurring photoactivated adenylyl cyclases containing a BLUF domain have been used to control cAMP levels in various models (Nagahama et al., 2007; Schroder-Lang et al., 2007; Stierl et al., 2011) and reengineered to function as guanylyl cyclases (Ryu et al., 2010). Other studies have taken advantage of LOV domains to confer photosensitivity to DNA-binding proteins, enzymes, and small GTPases (Lee et al., 2008; Moglich et al., 2009; Strickland et al., 2008; Wu et al., 2009) and more recently to activate endogenous calcium channels (Pham et al., 2011). One last exciting development consists in exploiting reversible light-dependent protein binding mechanisms found in plants (reviewed in Kami et al., 2010). These mechanisms involve identified partners such as phytochromes (Phy) and phytochrome-interacting factors which can be fused to proteins of interest to gain photocontrol over their association. The resulting "photoactivated dimerizers" can be used to investigate biological processes with exquisite spatiotemporal resolution or to create new molecular pathways. This strategy has already been used to achieve light-gated protein translocation, protein splicing, gene transcription, and DNA recombination (Kennedy et al., 2010; Levskaya et al., 2009; Shimizu-Sato et al., 2002; Toettcher et al., 2011; Tyszkiewicz and Muir, 2008; Yazawa et al., 2009) and is reviewed extensively in Chapter 6 of this issue.

Photosensitizers: using light to destroy proteins and cells

Interacting with cells and proteins in a rapid and reversible way is one key program of optogenetics. But light can also be used to produce targeted lesions, an approach which can be relevant for perturbing neural circuits and designing models of neurodegenerative disorders. One way of making these lesions specific is to target light-sensitive molecules called photosensitizers (PSs) to particular proteins or cells. When irradiated with light, PSs generate reactive oxygen species (ROS) which very rapidly react with any nearby biomolecule and can eventually kill cells through apoptosis or necrosis. The technique, called chromophore-assisted light inactivation (CALI), has been used extensively for the treatment of precancerous lesions and superficial tumors. Most available PSs are organic molecules which have to be introduced exogenously into living systems and offer very poor selectivity for particular cell types or proteins. One solution to this issue is to rely on peptide-like (peptoids) PSs which are resistant to proteolysis and can be designed to bind specifically to virtually any given protein (Lee et al., 2010). But a more definitive solution for protein- and cell-type-specific CALI was the design of the first genetically encoded PS (Bulina et al., 2006a). The protein called KillerRed was isolated by screening a collection of GFP homologs for phototoxic effects on *Escherichia coli* cells. Although its ROS-production capacity is still inferior to chemical PSs, KillerRed has been used successfully in zebrafish embryos to induce cell death (Teh et al., 2010) and in cell cultures to achieve target protein inactivation (Bulina et al., 2006b) and reversible blockage of cell division (Serebrovskaya et al., 2011).

A comprehensive definition of optogenetics

As pointed out before (Miesenböck, 2009), the term "optogenetics" is a bit of a misnomer as it does not involve any interaction between light and the genome. But coming up with this label was definitely a smart move judging from how quickly it was adopted by the research community. The term is now firmly established both in the scientific literature and in the popular media, but its usage has not yet crystallized around a common acceptation.

Etymologically, "optogenetics" simply refers to the combination of optical and genetic approaches and implicitly designates all strategies using genetically addressable light-sensitive tools to study biological systems. As a consequence, the term should seize on 20 years of utilization of FPs, including for simply labeling cells and proteins. More reasonably, optogenetics can designate the use of genetically addressable photosensitive elements not as inert dyes but as environmentally sensitive fluorophores (in which light emission is affected by identified factors) and/or as active agents (which can transduce optical energy into biophysical effects). This definition encompasses both monitoring and control strategies. We believe that a narrower acceptation of the word is unjustified. Just like "optoelectronics" designates the use of both light sources and detectors, "optogenetics" should encompass the use of both control tools and reporters.

Which control tools and which reporters should be included in this definition? In the broadest sense, optogenetic tools do not need to be fully genetically encoded but only genetically "addressable." This means that proteins requiring an exogenous cofactor to function can also be considered as "optogenetic" as long as their expression can be restricted to certain groups of cells. This definition includes a range of photochemical approaches where proteins are engineered to bind to a given photochromic ligand. It also encompasses the use of photoreceptor proteins in organisms lacking their specific chromophore. In such cases, the chromophore molecule has to be added exogenously (e.g., retinal in invertebrates for channelrhodopsin-based applications and bilin in nonplant organisms for phytochrome-based applications).

Finally, to what areas of biology should the term optogenetics apply? There is no valid reason to restrict its use to neuroscience only. Current developments even tend to show an accelerated expansion of optogenetic approaches toward general cellular and molecular biology. Overall, we wish to conclude that a comprehensive definition of optogenetics might be the following: optogenetics is the combination of optical and molecular strategies to monitor and control designated molecular and cellular activities in living tissues and cells using genetically addressable photosensitive tools.

Combining the tools

Optogenetic control tools clearly made their breakthrough in neuroscience with manipulations at the level of cell populations while observing the consequences at the systems and behavioral levels (Carter and de Lecea, 2011). However, it is also clear that a detailed mechanistic analysis and understanding of brain function will require simultaneous observation and/or manipulation of various neuronal types at the same cellular or circuit level. This can be partially implemented by using optogenetics in conjunction with existing techniques like electrical recordings. But we believe that this methodological challenge will be eventually more perfectly met by combining optogenetic tools within the same experiment, an important step which will unleash the full power of optogenetics.

"See it, block it, move it"

To understand and demonstrate how a biological phenomenon works ultimately requires using a canonical scientific methodology often summarized by the formula "see it, block it, move it." The first step is to identify the conditions for this phenomenon to occur (see it); the second step is trying to find out which of these conditions are necessary using loss-of-function experiments (block it); the third step is to test the sufficiency of one or more conditions through gain-of-function experiments (move it). The first step aims at establishing a correlation, while the two others aim at demonstrating causation. Correlation in neuroscience has been investigated in particular using invasive electrical recordings in order to match neuronal activity with behavior. Although such recordings can provide hints on possible causal relationships (e.g., when identified electrical events precede or follow behavioral events), causation is traditionally approached using genetic (KOs and overexpression), electrical (stimulations), surgical (lesions), or pharmacological (agonists and antagonists) interventions. None of these techniques alone provides both high temporal and spatial (cellular) specificity. Electrical stimulations and recordings of neuronal firing display exquisite microsecond-scale temporal resolution but are usually unable to discriminate between neurochemical cell types. In addition, electrical stimulations do not discriminate between axons and cell bodies, which seriously limits their interpretative value. Conversely, pharmacological interventions are hampered by their poor temporal resolution although they can provide very good neurochemical specificity.

Optogenetics is considered a true technological breakthrough because it makes it possible to implement the "see it, block it, move it" approach with both high temporal resolution and high cellular (even subcellular and molecular) resolution. Thus compared to standards of the past decade, modern optogenetic studies might bring more definitive answers and allow biologists to form stronger interpretations. More remarkably, combining optogenetic tools will offer the possibility of implementing this approach in the same experiment, which will dramatically increase the yield of individual studies.

Combining optogenetics with electrical recordings

Microelectrode recordings are still the golden standard for measuring neuronal firing, surpassing by far optical sensors at least at the single cell resolution and millisecond time scale. But electrical recordings can be very advantageously combined with optogenetic tools, in general, and light-gated actuators, in particular.

First of all, electrical recordings are and will probably remain the ultimate readout of the efficiency of optogenetic activation and inactivation protocols. Indeed, the reliability of optogenetic control depends on a series of important parameters which can be preparation specific, such as the electrophysiological properties of the target cell type, the optical properties of the tissue, or the expression level of the optogenetic tool (which depends on the time postinfection when using viral vectors). Assessing photoevoked changes in firing during optogenetic control experiments might even become systematic practice with the use of "optoelectrodes" which integrate light guides and electrodes in the same device.

Second, optogenetic control tools can elegantly replace stimulating electrodes in circuit mapping experiments. Classically, these experiments consist in probing functional connections between neuronal types and brain regions using pairs of recording and stimulating electrodes. Light-gated actuators can substitute for electrical stimulations in order to control efferent and afferent connections in isolation based on their origin, destination, or neurochemical identity. For example, labeling a group of neurons anterogradely with ChR2 or NpHR allows the experimenter to photoexcite or inhibit specifically its axonal projections in distant areas, even when cut from their soma (Atasoy et al., 2008; Cruikshank et al., 2010; Kaneda et al., 2011; Petreanu et al., 2007, 2009; Varga et al., 2009). Conversely, expressing these tools using retrograde transsynaptic activators (Gradinaru et al., 2010) or retrogradely transported viruses (Lima et al., 2009) offers the opportunity to control specifically cells projecting to a particular region. These dual electrical-optogenetic strategies can even be paired with targeted illumination and scanning techniques to refine and accelerate mapping processes.

Third, optogenetics can help overcome the fact that extracellular electrical recordings do not easily distinguish spikes from different neuronal populations. This has been a central issue and a great source of debates in the study of neuronal firing *in vivo*. Traditionally, neuronal types are identified during recording based on electrophysiological criteria like spike shape and firing patterns. But this approach fails in the case where two different populations have overlapping properties. Labeling a group of neurons with ChR2 offers the possibility to confirm their identity online. As shown by Lima et al. (2009), ChR2-tagged neurons can be identified *in vivo* by their reliable and short latency response to brief flashes of blue light, a strategy called PINP (photostimulation-assisted identification of neuronal populations).

Combining optogenetic tools: Toward multicolor interrogation of neural circuits

Multicolor control of neuronal populations

Combining several light-gated actuators in the same experiment requires the ability to recruit one with minimal cross-excitation of the others. Maybe because they were isolated from organisms living in very different ecosystems, microbial opsins display a great diversity of spectral sensitivities. A few of them can be excited almost separately using different wavelengths. The best example so far is the association of ChR2 and NpHR which allows bidirectional control of firing of the same cells using blue- and yellow light (Zhang et al., 2007) opening the possibility of performing loss-of-function and gain-of-function experiments (block it and move it) on the same preparation.

Other "optically compatible" opsin pairs include the blue- and yellow light-gated channelrhodopsin variants ChR2 and VChR1 (Zhang et al., 2008) and the blue- and red-light drivable ion pumps Mac and NpHR (Chow et al., 2010). In theory, these tools allow multicolor control of separate populations of neurons simultaneously. A new generation of red-shifted actuators includes novel channelrhodopsins such as MChR1 from *M. viride* (Govorunova et al., 2011) and C1V1s, a family of ChR1/VChR1 chimera displaying large photocurrents and minimal cross-activation with ChR2 (Yizhar et al., 2011a,b,c) as well as new light-driven pumps such as Halo57, a naturally occurring halorhodopsin displaying larger photocurrents than NpHR when excited in the far red (Klapoetke et al., 2010). These new opsins are expanding the catalog of compatible actuators for multicolor control of neural circuits. Optogenetic control tools can also be combined physically as a unique protein. Recently a tandem gene fusion strategy was proposed for co-localized and stoichiometric expression of opsin pairs (Kleinlogel et al., 2011). This approach has a number of potential applications. Precise bidirectional control of firing with low cell-to-cell variability of the excitation-to-inhibition ratios can be achieved by fusing a ChR variant and a light-driven pump. This strategy is also a useful way of creating new tools with new properties: for example, ChR variants with different excitation spectra can be combined to create a hybrid tool with a wider action spectrum.

Multicolor probing of neuronal activity

Contrary to microbial opsins, most genetically encoded reporters were not isolated from ecologically diverse species but were engineered based on a very limited number of FPs (GFP or YFP for single FP sensors and CFP and YFP for FRET sensors). However, the color palette of available FP variants has been continuously expanding for the past 10 years, and available FPs now span almost the entire visible spectrum (Chudakov et al., 2010; Day and Davidson, 2009). This opens the door for a new generation of genetically encoded probes with diversified and minimally overlapping spectral characteristics. These novel tools will be used to visualize the activity of distinct neuronal populations in parallel or to image multiple parameters in the same cells.

FRET sensors were the first category of optical reporters to be spectrally diversified. Indeed, grafting a new pair of FPs in an existing FRET sensor scaffold is relatively straightforward since it does not require major modifications of the FPs. In contrast, updating single FP sensors can require more work since they often incorporate modified versions of the FP (e.g., circularly permutated FPs). New blue- and red-shifted spectral variants were already produced for several FRET sensors including the voltage sensors VSFP2s (Akemann et al., 2010), sensors of cyclic nucleotides (Niino et al., 2009), reporters of enzymatic activities (Ai et al., 2008; Grant et al., 2008; Ouyang et al., 2010), or protein translocation (Piljic and Schultz, 2008). Some of these variants were used to demonstrate the feasibility of double and triple FRET measurements (Ai et al., 2008; Grant et al., 2008; Niino et al., 2009; Ouyang et al., 2010; Piljic and Schultz, 2008).

FRET sensors have the advantage of enabling ratiometric measurements but the inconvenience of using two FPs (one donor and one acceptor). For this reason, combining more than two or three spectrally nonoverlapping FRET sensors is very challenging. A smart workaround is to free up one color channel by using a nonfluorescent (dark) acceptor which acts as a dynamic quencher for the donor fluorescence (Ganesan et al., 2006; Niino et al., 2010). Still, single FP sensors provide a simpler and more flexible solution to the problem of spectral crossover. Single FP sensors are still almost exclusively based on GFP or YFP variants except for the blue-shifted kinase activity sensor Cyan Sinphos (Kawai et al., 2004) and the red-shifted monochromatic voltage sensors VSFP3s (Perron et al., 2009). However, the attractiveness of multicolor imaging should promote the construction of additional spectral variants of single FP sensors in the near future. Recent efforts have focused on mutating the calcium indicator scaffold introduced as GCaMP (Nakai et al., 2001) to obtain hue-shifted variants. Using a "molecular evolution strategy" (iterative rounds of mutagenesis and screening of bacterial colonies), Zhao et al. (2011) have engineered a new set of GCaMP mutants called GECOs, comprising blue and red variants. Another initiative which will accelerate the development of new calcium sensors is the GECI project from the HHMI Janelia Farm research campus (http://www.janelia.org/team-project/geci). This project uses a high-throughput, mammalian neuron-based imaging platform to screen through libraries of variants. Current lead variants include blue, cyan, and yellow versions of the GCaMP scaffold (BCaMP, CyCaMP, and YCaMP) as well as a red version (RCaMP) which was engineered from scratch using the red FP mRuby (Loren L. Looger, personal communication).

Combined optogenetic monitoring and control of neuronal activity

The next big step remains the association of optical reporters and control tools within the same experiment to allow all-optical interrogation of neural circuits. To date, only one study has employed this type of strategy: the work by the team of Sharad Ramanathan described how ChR2 and GCaMP can be combined to map functional connections between groups of neurons in *C. elegans* (Guo et al., 2009). Because ChR2 and GCaMP have highly overlapping excitation spectra, the authors had to separate the excitation channels of the two proteins both temporally and spatially. Similar experiments should be greatly simplified by the use of red-shifted activity reporters such as RCaMP and VSFP3s (Perron et al., 2009) or alternatively by the combination of red-shifted opsins with blue-shifted reporters.

Given the current rate of expansion of the optogenetic toolkit, the number of possible tool combinations might soon become overwhelming, giving unprecedented latitude for the experimenter's imagination. Most important, the analytical power of "all-optogenetic" approaches is potentially mind-blowing: combining monitoring and control will allow researchers to establish correlation and causation in the same experiment. This should increase the yield of individual experiments and raise the standards in many fields of neurobiological research.

New challenges for old technologies

Optogenetics did not evolve as a stand-alone approach but rather emerged at the crossroads of several independent technologies. These technologies include methods for gene delivery on the one hand and for light delivery and collection on the other. By constantly setting new technical requirements, optogenetics is regularly challenging these parent technologies and driving technical innovation. Several key techniques for optogenetics are reviewed in the following chapters. Here, our intention is to provide an overview of the emerging optogenetic know-how in neuroscience, with a strong focus on mammalian models.

Current and future challenges for gene delivery approaches

Optogenetic tools are genetically addressable, which means that all or parts of them are genetically encoded. Thus, the starting point of any optogenetic experiment consists in choosing a particular optogenetic tool and a method to deliver it to a target system. The main objective of this step is to achieve expression in a functionally and/or genetically well-defined set of neurons. Depending on the time and resources available as well as experimental requirements, one can chose to build transgenic lines (germline transgenesis) or to acutely transfer the gene of interest to a particular organ, region, or group of cells in individual animals (somatic gene delivery). Here, we review the current advantages and limitations of these strategies (see also Zhang et al., 2010).

Somatic gene transfer in the central nervous system

Acute gene transfer can be performed using viral vectors, the two most popular agents currently being retroviruses (which include lentiviruses) and adeno-associated viruses (AAVs) (Aronoff and Petersen, 2006; Davidson and Breakefield, 2003; Monahan and Samulski, 2000; Teschemacher et al., 2005; Wong et al., 2006). Virus injection into the brain can be performed at all stages of life through a simple surgical procedure (Cetin et al., 2006; Lowery and Majewska, 2010; Pilpel et al., 2009; Puntel et al., 2010). Electroporation is another method for quick gene delivery which works by forcing expression plasmids into single or groups of cells using an electric field (Judkewitz et al., 2009). When performed on mouse embryos *in utero*, this technique can provide large numbers of transgenic animals in a short time frame (Walantus et al., 2007). Electroporation techniques are reviewed in greater details in Chapter 9 of this issue. Acute gene transfer offers a number of advantages over transgenic lines. First, it offers the possibility to test new genetic constructs rapidly (in several weeks), allowing researchers to keep up with new optogenetic tools. Second, it can provide higher expression levels than transgenic lines, a feature which can be particularly important when working with actuators with low unitary photocurrents such as microbial opsins.

Viral vectors are currently the most popular method for rapid gene delivery mainly because of the versatility that they offer. Viral strategies can be designed to yield both high levels and high cell-type specificity of expression. Cell-type specificity can eventually be empirically achieved through viral serotype-specific tropism but most commonly relies on the use of specific gene promoters. Promoters can be included in the encapsidated transgene to allow autonomous specific expression. In case this approach yields insufficient expression, transcriptional amplification strategies can help enhancing the expression of the transgene (Liu et al., 2008). A popular alternative is to inject viruses containing a Cre-responsive expression cassette into the brain of a Cre-expressing line (Kuhlman and Huang, 2008), a technique which was perfected with the flip-excision (FLEX) switch system (Atasoy et al., 2008; this system is also referred to as DiO for doublefloxed inverse open-reading-frame). In the FLEX/DiO system, cell-type specificity is provided by the expression of the Cre recombinase, while high transcription rate of the optogenetic tool is guaranteed by a strong ubiquitous promoter present in the Cre-responsive cassette. This method was a godsend to optogenetics because it made the hundreds of well-characterized Cre-expressing mouse strains generated over the past decade amenable to optogenetics. In particular, FLEX/DiO constructs offer the possibility to quickly test several optogenetic tools on the same type of neurons using the same mouse strain. In theory, FLEX/DiO viruses can also be coinjected with custom Cre-expressing viruses to implement a transcriptional amplification strategy. This approach might reveal useful in species with still very limited catalogs of Cre-expressing transgenic lines such as rats (Witten et al., in preparation).

Viral approaches can also provide multiple levels of spatial specificity. First, stereotaxic viral injections can be optimized in order to restrict the expression of one or several optogenetic tools to one or more anatomically identified brain regions. Second, strategies for retrograde trans-synaptic expression can be used to target neurons projecting to a particular brain area. Such strategies can employ retrogradely transported viruses like the herpes simplex virus (Berges et al., 2007; Lima et al., 2009), rabies, and pseudo-rabies viruses (Osakada et al., 2011; Wickersham et al., 2007a,b) or certain AAV serotypes (Masamizu et al., 2011). An elegant alternative consists in using a dual-virus approach in which one virus expresses WGA-Cre, a fusion between the Cre and the transcellular tracer protein wheat germ agglutinin (WGA), while the other expresses an optogenetic tool under the control of a FLEX/DiO cassette. The method (described in Gradinaru et al., 2010) follows a three-step process: (1) the first virus is used to infect a particular brain region (region A), while the other is injected into an upstream structure (region B); (2) WGA-Cre is produced in neurons of region A and traffics transsynaptically into their presynaptic neurons; and (3) WGA-Cre activates the transcription of the tool of interest only neurons of region B projecting to region A.

Despite their appreciable flexibility, viral approaches have a number of limitations. First, infection efficiency is usually spatially inhomogeneous, with expression decreasing away from the injection point. Even within the site of injection not all potentially targeted cells express the same amount of the protein depending, for example, on the number of viral copies incorporated into the cells. Overall infection rates are also highly dependent on the quality/titer of the virus preparation which can vary from one batch to the other and introduce variability in the experiment outcome. This important issue has only been addressed and discussed on very few occasions (Aponte et al., 2011; Haubensak et al., 2010; Lin et al., 2011). Inhomogeneous expression can be partially overcome by performing viral injections in neonates, a method which can yield more widespread expression (Passini and Wolfe, 2001; Pilpel et al., 2009).

Another limitation is the potential toxicity of proteins expressed at high levels using viral gene delivery. High transcription rates may rapidly lead to toxic accumulation of the protein, thus reducing the time window for experimentation. This issue has not been clearly addressed yet in the literature. Other issues include potential immunogenicity of viral particles and DNA packaging limitations of viruses. Indeed, viral capsids can only accommodate exogenous DNA fragments up to a certain limit. This limit (around 5kb for AAVs and 10–15kb for lentiviruses) is not an absolute one in the sense that viral particles can still be produced with larger inserts but with lower titers. Viral gene delivery approaches are reviewed in Chapters 9 and 11 of this issue.

Germline transgenesis

Most of the drawbacks of viruses can be overcome by germline transgenesis. This approach aims at establishing lines of transgenic animals expressing the protein of interest stably and constitutively, eliminating the need of delivering the gene of interest on a single animal basis. Available methods for germline transgenesis are detailed in Chapter 9 of this issue. When compared to viral approaches, the main issue of "optogenetic" transgenic lines so far has been their lower expression levels. This limitation can be problematic for optogenetic tools requiring high expression levels such as microbial opsins. Nevertheless, a number of mouse strains expressing optical reporters (GCaMPs, VSFPs, synaptopHluorin and Clomeleon) or control tools (ChR2, VChR1, ChETA, NpHR, eNpHR3 and Arch) have been generated. These lines express the optogenetic tool under the control of either a specific promoter (Thy1, ChAT, VGAT,

TPH2, VGluT2, PV) or a Cre-activated cassette. The latter type of strain can be crossed with any existing Cre driver line to achieve targeted expression through cell-type-specific recombination. Available mouse lines for optogenetic applications are described in Chapter 10 of this issue.

Although transgenesis clearly saves time and money on the long run, its implementation can be costly and time consuming. While this is true for classical transgenesis techniques (pronuclear microinjection and ES integration into blastocysts), new techniques such as testis electroporation (Dhup and Majumdar, 2008), lentivirus-mediated, and zinc finger nucleases-mediated transgenesis (Le Provost et al., 2010) might hold the keys for rapid and efficient germline transgenesis in various mammalian species.

Cell-subtype specificity through intersectional genetic strategies

Current optogenetic approaches achieve cell-type specificity through the use of single promoters, but discrete cellular subtypes are often defined by the selective coexpression of several markers rather than just one. This is the case for cortical circuits in which functionally distinct subtypes of inhibitory interneurons express specific combinations of calcium-binding proteins, neuropeptides, enzymes, and receptors (Ascoli et al., 2008; Kubota et al., 2011). To target neuronal subpopulations, future optogenetic approaches might employ intersectional strategies to restrict the expression of a transgene to cells coexpressing a particular set of genes. In the mouse, intersectional gene activation was implemented using a dual-recombinase method in order to refine fate mapping studies (Dymecki et al., 2010). In this method, the transcription of a transgene is dependent on the removal of two STOP cassettes by two independent recombinases (e.g., Cre and FLPe) expressed under the control of different promoters. A similar intersectional strategy could easily be transposed to the FLEX switch system, which already requires two recombination events to produce stable transgene inversion (Atasoy et al., 2008). Intersectional approaches might become more and more attractive with the increasing number of FLPe driver lines and the use of novel site-specific recombinases (Nern et al., 2011). Intersectional expression strategies are reviewed in detail in Chapters 9 and 10 of this issue.

Light delivery and collection

Optogenetics builds on an experimental hardware that blends standard technologies and recent innovations in optical imaging, digital microscopy, and photonics. While much of today's instrumentation is inspired by widespread applications of light microscopy in biology and other disciplines, optogenetics poses new challenges that are likely to expand the technical platform in bioimaging and biophotonics.

Breaking new grounds in microscopic imaging of neuronal activity

Before the advent of modern genetically encoded optical sensors, the use of nonprotein reporters of neuronal activity had already prompted significant advances in microscopic imaging techniques. Optogenetic probes should prolong this momentum by opening new possibilities such as deeper imaging over longer periods of time.

Classically, single cell-resolved fluorescence images are obtained using conventional mono- or multiphoton microscopy combined with laser-scanning techniques. This approach has been used extensively in combination with organic dyes to image neuronal activity in thin preparations (small animals, cultured cells, or brain slices) or superficial brain structures in head-fixed animals. Because dyes can report neuronal activity with relatively high temporal precision (several milliseconds), one major improvement in the past decade was to achieve high scan rates using new scanning schemes (Saggau, 2006). Fast-scan optical imaging allowed researchers to follow the activity of neuronal networks with combined high temporal and spatial resolution. Genetically encoded optical probes offer a new ground for further developments in the field by allowing multiscale imaging (from large cortical areas to subcellular compartments) over longer periods of time (weeks vs. hours in the case of organic dyes).

Today's challenges for light microscopy consist in accessing deep structures (>1 mm) and imaging neuronal activity in unrestrained animals, with the long-term goal of combining the two. By eliminating the need of a dye-loading step and allowing long-term imaging, genetically encoded activity reporters have dramatically increased the attractiveness of such approaches. Several options have already been investigated. On the side of deep-brain imaging, thin (<1 mm of diameter) gradient refractive index microlenses can be inserted into the brain to increase the reach of laser-scanning microscopy, acting as an optical relay or "microendoscope." Microendoscopy can also be implemented by scanning through high-density optical fiber bundles implanted into deep-brain structures, with each optical fiber behaving as an individual pixel (Vincent et al., 2006). On the side of freely behaving animals, prototypes of lightweight (1–4 g) portable optical microscopes have been designed but their usability is still limited (see Wilt et al., 2009 for review).

Optogenetics might also influence the evolution of light sources and detectors. Because the sensitivity of genetically encoded reporters still lags behind the one of nonprotein sensors (Knöpfel et al., 2006), increasing the quantum yields of detectors and the stability of light sources will represent critical improvements. One solution for low-noise illumination consists in adapting semiconductor light sources like light-emitting diodes (LEDs) to optical microscopy (Albeanu et al., 2008).

Light delivery techniques for optogenetic control of brain activity

Compared to the challenges of optical microscopy, delivering light into the brain to control neuronal activity can seem almost trivial. But this is probably the field where optogenetics is driving the strongest innovation. The main concern is to supply light at a sufficient intensity to a defined volume of brain tissue. Illuminating spatially restricted areas in superficial brain structures or thin preparations can be done using conventional laser-scanning techniques, digital micromirror devices (Jerome et al., 2011), holographic patterned light (Papagiakoumou et al., 2010, this technique is reviewed in Chapter 7), or LED microarrays (Grossman et al., 2010). Alternatively, head-mounted LEDs offer a simple way of delivering light to the surface of the brain in unrestrained animals (Huber et al., 2008; Iwai et al., 2011). Along this line of work, the team of Ed Boyden recently engineered a wirelessly powered and controlled LED system for brain surface illumination weighting only 2 g (Wentz et al., 2011).

Illuminating deep brain areas requires the use of light guides such as optical fibers, an approach which has become standard in the past couple of years. Fiberoptic light delivery can be implemented easily in freely behaving animals. A number of accessories have been developed for this purpose, including implantable optical fiber pieces with miniaturized connectors as well as fiberoptic rotary joints to allow free rotations of the fiber connected to the animal (Gradinaru et al., 2007; Kravitz and Kreitzer, 2011; Yizhar et al., 2011a,b,c). Recent commercial versions of these products can incorporate an independent channel for liquid delivery. Laser beams offer convenient light sources which can be easily manipulated and focused (launched) into the fiber core (typically <200 μm of diameter). LEDs offer a cheaper alternative and can also be coupled to optical fibers (pigtailed) but with lower coupling efficiency than lasers. Alternatives to optical fibers are under investigation such as waveguide materials which can be deposited onto thin layers of silicon to create multipoint light delivery probes (Zorzos et al., 2010). Chapter 11 of this issue reviews some important technical considerations to take into account for fiberoptic light delivery into the brain.

Optoelectrodes: New devices for combined light delivery and electrical recording

One new requirement of optogenetics is the possibility of delivering light and recording electrical activity with the same implantable device. These "optoelectrodes" (or optrodes) are particularly useful to assess the efficiency of photostimulation and -inhibition *in vivo*. Many different optoelectrode designs have been introduced in the past 5 years. Optical fibers can be simply glued onto or bundled with existing single and multielectrode systems (Anikeeva et al., 2011; Diester et al., 2011; Gradinaru et al., 2007, 2009; Halassa et al., 2011; Kravitz and Kreitzer, 2011; Royer et al., 2010; Zhang et al., 2010) or gold metalized to behave as electrodes (Zhang et al., 2009). Alternatively, 2D multielectrode silicon probes can be modified to integrate waveguide materials (Cho et al., 2010). In the near future, electrical wires might be embedded into the structure of optical fibers. Wires can be added at the fabrication stage by inserting them in a large-diameter optical fiber "preform" and then pulling out a thin string from the heated preform.

One problem facing the use of optoelectrodes is direct interaction between light and metal electrodes when immersed in brain tissue (or saline), a phenomenon which causes light-induced electrical artifacts that can obscure local field potential and spike recordings. These artifacts are most likely due to a photovoltaic effect (also referred to as photogalvanic or Becquerel effect) but have not been investigated and discussed in detail, except on a few occasions (Ayling et al., 2009; Cardin et al., 2010; Han et al., 2009). Efforts are underway to develop "light-proof" electrodes using, for example, indium tin oxide coating (Zorzos et al., 2009). Glass electrodes are devoid of such light-induced artifacts but are mostly suited for recordings in isolated preparations or immobilized animals. Recently, LeChasseur et al. (2011) introduced a thin ($<20\mu m$ at the tip) fiberoptic microprobe containing one core for light delivery and one electrolyte-filled hollow core for neuronal recording. This strategy allows combined artifact-free single unit recording and photocontrol with minimal damage to the tissue. The issue of light artifacts on metal electrodes is covered in Chapter 11.

Collecting bulk fluorescence in vivo

Monitoring neuronal activity optically often rhymes with imaging the activity of single neurons. But a simpler approach consists in retrieving global (bulk) fluorescence signals rather than single cell-resolved microscopic images, in order to obtain information on population activity. This approach was implemented on anesthetized mice expressing optical reporters (GCaMP2 or synaptopHluorin) in order to map responses elicited by sensory or local electrical stimulations (Diez-Garcia et al., 2007; Fletcher et al., 2009; Petzold et al., 2009). Bulk fluorescence measures are also easy to implement in freely behaving animals using the same fiberoptic hardware used for light delivery. This approach has been pioneered in particular by the team of Matthew E. Larkum using nonprotein calcium-sensitive dyes (Murayama and Larkum, 2009; Murayama et al., 2007). Since genetically encoded probes have raised the relevance of bulk fluorescence measurements by providing cell-type-specific signals, this methodology might soon become more widely adopted. As an example, Lütcke et al. (2010) recently reported the use of optical fibers to measure large-scale sensory-evoked cortical activity in GECI-expressing freely moving mice.

Another potential application of bulk fluorescence measurements is to map and assess transgene expression *in vivo* following viral infection. Most optogenetic tools are indeed either fluorescent or fused with FPs in order to facilitate histological examination. Optical fibers can be used to localize the core of an infection, where optogenetic signals and effects are expected to be higher, prior to recording or stimulation. Chronic measurements of fluorescence signals can also track the expression of a transgene and help define optimal time windows for experimentation. This approach was used recently to guide optogenetic stimulations in nonhuman primates (Diester et al., 2011). In the future, fiberoptic deep-brain fluorescence measurements might even be adopted to assess transgene expression in viral therapy approaches in humans, including potential optogenetic therapies.

Spreading the knowledge and tools

Optogenetics is a field with nearly unprecedented momentum. Optogenetic approaches are becoming standard practice in neuroscience and optogenetic tools are evolving at a frenetic pace. While the vitality and creativity of the field can be nothing but warmly acclaimed, it has its flip side: it is simply becoming challenging for researchers to keep up with it.

Acknowledging this glaring situation naturally leads to the question of how scientific and technical knowledge should be optimally spread among researchers. Review articles, which are the traditional way of summarizing the current state of the art on a particular topic, are intrinsically not adapted to a rapidly evolving field because of their relatively long periodicity, limited article lengths, and incompressible publication delays. The snapshot that they provide, although potentially comprehensive at the time of publication, often loses some of its representativity within the next months.

Keeping up with optogenetics through Web collaboration

On the use of wikis in science

Collaborative Web-based solutions are a weapon of choice to tame a rapid flow of information and hold a constantly up-to-date body of knowledge. Among them, wikis have proven their efficiency for sharing and organizing many bits of information across large communities of users. They provide key advantages which are worth enumerating. First, they are extremely reactive since they can be rapidly edited from anywhere and by anyone. Second, they are extraordinarily dynamic since their content can be modified and corrected at will by a potentially unlimited number of users. Third, the content of a wiki is very rich since it can incorporate much more than just text and pictures by storing virtually any type of file. And last but not least, the access to the large majority of them is totally free. However, the success of a wiki is never guaranteed from start and resides in its ability to attain the critical mass of users necessary for its effectiveness and durability. Without this critical mass, it usually falls into disuse.

Inspired by the remarkable success of the online encyclopedia Wikipedia, scientific wikis have started to proliferate within the past few years with the common aim of sharing protocols, tricks, and ideas (Butler, 2005; Pearson, 2006; Waldrop, 2008). As a result, many different fields of biological research of various sizes and focuses have been "wiki-fied." For example, OpenWetWare (http://www.openwetware.org), created in 2005 by students from the Massachusetts Institute of Technology, is hosting information about protocols in biochemistry and molecular genetics, gathering around 9000 users from more than 100 laboratories. These efforts participate in promoting the general concept of "open research" where researchers make clear and exhaustive accounts of their methodology unlike in many peer-reviewed articles.

As new genetically encoded optical tools are greeted by an irresistible wind of enthusiasm and excitement among neurobiologists, it seems that a similar effort for sharing the emerging optogenetic know-how is very timely. Such a project exists: a wiki platform called OpenOptogenetics was launched during the summer of 2010 (http://www.openoptogenetics.org). This platform is fully supported by researchers and aims at providing up-to-date technical information about all aspects of optogenetics.

OpenOptogenetics, an open wiki about optogenetics

Pioneer laboratories in optogenetics have started to share their protocols and experience online through dedicated Web pages such as the Optogenetic Resource Center of the Deisseroth lab (http://www.stanford.edu/group/dlab/optogenetics/) or the section "Protocols and Reagents" of Edward Boyden's Synthetic Neurobiology Group (http://www.syntheticneurobiology.org/protocols). By adopting a fully open framework, OpenOptogenetics aims at pushing the collaboration a step further. The wiki is fully supported and administered by researchers and allows anyone to create and edit pages through a simple registration step. OpenOptogenetics is committed to a number of missions: (1) maintaining an inventory of all available optogenetic tools, their properties, and where to obtain them or the transgenic lines expressing them; (2) providing detailed protocols related to common procedures (e.g., gene or light delivery into the brain); (3) describing and comparing the available hardware (light sources, fiberoptic components, optoelectrodes), as well as guides on how to build economical setups for most common applications; (4) linking to books, reviews, or other documentations about optogenetics; (5) keeping an up-to-date list of scientific events (workshops, conferences) related to optogenetics; (6) providing a news feed tracking the latest technical developments in the field.

Promoting the diffusion of an emerging technical know-how is especially meaningful. Facilitating and democratizing the implementation of new techniques has a positive impact on how quickly and how widely they are adopted. OpenOptogenetics will help newcomers take the right decisions on how to best set up their experiments according to their needs and resources. The information available in the wiki should, for example, help people pick the appropriate channelrhodopsin variant or decide which light source to buy, a key question for laboratories with tight budgets. OpenOptogenetics should not just benefit to the end users. By channelizing the demand for information and guidance, it should help technology makers deal with an increasing number of inquiries about their resources and protocols. Finally, an efficient horizontal transfer of information can be expected to raise the profile of methods by spreading subtle and usually unpublished (but nonetheless important) methodological tricks.

Distributing the tools

Sharing sequences

Knowledge is not the only thing that can be shared. One of the pillars of modern molecular biology is the management and sharing of the millions of nucleotide sequences isolated every year across the world. These sequences are made publicly available and searchable by institutional databases such as the International Nucleotide Sequence Database Collaboration (http://www.insdc.org/). At a smaller scale, biologists also share recombinant DNA under "ready-to-use" forms such as bacterial plasmids. Synthetic plasmids used routinely in genetic engineering are called vectors. These vectors have become essential tools to store and multiply genes and to perform common cloning procedures. Several initiatives are exploiting the power of Web-based strategies not only to better share vector information but also to physically distribute them. These existing solutions are most appropriate for a rapid and fair diffusion of optogenetic tools.

Several laboratories and companies are hosting and maintaining their own online vector database but only two platforms have emerged as popular and potentially comprehensive vector repositories: PlasmID (http://plasmid.med.harvard.edu/PLASMID/) and Addgene (http://www.addgene.org). PlasmID was established in 2004 to facilitate the search and request of plasmids from the DNA Resource Core of the Dana-Farber/Harvard Cancer Center and is currently supported by several institutions (Zuo et al., 2007). Addgene was started the same year as a nonprofit organization and was initially funded by private donations and small business loans (Fan et al., 2005). Both platforms are backed up by solid plasmid storage and sequencing facilities and distribute their samples internationally at low cost.

PlasmID's strategy is oriented toward gathering large existing collections, whereas Addgene is very proactively working with individual laboratories. Addgene was created with the objective of reducing the burden on researchers to store, maintain, and distribute plasmid clones and supporting information. Its financial model is to set prices such that revenues from requests (around $65 per plasmid) are just enough to cover the operating costs. This point of financial self-sustainability was reached in 2007, only 3 years after the creation of the organization. Addgene's catalog currently has over 15,000 plasmids from around 1000 laboratories and keeps on growing. Searching this catalog online is particularly user-friendly: most plasmids are linked to the corresponding articles, allowing scientists to quickly find the actual experiments performed with them, and a color code tracks their popularity (how many times they were requested). Addgene was born just on time to catch on with the recent boom of optogenetics. The major players in optogenetic engineering have already deposited over 60 constructs in only a few years. It is thus relatively safe to predict that Addgene will become a major platform for the distribution of optogenetic tools.

Sharing transgenic animals

Distribution channels similar to Addgene also exist for transgenic animals. The Jackson Laboratories (http://www.jax.org/), one of the biggest world's source of genetically modified mouse strains, already distributes more than 20 lines expressing optogenetic tools under the control of specific promoters or a Cre-activated cassette (http://research.jax.org/grs/optogenetics.html). Both reporter strains and FLEX/DiO viruses allow the use of the hundreds of available Cre driver mouse lines for optogenetic applications. Data repositories for Cre transgenics are starting to emerge, such as Cre-X-Mice (http://nagy.mshri.on.ca/cre_new/index.php), and large-scale efforts are under way to expand the catalog of Cre lines through the GENSAT project (Geschwind, 2004; Heintz, 2004, http://www.gensat.org), the Cre Driver Network (http://www.credrivermice.org/), and the Allen Institute for Brain Science (http://www.brain-map.org/).

Concluding remarks

From the beginning, optogenetics has been more than just a toolbox. It represents the answer to the experimental difficulties in bridging the complexity and diversity of molecular and cellular neurophysiology with systemic functions. This is a necessary step for several large-scale projects aimed at reverse engineering the cerebral cortex, including the Blue Brain Project (http://bluebrain.epfl.ch) at EPFL (Ecole Polytechnique Fédérale de Lausanne) that is currently running for a very large European Commission research grant and the ambitious goal to deciphering the neuronal code at the Allen Institute for Brain Science (www.alleninstitute.org). Optogenetics can be seen as a revolutionary new field, but at the same time, it evolved over decades propelled by visionary expectations and a pedigree of scientific discoveries and inventions. The perspective presented in this introductory chapter of the *Progress in Brain Research* issue on optogenetics sets the scene for the subsequent chapters dealing with specific optogenetic tools and their application to brain research.

References

Ai, H. W., Hazelwood, K. L., et al. (2008). Fluorescent protein FRET pairs for ratiometric imaging of dual biosensors. *Nature Methods*, *5*(5), 401–403.

Airan, R. D., Thompson, K. R., et al. (2009). Temporally precise in vivo control of intracellular signalling. *Nature*, *458*(7241), 1025–1029.

Akemann, W., Mutoh, H., et al. (2010). Imaging brain electric signals with genetically targeted voltage-sensitive fluorescent proteins. *Nature Methods*, *7*(8), 643–649.

Albeanu, D. F., Soucy, E., et al. (2008). LED arrays as cost effective and efficient light sources for widefield microscopy. *PLoS One*, *3*(5), e2146.

Anikeeva, P., Andalman, A. S., Witten, I., Warden, M., Goshen, I., Grosenick, L., et al. (2011). Optetrode: a multichannel readout for optogenetic control in freely moving mice. *Nature Neuroscience*, advance online publication.

Aponte, Y., Atasoy, D., et al. (2011). AGRP neurons are sufficient to orchestrate feeding behavior rapidly and without training. *Nature Neuroscience*, *14*(3), 351–355.

Aronoff, R., & Petersen, C. C. (2006). Controlled and localized genetic manipulation in the brain. *Journal of Cellular and Molecular Medicine*, *10*(2), 333–352.

Ascoli, G. A., Alonso-Nanclares, L., et al. (2008). Petilla terminology: Nomenclature of features of GABAergic interneurons of the cerebral cortex. *Nature Reviews Neuroscience*, *9*(7), 557–568.

Atasoy, D., Aponte, Y., et al. (2008). A FLEX switch targets Channelrhodopsin-2 to multiple cell types for imaging and long-range circuit mapping. *The Journal of Neuroscience*, *28*(28), 7025–7030.

Ayling, O. G., Harrison, T. C., et al. (2009). Automated light-based mapping of motor cortex by photoactivation of channelrhodopsin-2 transgenic mice. *Nature Methods*, *6*(3), 219–224.

Banghart, M., Borges, K., et al. (2004). Light-activated ion channels for remote control of neuronal firing. *Nature Neuroscience*, *7*(12), 1381–1386.

Bartels, E., Wassermann, N. H., et al. (1971). Photochromic activators of the acetylcholine receptor. *Proceedings of the National Academy of Sciences of the United States of America*, *68*(8), 1820–1823.

Beja, O., Aravind, L., et al. (2000). Bacterial rhodopsin: Evidence for a new type of phototrophy in the sea. *Science*, *289*(5486), 1902–1906.

Berges, B. K., Wolfe, J. H., et al. (2007). Transduction of brain by herpes simplex virus vectors. *Molecular Therapy*, *15*(1), 20–29.

Boyden, E. S. (2011). A history of optogenetics: The development of tools for controlling brain circuits with light. *F1000 Biology Reports*, *3*, 11.

Boyden, E. S., Zhang, F., et al. (2005). Millisecond-timescale, genetically targeted optical control of neural activity. *Nature Neuroscience*, *8*(9), 1263–1268.

Bulina, M. E., Chudakov, D. M., et al. (2006). A genetically encoded photosensitizer. *Nature Biotechnology*, *24*(1), 95–99.

Bulina, M. E., Lukyanov, K. A., et al. (2006). Chromophore-assisted light inactivation (CALI) using the phototoxic fluorescent protein KillerRed. *Nature Protocols*, *1*(2), 947–953.

Butler, D. (2005). Science in the web age: Joint efforts. *Nature*, *438*, 548–549.

Cardin, J. A., Carlen, M., et al. (2010). Targeted optogenetic stimulation and recording of neurons in vivo using cell-type-specific expression of Channelrhodopsin-2. *Nature Protocols*, *5*(2), 247–254.

Carter, M. E., & de Lecea, L. (2011). Optogenetic investigation of neural circuits in vivo. *Trends in Molecular Medicine*, *17*(4), 197–206.

Cetin, A., Komai, S., et al. (2006). Stereotaxic gene delivery in the rodent brain. *Nature Protocols*, *1*(6), 3166–3173.

Chambers, J. J., Banghart, M. R., et al. (2006). Light-induced depolarization of neurons using a modified Shaker K(+) channel and a molecular photoswitch. *Journal of Neurophysiology*, *96*(5), 2792–2796.

Cho, J., Won Baac, H., et al. (2010). A 16-site neural probe integrated with a waveguide for optical stimulation. In: *IEEE 23rd International Conference on Micro Electro Mechanical Systems*, (pp. 995–998). Wanchai, Hong Kong, IEEE.

Chow, B. Y., Chuong, A. S., et al. (2011). Synthetic physiology strategies for adapting tools from nature for genetically targeted control of fast biological processes. *Methods in Enzymology*, *497*, 425–443.

Chow, B. Y., Han, X., et al. (2010). High-performance genetically targetable optical neural silencing by light-driven proton pumps. *Nature*, *463*(7277), 98–102.

Chudakov, D. M., Matz, M. V., et al. (2010). Fluorescent proteins and their applications in imaging living cells and tissues. *Physiological Reviews*, *90*(3), 1103–1163.

Cohen, L. B., Keynes, R. D., et al. (1968). Light scattering and birefringence changes during nerve activity. *Nature*, *218*(5140), 438–441.

Crick, F. H. (1979). Thinking about the brain. *Scientific American*, *241*(3), 219–232.

Crick, F. (1999). The impact of molecular biology on neuroscience. *Philosophical Transactions of the Royal Society of London. Series B, Biological Sciences*, *354*(1392), 2021–2025.

Cruikshank, S. J., Urabe, H., et al. (2010). Pathway-specific feedforward circuits between thalamus and neocortex revealed by selective optical stimulation of axons. *Neuron*, *65* (2), 230–245.

Davidson, B. L., & Breakefield, X. O. (2003). Viral vectors for gene delivery to the nervous system. *Nature Reviews Neuroscience*, *4*(5), 353–364.

Day, R. N., & Davidson, M. W. (2009). The fluorescent protein palette: Tools for cellular imaging. *Chemical Society Reviews*, *38*(10), 2887–2921.

Deisseroth, K. (2010). Controlling the brain with light. *Scientific American*, *303*(5), 48–55.

Deisseroth, K. (2011). Optogenetics. *Nature Methods*, *8*(1), 26–29.

Deisseroth, K., Feng, G., et al. (2006). Next-generation optical technologies for illuminating genetically targeted brain circuits. *The Journal of Neuroscience*, *26*(41), 10380–10386.

Denk, W., Strickler, J. H., et al. (1990). Two-photon laser scanning fluorescence microscopy. *Science*, *248*(4951), 73–76.

Deuschle, K., Okumoto, S., et al. (2005). Construction and optimization of a family of genetically encoded metabolite sensors by semirational protein engineering. *Protein Science*, *14*(9), 2304–2314.

Dhup, S., & Majumdar, S. S. (2008). Transgenesis via permanent integration of genes in repopulating spermatogonial cells in vivo. *Nature Methods*, *5*(7), 601–603.

Diester, I., Kaufman, M. T., et al. (2011). An optogenetic toolbox designed for primates. *Nature Neuroscience*, *14*(3), 387–397.

Diez-Garcia, J., Akemann, W., et al. (2007). In vivo calcium imaging from genetically specified target cells in mouse cerebellum. *Neuroimage*, *34*(3), 859–869.

Dymecki, S. M., Ray, R. S., et al. (2010). Mapping cell fate and function using recombinase-based intersectional strategies. *Methods in Enzymology*, *477*, 183–213.

Fan, M., Tsai, J., et al. (2005). A central repository for published plasmids. *Science*, *307*(5717), 1877.

Fenno, L., Yizhar, O., et al. (2011). The development and application of optogenetics. *Annual Review of Neuroscience*, *34*, 389–412.

Fletcher, M. L., Masurkar, A. V., et al. (2009). Optical imaging of postsynaptic odor representation in the glomerular layer of the mouse olfactory bulb. *Journal of Neurophysiology*, *102*(2), 817–830.

Fortin, D. L., Dunn, T. W., et al. (2011). Optogenetic photochemical control of designer K+ channels in mammalian neurons. *Journal of Neurophysiology*, *106*(1), 488–496.

Frommer, W. B., Davidson, M. W., et al. (2009). Genetically encoded biosensors based on engineered fluorescent proteins. *Chemical Society Reviews*, *38*(10), 2833–2841.

Ganesan, S., Ameer-Beg, S. M., et al. (2006). A dark yellow fluorescent protein (YFP)-based Resonance Energy-Accepting Chromoprotein (REACh) for Forster resonance energy transfer with GFP. *Proceedings of the National Academy of Sciences of the United States of America*, *103*(11), 4089–4094.

Geschwind, D. (2004). GENSAT: A genomic resource for neuroscience research. *Lancet Neurology*, *3*(2), 82.

Göbel, W., & Helmchen, F. (2007). In vivo calcium imaging of neural network function. *Physiology (Bethesda)*, *22*, 358–365.

Gorostiza, P., & Isacoff, E. (2007). Optical switches and triggers for the manipulation of ion channels and pores. *Molecular BioSystems*, *3*(10), 686–704.

Gorostiza, P., & Isacoff, E. Y. (2008). Optical switches for remote and noninvasive control of cell signaling. *Science*, *322*(5900), 395–399.

Govorunova, E. G., Spudich, E. N., et al. (2011). New channelrhodopsin with a red-shifted spectrum and rapid kinetics from Mesostigma viride. *MBio*, *2*(3), e00115–11.

Gradinaru, V., Mogri, M., et al. (2009). Optical deconstruction of parkinsonian neural circuitry. *Science*, *324*(5925), 354–359.

Gradinaru, V., Thompson, K. R., et al. (2007). Targeting and readout strategies for fast optical neural control in vitro and in vivo. *The Journal of Neuroscience*, *27*(52), 14231–14238.

Gradinaru, V., Zhang, F., et al. (2010). Molecular and cellular approaches for diversifying and extending optogenetics. *Cell*, *141*(1), 154–165.

Grant, D. M., Zhang, W., et al. (2008). Multiplexed FRET to image multiple signaling events in live cells. *Biophysical Journal*, *95*(10), L69–L71.

Grossman, N., Poher, V., et al. (2010). Multi-site optical excitation using ChR2 and micro-LED array. *Journal of Neural Engineering*, *7*(1), 16004.

Guo, Z. V., Hart, A. C., et al. (2009). Optical interrogation of neural circuits in Caenorhabditis elegans. *Nature Methods*, *6* (12), 891–896.

Gutierrez, D. V., Mark, M. D., et al. (2011). Optogenetic control of motor coordination by Gi/o protein-coupled vertebrate rhodopsin in cerebellar Purkinje cells. *Journal of Biological Chemistry*, *286*(29), 25848–25858.

Halassa, M. M., Siegle, J. H., et al. (2011). Selective optical drive of thalamic reticular nucleus generates thalamic bursts and cortical spindles. *Nature Neuroscience*, *14*(9), 1118–1120.

Han, X., Chow, B. Y., et al. (2011). A high-light sensitivity optical neural silencer: Development, and application to optogenetic control of nonhuman primate cortex. *Frontiers in Systems Neuroscience*, *5*, 18.

Han, X., Qian, X., et al. (2009). Millisecond-timescale optical control of neural dynamics in the nonhuman primate brain. *Neuron*, *62*(2), 191–198.

Haubensak, W., Kunwar, P. S., et al. (2010). Genetic dissection of an amygdala microcircuit that gates conditioned fear. *Nature*, *468*(7321), 270–276.

Hegemann, P., & Moglich, A. (2011). Channelrhodopsin engineering and exploration of new optogenetic tools. *Nature Methods*, *8*(1), 39–42.

Heintz, N. (2004). Gene expression nervous system atlas (GENSAT). *Nature Neuroscience*, *7*(5), 483.

Huber, D., Petreanu, L., et al. (2008). Sparse optical microstimulation in barrel cortex drives learned behaviour in freely moving mice. *Nature*, *451*(7174), 61–64.

Inouye, S., Noguchi, M., et al. (1985). Cloning and sequence analysis of cDNA for the luminescent protein aequorin. *Proceedings of the National Academy of Sciences of the United States of America*, *82*(10), 3154–3158.

Iwai, Y., Honda, S., et al. (2011). A simple head-mountable LED device for chronic stimulation of optogenetic molecules in freely moving mice. *Neuroscience Research*, *70*(1), 124–127.

Janovjak, H., Szobota, S., et al. (2010). A light-gated, potassium-selective glutamate receptor for the optical inhibition of neuronal firing. *Nature Neuroscience*, *13*(8), 1027–1032.

Jerome, J., Foehring, R. C., et al. (2011). Parallel optical control of spatiotemporal neuronal spike activity using high-speed digital light processing. *Frontiers in Systems Neuroscience*, *5*, 70.

Johnson, I., & Spence, M. T. Z. (2010). *The Molecular Probes Handbook: A guide to fluorescent probes and labeling technologies* (11th ed.). Eugene, Oregon: Life Technologies.

Judkewitz, B., Rizzi, M., et al. (2009). Targeted single-cell electroporation of mammalian neurons in vivo. *Nature Protocols*, *4*(6), 862–869.

Kami, C., Lorrain, S., et al. (2010). Light-regulated plant growth and development. *Current Topics in Developmental Biology*, *91*, 29–66.

Kaneda, K., Kasahara, H., et al. (2011). Selective optical control of synaptic transmission in the subcortical visual pathway by activation of viral vector-expressed halorhodopsin. *PLoS One*, *6*(4), e18452.

Kawai, Y., Sato, M., et al. (2004). Single color fluorescent indicators of protein phosphorylation for multicolor imaging of intracellular signal flow dynamics. *Analytical Chemistry*, *76*(20), 6144–6149.

Kennedy, M. J., Hughes, R. M., et al. (2010). Rapid blue-light-mediated induction of protein interactions in living cells. *Nature Methods*, *7*(12), 973–975.

Khorana, H. G., Knox, B. E., Nasi, E., Swanson, R., & Thompson, D. A. (1988). Expression of a bovine rhodopsin gene in Xenopus oocytes: demonstration of light-dependent ionic currents. *PNAS*, *85*(21), 7917–7921.

Kim, J. M., Hwa, J., et al. (2005). Light-driven activation of beta 2-adrenergic receptor signaling by a chimeric rhodopsin containing the beta 2-adrenergic receptor cytoplasmic loops. *Biochemistry*, *44*(7), 2284–2292.

Klapoetke, N., Chuong, A., et al. (2010). Novel classes of optogenetic reagent derived from screening genomic and ecological diversity. In: *2010 Society for Neuroscience Meeting*, San Diego, CA: Society for Neuroscience Program No. 106.1/MMM8.

Kleinlogel, S., Terpitz, U., Legrum, B., Gökbuget, D., Boyden, E. S., Bamann, C., et al. (2011). A gene-fusion strategy for stoichiometric and co-localized expression of light-gated membrane proteins. *Nature Methods*, *8*(12), 1083–1088.

Knöpfel, T., Diez-Garcia, J., et al. (2006). Optical probing of neuronal circuit dynamics: Genetically encoded versus classical fluorescent sensors. *Trends in Neurosciences*, *29*(3), 160–166.

Kravitz, A. V., & Kreitzer, A. C. (2011). Optogenetic manipulation of neural circuitry in vivo. *Current Opinion in Neurobiology*, *21*(3), 433–439.

Kubota, Y., Shigematsu, N., et al. (2011). Selective coexpression of multiple chemical markers defines discrete populations of neocortical GABAergic neurons. *Cerebral Cortex*, *21*(8), 1803–1817.

Kuhlman, S. J., & Huang, Z. J. (2008). High-resolution labeling and functional manipulation of specific neuron types in mouse brain by Cre-activated viral gene expression. *PLoS One*, *3*(4), e2005.

Lalonde, S., Ehrhardt, D. W., et al. (2005). Shining light on signaling and metabolic networks by genetically encoded biosensors. *Current Opinion in Plant Biology*, *8* (6), 574–581.

Le Provost, F., Lillico, S., et al. (2010). Zinc finger nuclease technology heralds a new era in mammalian transgenesis. *Trends in Biotechnology*, *28*(3), 134–141.

LeChasseur, Y., Dufour, S., et al. (2011). A microprobe for parallel optical and electrical recordings from single neurons in vivo. *Nature Methods*, *8*(4), 319–325.

Lee, J., Natarajan, M., et al. (2008). Surface sites for engineering allosteric control in proteins. *Science*, *322*(5900), 438–442.

Lee, J., Udugamasooriya, D. G., et al. (2010). Potent and selective photo-inactivation of proteins with peptoid-ruthenium conjugates. *Nature Chemical Biology*, *6*(4), 258–260.

Levskaya, A., Weiner, O. D., et al. (2009). Spatiotemporal control of cell signalling using a light-switchable protein interaction. *Nature*, *461*(7266), 997–1001.

Li, X., Gutierrez, D. V., et al. (2005). Fast noninvasive activation and inhibition of neural and network activity by vertebrate rhodopsin and green algae channelrhodopsin. *Proceedings of the National Academy of Sciences of the United States of America*, *102*(49), 17816–17821.

Lima, S. Q., Hromadka, T., et al. (2009). PINP: A new method of tagging neuronal populations for identification during in vivo electrophysiological recording. *PLoS One*, *4*(7), e6099.

Lima, S. Q., & Miesenböck, G. (2005). Remote control of behavior through genetically targeted photostimulation of neurons. *Cell*, *121*(1), 141–152.

Lin, J. Y. (2011). A user's guide to channelrhodopsin variants: Features, limitations and future developments. *Experimental Physiology*, *96*(1), 19–25.

Lin, D., Boyle, M. P., et al. (2011). Functional identification of an aggression locus in the mouse hypothalamus. *Nature*, *470*(7333), 221–226.

Liu, B., Paton, J. F., et al. (2008). Viral vectors based on bidirectional cell-specific mammalian promoters and transcriptional amplification strategy for use in vitro and in vivo. *BMC Biotechnology*, *8*, 49.

Losi, A., & Gärtner, W. (2012). The Evolution of Flavin-Binding Photoreceptors: An Ancient Chromophore Serving Trendy Blue-Light Sensors. *Annual Review of Plant Biology*, *63*, 1.1–1.24.

Lowery, R. L., & Majewska, A. K. (2010). Intracranial injection of adeno-associated viral vectors. *Journal of Visualized Experiments*, (45), e2140.

Lütcke, H., Murayama, M., et al. (2010). Optical recording of neuronal activity with a genetically-encoded calcium indicator in anesthetized and freely moving mice. *Front Neural Circuits*, *4*, 9.

Markvicheva, K. N., Bilan, D. S., et al. (2011). A genetically encoded sensor for H2O2 with expanded dynamic range. *Bioorganic & Medicinal Chemistry*, *19*(3), 1079–1084.

Masamizu, Y., Okada, T., et al. (2011). Local and retrograde gene transfer into primate neuronal pathways via adeno-associated virus serotype 8 and 9. *Neuroscience*, *193*, 249–258.

Masseck, O. A., Rubelowski, J. M., et al. (2011). Light- and drug-activated G-protein-coupled receptors to control intracellular signalling. *Experimental Physiology*, *96*(1), 51–56.

Melyan, Z., Tarttelin, E. E., et al. (2005). Addition of human melanopsin renders mammalian cells photoresponsive. *Nature*, *433*(7027), 741–745.

Miesenböck, G. (2009). The optogenetic catechism. *Science*, *326*(5951), 395–399.

Miesenböck, G. (2011). Optogenetic control of cells and circuits. *Annual Review of Cell and Developmental Biology*, *27*, 731–758.

Möglich, A., Ayers, R. A., et al. (2009). Design and signaling mechanism of light-regulated histidine kinases. *Journal of Molecular Biology*, *385*(5), 1433–1444.

Möglich, A., & Moffat, K. (2010). Engineered photoreceptors as novel optogenetic tools. *Photochemical and Photobiological Sciences*, *9*(10), 1286–1300.

Monahan, P. E., & Samulski, R. J. (2000). Adeno-associated virus vectors for gene therapy: More pros than cons? *Molecular Medicine Today*, *6*(11), 433–440.

Mukohata, Y., Ihara, K., et al. (1999). Halobacterial rhodopsins. *Journal of Biochemistry*, *125*(4), 649–657.

Muller, V., & Oren, A. (2003). Metabolism of chloride in halophilic prokaryotes. *Extremophiles*, *7*(4), 261–266.

Murayama, M., & Larkum, M. E. (2009). In vivo dendritic calcium imaging with a fiberoptic periscope system. *Nature Protocols*, *4*(10), 1551–1559.

Murayama, M., Perez-Garci, E., et al. (2007). Fiberoptic system for recording dendritic calcium signals in layer 5 neocortical pyramidal cells in freely moving rats. *Journal of Neurophysiology*, *98*(3), 1791–1805.

Nagahama, T., Suzuki, T., et al. (2007). Functional transplant of photoactivated adenylyl cyclase (PAC) into Aplysia sensory neurons. *Neuroscience Research*, *59*(1), 81–88.

Nagel, G., Ollig, D., et al. (2002). Channelrhodopsin-1: A light-gated proton channel in green algae. *Science*, *296* (5577), 2395–2398.

Nagel, G., Szellas, T., et al. (2003). Channelrhodopsin-2, a directly light-gated cation-selective membrane channel. *Proceedings of the National Academy of Sciences of the United States of America*, *100*(24), 13940–13945.

Nagel, G., Szellas, T., et al. (2005). Channelrhodopsins: Directly light-gated cation channels. *Biochemical Society Transactions*, *33*(Pt. 4), 863–866.

Nakai, J., Ohkura, M., et al. (2001). A high signal-to-noise Ca (2+) probe composed of a single green fluorescent protein. *Nature Biotechnology*, *19*(2), 137–141.

Nerbonne, J. M. (1996). Caged compounds: Tools for illuminating neuronal responses and connections. *Current Opinion in Neurobiology*, *6*(3), 379–386.

Nern, A., Pfeiffer, B. D., Svoboda, K., & Rubin, G. M. (2011). Multiple new site-specific recombinases for use in manipulating animal genomes. *PNAS*, *108*(34), 14198–14203.

Niino, Y., Hotta, K., et al. (2009). Simultaneous live cell imaging using dual FRET sensors with a single excitation light. *PLoS One*, *4*(6), e6036.

Niino, Y., Hotta, K., et al. (2010). Blue fluorescent cGMP sensor for multiparameter fluorescence imaging. *PLoS One*, *5* (2), e9164.

Oh, E., Maejima, T., et al. (2010). Substitution of 5-HT1A receptor signaling by a light-activated G protein-coupled receptor. *Journal of Biological Chemistry*, *285*(40), 30825–30836.

Okumoto, S. (2010). Imaging approach for monitoring cellular metabolites and ions using genetically encoded biosensors. *Current Opinion in Biotechnology*, *21*(1), 45–54.

Okumoto, S., Takanaga, H., et al. (2008). Quantitative imaging for discovery and assembly of the metabo-regulome. *New Phytologist*, *180*(2), 271–295.

Osakada, F., Mori, T., et al. (2011). New rabies virus variants for monitoring and manipulating activity and gene expression in defined neural circuits. *Neuron*, *71*(4), 617–631.

Ouyang, M., Huang, H., et al. (2010). Simultaneous visualization of protumorigenic Src and MT1-MMP activities with fluorescence resonance energy transfer. *Cancer Research, 70* (6), 2204–2212.

Papagiakoumou, E., Anselmi, F., et al. (2010). Scanless two-photon excitation of channelrhodopsin-2. *Nature Methods*, *7*(10), 848–854.

Passini, M. A., & Wolfe, J. H. (2001). Widespread gene delivery and structure-specific patterns of expression in the brain after intraventricular injections of neonatal mice with an adeno-associated virus vector. *Journal of Virology*, *75*(24), 12382–12392.

Pearson, H. (2006). Online methods share insider tricks. *Nature*, *441*, 678.

Perron, A., Mutoh, H., et al. (2009). Red-shifted voltage-sensitive fluorescent proteins. *Chemical Biology*, *16*(12), 1268–1277.

Peterka, D. S., Takahashi, H., et al. (2011). Imaging voltage in neurons. *Neuron*, *69*(1), 9–21.

Petreanu, L., Huber, D., et al. (2007). Channelrhodopsin-2-assisted circuit mapping of long-range callosal projections. *Nature Neuroscience*, *10*(5), 663–668.

Petreanu, L., Mao, T., et al. (2009). The subcellular organization of neocortical excitatory connections. *Nature*, *457* (7233), 1142–1145.

Petzold, G. C., Hagiwara, A., et al. (2009). Serotonergic modulation of odor input to the mammalian olfactory bulb. *Nature Neuroscience*, *12*(6), 784–791.

Pham, E., Mills, E., et al. (2011). A synthetic photoactivated protein to generate local or global ca(2+) signals. *Chemical Biology*, *18*(7), 880–890.

Piljic, A., & Schultz, C. (2008). Simultaneous recording of multiple cellular events by FRET. *ACS Chemical Biology*, *3*(3), 156–160.

Pilpel, N., Landeck, N., et al. (2009). Rapid, reproducible transduction of select forebrain regions by targeted recombinant virus injection into the neonatal mouse brain. *Journal of Neuroscience Methods*, *182*(1), 55–63.

Prasher, D. C., Eckenrode, V. K., et al. (1992). Primary structure of the Aequorea victoria green-fluorescent protein. *Gene*, *111*(2), 229–233.

Prasher, D., McCann, R. O., et al. (1985). Cloning and expression of the cDNA coding for aequorin, a bioluminescent calcium-binding protein. *Biochemical and Biophysical Research Communications*, *126*(3), 1259–1268.

Puntel, M., Kroeger, K. M., et al. (2010). Gene transfer into rat brain using adenoviral vectors. *Current Protocols in Neuroscience*, Chapter 4: Unit 4. 24. See: http://onlinelibrary.wiley.com/doi/10.1002/0471142301.ns0424s50/abstract.

Ridgway, E. B., & Ashley, C. C. (1967). Calcium transients in single muscle fibers. *Biochemical and Biophysical Research Communications*, *29*(2), 229–234.

Royer, S., Zemelman, B. V., et al. (2010). Multi-array silicon probes with integrated optical fibers: Light-assisted perturbation and recording of local neural circuits in the behaving animal. *The European Journal of Neuroscience*, *31*(12), 2279–2291.

Ryu, M. H., Moskvin, O. V., et al. (2010). Natural and engineered photoactivated nucleotidyl cyclases for optogenetic applications. *Journal of Biological Chemistry*, *285*(53), 41501–41508.

Saggau, P. (2006). New methods and uses for fast optical scanning. *Current Opinion in Neurobiology*, *16*(5), 543–550.

Schroder-Lang, S., Schwarzel, M., et al. (2007). Fast manipulation of cellular cAMP level by light in vivo. *Nature Methods*, *4*(1), 39–42.

Serebrovskaya, E. O., Gorodnicheva, T. V., et al. (2011). Light-induced blockage of cell division with a chromatin-targeted phototoxic fluorescent protein. *Biochemistry Journal*, *435*(1), 65–71.

Sherrington, C. S. S. (1940). *Man on his nature, O.M. The Gifford lectures, Edinburgh, 1937–8.* Cambridge: Cambridge University Press.

Shimizu-Sato, S., Huq, E., et al. (2002). A light-switchable gene promoter system. *Nature Biotechnology*, *20*(10), 1041–1044.

Souslova, E. A., & Chudakov, D. M. (2007). Genetically encoded intracellular sensors based on fluorescent proteins. *Biochemistry (Mosc)*, *72*(7), 683–697.

Stierl, M., Stumpf, P., et al. (2011). Light modulation of cellular cAMP by a small bacterial photoactivated adenylyl cyclase, bPAC, of the soil bacterium Beggiatoa. *Journal of Biological Chemistry*, *286*(2), 1181–1188.

Strickland, D., Moffat, K., et al. (2008). Light-activated DNA binding in a designed allosteric protein. *Proceedings of the National Academy of Sciences of the United States of America*, *105*(31), 10709–10714.

Strickland, D., Yao, X., et al. (2010). Rationally improving LOV domain-based photoswitches. *Nature Methods*, *7*(8), 623–626.

Teh, C., Chudakov, D. M., et al. (2010). Optogenetic in vivo cell manipulation in KillerRed-expressing zebrafish transgenics. *BMC Developmental Biology*, *10*, 110.

Teschemacher, A. G., Wang, S., et al. (2005). Targeting specific neuronal populations using adeno- and lentiviral vectors: Applications for imaging and studies of cell function. *Experimental Physiology*, *90*(1), 61–69.

Tian, L., & Looger, L. L. (2008). Genetically encoded fluorescent sensors for studying healthy and diseased nervous systems. *Drug Discovery Today: Disease Models*, *5*(1), 27–35.

Toettcher, J. E., Gong, D., et al. (2011). Light-based feedback for controlling intracellular signaling dynamics. *Nature Methods*, *8*(10), 837–839.

Tyszkiewicz, A. B., & Muir, T. W. (2008). Activation of protein splicing with light in yeast. *Nature Methods*, *5*(4), 303–305.

Varga, V., Losonczy, A., et al. (2009). Fast synaptic subcortical control of hippocampal circuits. *Science*, *326*(5951), 449–453.

Vincent, P., Maskos, U., et al. (2006). Live imaging of neural structure and function by fibred fluorescence microscopy. *EMBO Reports*, *7*(11), 1154–1161.

Volgraf, M., Gorostiza, P., et al. (2006). Allosteric control of an ionotropic glutamate receptor with an optical switch. *Nature Chemical Biology*, *2*(1), 47–52.

Walantus, W., Castaneda, D., et al. (2007). In utero intraventricular injection and electroporation of E15 mouse embryos. *Journal of Visualized Experiments*, *6*, 239.

Waldrop, M. (2008). Science 2.0—Is open access science the future? *Scientific American*, May 2008, 68–73.

Waschuk, S. A., Bezerra, A. G. Jr., et al. (2005). Leptosphaeria rhodopsin: Bacteriorhodopsin-like proton pump from a eukaryote. *Proceedings of the National Academy of Sciences of the United States of America*, *102*(19), 6879–6883.

Wentz, C. T., Bernstein, J. G., et al. (2011). A wirelessly powered and controlled device for optical neural control of freely-behaving animals. *Journal of Neural Engineering*, *8*(4), 046021.

Wickersham, I. R., Finke, S., et al. (2007). Retrograde neuronal tracing with a deletion-mutant rabies virus. *Nature Methods*, *4*(1), 47–49.

Wickersham, I. R., Lyon, D. C., et al. (2007). Monosynaptic restriction of transsynaptic tracing from single, genetically targeted neurons. *Neuron*, *53*(5), 639–647.

Wilt, B. A., Burns, L. D., et al. (2009). Advances in light microscopy for neuroscience. *Annual Review of Neuroscience*, *32*, 435–506.

Witten, I. B., Steinberg, E. E., Davidson, T., Tye, K. M., Zalocusky, K., Brodsky, M., et al. (2011). Recombinase-driver rat lines for optogenetics: tools, techniques, and application to dopamine-mediated positive reinforcement. Submitted.

Wong, L. F., Goodhead, L., et al. (2006). Lentivirus-mediated gene transfer to the central nervous system: Therapeutic and research applications. *Human Gene Therapy*, *17*(1), 1–9.

Wu, Y. I., Frey, D., et al. (2009). A genetically encoded photoactivatable Rac controls the motility of living cells. *Nature*, *461*(7260), 104–108.

Yazawa, M., Sadaghiani, A. M., et al. (2009). Induction of protein-protein interactions in live cells using light. *Nature Biotechnology*, *27*(10), 941–945.

Ye, H., Daoud-El Baba, M., et al. (2011). A synthetic optogenetic transcription device enhances blood-glucose homeostasis in mice. *Science*, *332*(6037), 1565–1568.

Yizhar, O., Fenno, L., et al. (2011a). Microbial opsins: A family of single-component tools for optical control of neural activity. *Cold Spring Harbor Protocols*, *2011(1)*, top102.

Yizhar, O., Fenno, L. E., et al. (2011b). Optogenetics in neural systems. *Neuron*, *71*(1), 9–34.

Yizhar, O., Fenno, L. E., et al. (2011c). Neocortical excitation/inhibition balance in information processing and social dysfunction. *Nature*, *477*, 171–178.

Zemelman, B. V., Lee, G. A., et al. (2002). Selective photostimulation of genetically chARGed neurons. *Neuron*, *33*(1), 15–22.

Zemelman, B. V., Nesnas, N., et al. (2003). Photochemical gating of heterologous ion channels: Remote control over genetically designated populations of neurons. *Proceedings of the National Academy of Sciences of the United States of America*, *100*(3), 1352–1357.

Zhang, F., Gradinaru, V., et al. (2010). Optogenetic interrogation of neural circuits: Technology for probing mammalian brain structures. *Nature Protocols*, *5*(3), 439–456.

Zhang, J., Laiwalla, F., et al. (2009). Integrated device for optical stimulation and spatiotemporal electrical recording of neural activity in light-sensitized brain tissue. *Journal of Neural Engineering*, *6*(5), 055007.

Zhang, F., Prigge, M., et al. (2008). Red-shifted optogenetic excitation: A tool for fast neural control derived from Volvox carteri. *Nature Neuroscience*, *11*(6), 631–633.

Zhang, F., Wang, L. P., et al. (2007). Multimodal fast optical interrogation of neural circuitry. *Nature*, *446*(7136), 633–639.

Zhao, Y., Araki, S., et al. (2011). An expanded palette of genetically encoded Ca(2) indicators. *Science*, *333*(6051), 1888–1891.

Zorzos, A. N., Boyden, E. S., et al. (2010). Multiwaveguide implantable probe for light delivery to sets of distributed brain targets. *Optics Letters*, *35*(24), 4133–4135.

Zorzos, A. N., Dietrich, A., et al. (2009). Light-proof neural recording electrodes. In: *2009 Society for Neuroscience Meeting*, Chicago, Illinois: Society for Neuroscience Program No. 388.12/GG107.

Zuo, D., Mohr, S. E., et al. (2007). PlasmID: A centralized repository for plasmid clone information and distribution. *Nucleic Acids Research*, *35*(Database issue), D680–D684.

T. Knöpfel and E. Boyden (Eds.)
Progress in Brain Research, Vol. 196
ISSN: 0079-6123

CHAPTER 2

Optogenetic excitation of neurons with channelrhodopsins: Light instrumentation, expression systems, and channelrhodopsin variants

John Y. Lin*

Department of Pharmacology, University of California at San Diego, La Jolla, CA, USA

Abstract: Classically, temporally precise excitation of membrane potential in neurons within intact tissue can be achieved by direct electrical stimulation or indirect electrical stimulation induced by changing magnetic fields. Both of these approaches have a predetermined selectivity based on the biophysical properties of the nervous tissue and membrane in the region of the stimulation. A recent advance in selective excitation of neurons is the "optogenetic" approach utilizing channelrhodopsins (ChRs). By expressing the light-responsive ChR in neurons using cell-type selective promoters or other methods, specific neurons can be depolarized by light in a temporally precise manner with millisecond resolution even if their membrane biophysical properties are less favorable for electrical stimulation. In addition, ChRs can be used to depolarize nonneuronal cells in the nervous tissue, and to sustain depolarization over a prolonged period of time, both of which cannot be achieved with electrical or magnetic stimulations. To conduct an experiment with ChR, experimenters need to make the correct choices on the three main components to such an experiment: the expression system, the illumination source, and the ChR variant used. This chapter aims to provide some discussions on the current developments of these aspects of the experiments.

To express ChR in neurons, the common expression systems include viral vectors, *in utero* electroporation, and transgenic animals, each with their advantages and limitations regarding the cost, expression pattern, and the required effort. In terms of the instrumentation, an illumination source that is capable of providing the desired wavelength with high intensity is crucial for the success of the experiment. The important factors regarding the light source used include the cost, light density output, efficiency for fiber coupling for *in vivo* rodent experiments, and the available methods to control light intensity and onset/termination. The third component of the experiment is the choice of the appropriate variants of ChR. Many novel ChR variants with unique properties have been

*Corresponding author.
Tel.: +1-858-534-5268; Fax: +1-858-534-5270
E-mail: j8lin@ucsd.edu

DOI: 10.1016/B978-0-444-59426-6.00002-1

engineered, and it can be difficult for the experimenters to choose the right variant with the desired properties for their experiments, as some information necessary for the experimenter to make the right selection is often incomplete or unavailable. Currently, the available variants for neuroscientific research are wild-type ChR2, ChR2+H134R, ChETA, VChR1, SFO, ChD, ChEF, ChIEF, ChRGR, CatCh, and TC. The features and limitations of these different variants are presented here.

Lastly, this chapter will provide some suggestion for the future development of the light source, expression system, and the development of the "next" generation of ChRs.

Keywords: channelrhodopsin variants; optogenetics; channel biophysical properties; optical techniques; transgenetic expression; recombinant viral technology.

Information in the nervous systems is encoded in the pattern of action potentials and the subsequent release of neurotransmitters from presynaptic terminals. The ability to manipulate these activities in an effective and minimally intrusive manner has great potential to increase our understandings of the function and organization of circuitry of the nervous system. Classic approaches to achieve superthreshold excitation of neuronal membrane by external manipulation are based on direct electrical excitation with electrodes, indirect electrical excitation induced by changing magnetic field, and pharmacological and chemical manipulation. Pharmacological and chemical manipulations usually lack temporal resolution, especially *in vivo*. Both electrical approaches are highly effective and temporally precise in the induction of superthreshold excitation of neuronal membrane or the direct release of neurotransmitters. In addition, the extracellular fluid is a relative effective conductive medium for electric field. The major drawback of the electrical approaches is the inability to selectively excite specific neurons in the brain area stimulated, as the susceptibility of nervous tissue to electrical stimulation is predetermined by the biophysical properties and structures of the tissue in the region (e.g., myelination, channel expression, branching pattern, and fiber orientation and fiber termination relative to the electric field). It is unlikely that these properties can be altered without disrupting their physiological functions.

The optogenetic approach has provided an alternative way to selectively excite a genetically defined group of neurons with light regardless of their biophysical membrane properties by rendering these neurons light responsive, a property not observed in most native neurons. This approach of exciting neuronal membrane requires the expression of genetically encoded components, typically an exogenous membrane channel, and the incorporation of an engineered or endogenous light-responsive chemical to make these exogenous channels light sensitive, leading to light-induced depolarization of membrane potential. By using light to depolarize neurons, experimenters can control the intensity and location of the stimulating light with existing optical techniques and manipulate subcellular structures or populations of neurons with relative high temporal resolution.

Earlier attempts at optogenetic excitation use engineered membrane channels that incorporate designed chemicals that respond to light either through uncaging (Zemelman et al., 2003) or through photoisomerization (Volgraf et al., 2006). These approaches require the introduction of both the transgene and the exogenous chemicals, which increases the complexity of the experiments, as some of these chemicals are difficult to handle or introduce into the tissue. A more recent advance in optogenetic excitation of neuron is the use of microbial opsin channelrhodopsin-2 (ChR2; Boyden et al., 2005; Nagel et al., 2003), which incorporates all-*trans* retinal, a chemical that is endogenous to the mammalian nervous system, removing the need to introduce exogenous chemicals. Unlike most known microbial opsins that are vectorial membrane pumps that transport hydrogen or chloride ion across the membrane unidirectionally (Lanyi and Luecke, 2001), ChRs have membrane channel properties that allow the conduction of ions through the channel pore. Due to their channel-like properties, they are much more efficient at altering the membrane excitability of expressing cells than the microbial opsin pumps. As a channel, ChR is nonselectively permeable to sodium, potassium, proton, and calcium ions, and its activation in cerebrospinal fluid or physiological saline leads to depolarization of the membrane potential in most cells tested to date (Boyden et al., 2005; Nagel et al., 2003). When expressed in sufficient numbers on neuronal membrane, this can lead to superthreshold depolarization and triggering of an action potential when illuminated by light (Boyden et al., 2005). It is also possible to directly stimulate the release of neurotransmitters from presynaptic terminals (Petreanu et al., 2009). In most general experiments using ChRs, the experimenters would like to be able to induce rapid and strong depolarization of the membrane with minimal delay after the onset of illumination and terminate the depolarization rapidly after cessation of illumination. In addition, the response to light should be consistent with subsequent repetitive stimulation, with little desensitization or rundown (Lin, 2011). These are determined by the biophysical membrane properties of expressing cells, the expression level of the protein, the properties of the ChR variants used, and the properties of the stimulation light source used. As it is undesirable to change the biophysical properties of the cells, and to overexpress the protein at a high level that may lead to toxicity or altered membrane properties, the factors that the experimenters can control are the ChR variant used, expression system, expression level of the protein, and the hardware used to deliver light in the system of choice. In the following sections, I will provide some discussions on these aspects of optogenetic experiments.

Light-delivery instrumentation

The suitable light-delivery instrumentation for the optogenetic excitation would much depend on the experiment and vary greatly from *in vivo* to *ex vivo* experiments in different organisms. The typical factors are the physical size of the light source, light intensity that can be delivered to the specimen, size of illumination volume, methods to manipulate the intensity and onset/termination, and cost. The three common types of light sources are arc lamp, laser, and more recently LED, which all have their advantages and limitations. The following description is summarized in Table 1.

Arc lamp

Arc lamps are bright light sources with high radiant intensity and provide continuous wavelength output across the UV and visible spectra. However, experimenters should pay attention to the spectral peaks of the arc lamp used, as the different arc lamps (xenon, mercury, mercury-xenon, and metal halide) have distinct spectral profiles, with xenon arc lamp giving the most uniform spectral output across the visible light spectrum. The selection of desired wavelength can be achieved with interference filters or prism/diffraction grating systems. The control of light illumination is typically achieved with a mechanical shutter or filter wheel, with the opening/closing time of the shutter typically >3 ms, slower than the kinetics of most ChR variants. Galvanometer-based shuttering is also available for faster speed, but at higher cost. The slow speed of a mechanical shutter typically limits the frequency of reliable stimulation to <25 Hz. To control the intensity of the light output, neutral density filters can be used and continuous intensity control can

Table 1. The features and limitations of common light sources used in optogenetic excitation experiments with channelrhodopsin

Light source	Features	Limitations
Arc lamp	– Continuous light across visible wavelengths – Availability in most laboratories with fluorescence microscope	– Typically slower shuttering – Intensity control harder to achieve – Inefficient coupling into a submillimeter fiber – Large housing – Shorter lifetime
Laser	– Collimated – High brightness per area – Can be coupled efficiently into a submillimeter fiber	– Higher cost both for the laser and the accessories – One wavelength per laser – Limited wavelengths selection (some wavelength range not available) – Shuttering and intensity can be hard to control with some lasers
Light-emitting diode (LED)	– Low cost per LED – Long lifetime – Rapid "shuttering" through controlled current input – Intensity control through current input – LED coupled into a submillimeter optical fiber commercially available – Small size	– Require a good external triggered constant current supply – May require a heat sink (depending on experiment) – One wavelength per LED – Currently weaker light intensity compared to arc lamps and lasers when collimated

be achieved with a neutral density wheel. The lifetime of arc lamp is typically 200 (mercury) to 2000h (metal halide). Arc lamps are readily available in laboratories that routinely perform fluorescence microscopic imaging experiments and can be modified to excite ChR-expressing cells by adding a triggerable shutter (Zhang and Oertner, 2007). Patterned and rapid illumination of the specimen with an arc lamp under the microscope can be achieved by inserting a digital mirror device in an optical plane that is conjugate with the specimen plane (Leifer et al., 2011; Levskaya et al., 2009). The high radiant intensity and ease of coupling to focal lens makes arc lamp ideal light source to illuminate big region of interest, such as the whole organ expressing ChR. A limitation of the arc lamp is the large size of the housing and the inefficient light coupling into a submillimeter flexible fiber needed for behavioral experiments in a freely moving rodent. Overall, the arc lamp provides a bright light source for optogenetic excitation for multiple wavelengths but currently is more suitable for *ex vivo* preparations under the microscope.

Laser-based system

Lasers provide bright and highly collimated light that can be coupled efficiently into a small optical fiber due to its coherent nature (Adamantidis et al., 2007). High intensity over small illumination region is a major advantage over other light sources for photostimulation. The major limitation for the laser-based illumination system is its limited wavelength and higher cost, although recent advances in solid-state lasers have made lasers more affordable. Light can be shuttered mechanically or acousto-optically, depending on one's speed and budget requirements. Even with a mechanical shutter, the speed that can be achieved is faster than the corresponding shutter used with an arc lamp due to the smaller physical size of the shutter. With the use of a scanning galvanometer mirror, the laser light can be used to conduct point photostimulation of ChR for mapping experiments (Petreanu et al., 2009). The Ti-sapphire laser used in multiphoton imaging is capable stimulating ChRs (Andrasfalvy et al., 2010; Papagiakoumou et al., 2010; Rickgauer and Tank, 2009); however, the high cost of this laser and the modification required makes it difficult for most laboratories to establish a system dedicated to such optogenetic experiments. When the laser light is coupled into a lightweight submillimeter fiber, it is possible to conduct optogenetic experiments in freely moving rodents (Adamantidis et al., 2007; Aravanis et al., 2007).

Light-emitting diode-based system

Light-emitting diode (LED) technology has advanced rapidly in the past few years in the commercial sector, with increased brightness and efficiency of the LEDs engineered for lighting and electronic industry. The physical size of the LED can be very small (<1 mm × 1 mm for a surface emitting LED) compared to the arc lamp or laser and can be easily mounted directly on top of the brain of a freely moving rodent (Huber et al., 2008). To utilize the LED in an optogenetic experiment, a stable constant current supply that can achieve microsecond scale onset and termination in response to an external trigger input is desirable. The light intensity of an LED can also be easily controlled through changing the current amplitude, although the light output is not linear with the current input. LEDs are low cost compared to arc lamps and lasers, with a long lifetime if used correctly (10,000–100,000 h). Different wavelengths can be achieved easily and cost-effectively with different LEDs. One limitation of the LED is its susceptibility to overheating at sustained high current intensity. Commercial vendors supplying LED with heat sink, collimator, microscope adaptor, or even direct coupling to fiber are available; unfortunately, these components also increase the cost and size of the LED system. Currently, the light intensity of a LED is typically weaker than an arc lamp or laser at the same wavelength (often 3× to 10× weaker than a Xenon arc lamp with corresponding interference filter); however, the rapid development of LED technology may overcome this limitation in the near future.

Expression system

The different methods for expressing ChRs in particular neurons have varying efficiencies and limitations. In the cell culture model, there are many established methods of introducing transgenes into neurons of interest, whereas in the whole animal, the options are more limited, and the common approaches for ChR expression in animals are transgenic animals, viral vectors, and *in utero* electroporation. The key properties of each expression systems are cost, time, expression pattern required, targeting of cell type and the amount of labor required to generate these results.

Transgenic animals

In *Caenorhabditis elegans* (Nagel et al., 2005), zebrafish (Schoonheim et al., 2010), and the fruit fly (*Drosophila melanogaster*) (Schroll et al., 2006), transgenic animals are the most direct method of introducing ChRs into the neurons of interest. In *Drosophila* and the zebrafish, the Gal 4/UAS system allows for the introduction of ChRs under specific promoters in the cells (Schoonheim et al., 2010; Schroll et al., 2006). In the *C. elegans*, constructs with cell-type-specific promoters are available to target specific cell types (Nagel et al., 2005). It should be noted that all-*trans* retinal needs to be added externally in the *C. elegans* model and supplementation of all-*trans* retinal is also recommended in *Drosophila* to achieve sufficient depolarization.

In the mammalian system, both transgenic mice and rats expressing ChR2 under the Thy-1.2 promoter have been generated (Arenkiel et al., 2007; Tomita et al., 2009). Thy-1.2-driven ChR2 is randomly expressed in neurons of different lines, including layer V cortical neurons (Arenkiel et al., 2007). One feature of the Thy-1.2-driven expression of transgene is the increase in expression level as the animal matures (Feng et al., 2000), leading to very high expression of ChR2 in the adult mouse that allows for the transcranial stimulation with blue light (Drew et al., 2010). With transgenic animals, it is possible to achieve widespread expression of ChR in different locations in the brain. Generating transgenic mice and rats is typically slow, labor-intensive, and costly, and transgenic mice are more common than the corresponding transgenic rats. It is also difficult to predict whether the expression level of ChR under the specific promoter will be sufficient, making such projects a costly and risky investment. Another approach is to utilize Cre recombinase-expressing mice with a virus containing lox site flanked reversed ChR for targeted expression (Atasoy et al., 2008; Cardin et al., 2009). This technique allows the usage of a strong pan-neuronal promoter for ChR expression, while the selectivity is achieved by Cre-recombinase expression in the knock-in or BAC transgenic animal. This technique is currently only available in the mouse model.

Viral vector

With the viral vector approach, the recombinant virus carrying the ChR construct is injected into the region of interest, resulting in the expression of the protein in a selective or broad range of cells. The tropism of different viruses and serotypes should be noted when choosing the virus, as identical promoter and ChR variants with different viruses may result in different expression patterns (Nathanson et al., 2009). The two common types of virus used for the ChR experiments are HIV-based lentivirus and recombinant adeno-associated virus (rAAV). The generation and purification of both viral vectors typically takes 1–2 weeks, and the expression of the ChR typically >1 week, and often 3–4 weeks.

Lentivirus has the advantage of larger packaging capacity (6–10 kbp) compared to rAAV (Kumar et al., 2001). The larger packaging capacity suggests a longer cell-type specific promoter can be used compared to rAAV (Adamantidis et al., 2007). In our experience, achieving widespread uniform expression with lentivirus is difficult, as the expression level at the site of injection is usually the highest, and the expression level decreases away from the site of injection (Aronoff et al., 2010; Nathanson et al., 2009). Depending on the experiment, this localized infection can be used as a feature instead of a limitation. Damage to the core within the site of injection has been reported (Nathanson et al., 2009). Retrograde transport has been reported and can be improved with rabies virus glycoprotein (Kato et al., 2011; Wong et al., 2004).

rAAV is a safe, well-characterized virus due to its use in gene therapy. It is also suitable for long-term expression of transgenes. Its main limitation is its small packaging size (~4.7 kbp total), which limits the use of many cell-type-specific promoters. It is possible to achieve strong and widespread infection of neurons with one single injection of rAAV, possibly due to the smaller size of rAAV and higher titer that can be achieved (Aronoff et al., 2010; Nathanson et al., 2009). Recent methods to incorporate larger insert sizes into rAAV have been developed (Ghosh and Duan, 2007; Ghosh et al., 2008). Retrograde uptake and infection have been reported (Franich et al., 2008).

In utero ***electroporation***

In utero electroporation is a relatively safe approach, as it eliminates the use of biological infectious agents. *In utero* electroporation also bypasses the size limitation of the constructs associated with the viral approach, as most vectors designed for neuron transgene expression can be used. *In utero* electroporation is relatively low cost (low cost of consumable after initial instrumentation purchase). The limitations of the approach are the requirement of a higher level of surgical skill, and also the lower yield of surviving and expressing animals (Tabata and Nakajima, 2008). Targeted expression can be achieved by use of a cell-type-specific promoter, the placement of the electrodes (Borrell et al., 2005), and the choice of the age of the animal for the electroporation (Ajioka and Nakajima, 2005).

As described earlier, there is no single "perfect" expression system, and the optimal expression system is determined by the experiments.

Selection of ChR variants

Selection of correct ChR variants with the right properties influences the precision and efficiency of light-induced depolarization in the neurons. The recent engineering of several new variants has addressed some of the limitations of the earlier versions. The following seven properties of ChRs are crucial parameters for the experimenters to choose the most appropriate variant for their experiments.

1. *Channel conductance*: ChRs have low single-channel conductance compared to classical membrane channels (Feldbauer et al., 2009), making it necessary to express ChRs at high levels to achieve superthreshold membrane depolarization. Higher conductances would increase the level of depolarization per expression unit of protein.
2. *Ion selectivity*: All characterized ChRs are nonselectively permeant toward H^+, Na^+, K^+, and Ca^{2+} and have a reversal potential near 0 mV at physiological pH (see Fig. 1a; Berthold et al., 2008; Gunaydin et al., 2010; Lin et al., 2009a; Nagel et al., 2003; Tsunoda and Hegemann, 2009; Zhang et al., 2008). The maximal change in membrane potential that can be achieved with ChRs is its reversal potential of 0 mV if the cell has no basal "leak" current.
3. *Channel kinetics*: The opening and closing rates of ChRs are critical factors in achieving temporally precise manipulation of membrane potentials, as the speed of membrane potential change induced by ChRs is associated with both the channel kinetics and the intrinsic membrane properties of the cell (Grossman et al., 2011; Lin, 2011). Although the ideal kinetics of a ChR should be as fast as is physically allowed, it is impossible to increase the rate of channel opening and closing without sacrificing light sensitivity (Lin et al., 2009a). The opening rate of the channel is light intensity dependent, whereas the channel closure rate is light intensity independent (Fig. 1b; Ishizuka et al., 2006; Lin et al., 2009a). In addition to opening and closing rates of the channel, the rate of ChR desensitization in response to continuous or repetitive illumination is another kinetic process of ChRs that is light intensity dependent (Ishizuka et al., 2006). Recent literatures have also indicated that the kinetics is temperature, voltage, and illumination duration dependent (Chater et al., 2010; Kleinlogel et al., 2011).
4. *Desensitization and the recovery of the desensitized component in the dark*: In the presence of continuous strong illumination, the response of most ChRs decays from a peak response to a steady-state (or stationary) response (Berndt et al., 2009, 2011; Gunaydin et al., 2010; Kleinlogel et al., 2011; Lin et al., 2009a; Nagel et al., 2002, 2003; Zhang et al., 2008). This desensitization of photocurrent is also observed with repetitive pulsed stimulation at high frequency (Lin et al., 2009a). The ideal ChR response would be consistent with no desensitization to both repetitive and prolonged light stimulation. Most of the recent "blue" ChR variants (ChEF, ChIEF, ChRGR, and CatCh) have focussed on the reduction of desensitization (Kleinlogel et al., 2011; Lin

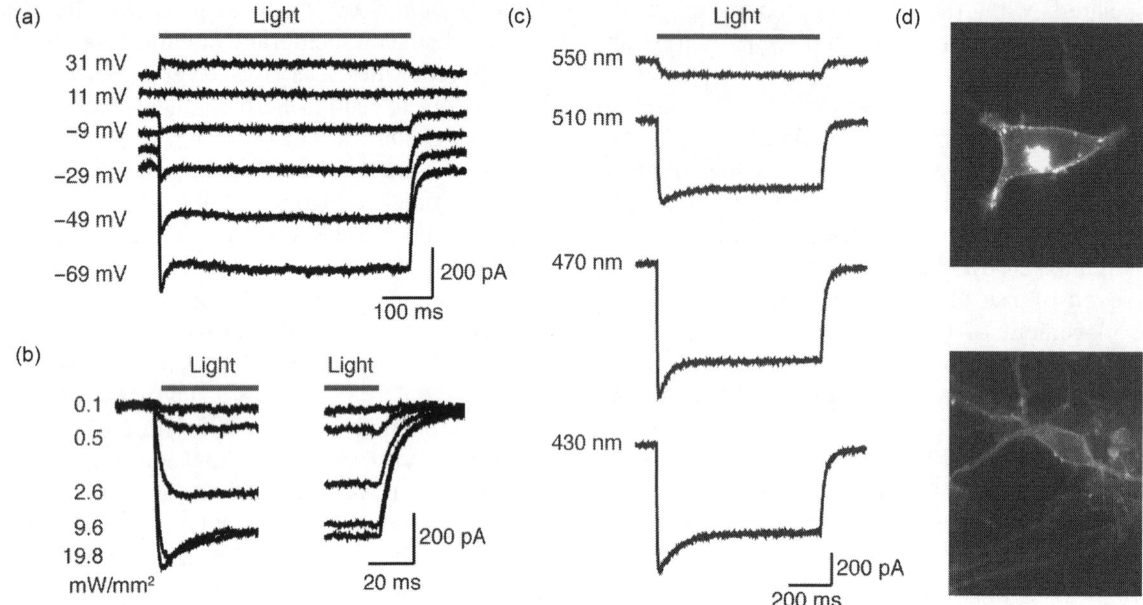

Fig. 1. Basic properties of channelrhodopsin illustrated with the ChIEF variant. (a) The ChIEF response at different holding potentials (−69, −49, −29, −9, 11, and 31mV) showing the reversal potential in physiological saline is ~10mV. (b) Light sensitivity and kinetics of ChIEF when activated with a 470-nm LED light source of varying light intensity (0.1, 0.5, 2.6, 9.6, and 19.8mW/mm^2). (c) The response of ChIEF to different wavelengths (550, 510, 470, and 430nm) of light with the same photon flux. (d) Expression and membrane trafficking of ChIEF in HEK293 cells (top) and cultured neuron (bottom). ChIEF has very efficient membrane trafficking and strong expression as observed with mCitrine (YFP) fluorescence fused to the C-terminal of the ChIEF. Blue (gray in print) bars indicate the light pulses. (For interpretation of the references to color in this figure legend, the reader is referred to the Web version of this chapter.)

et al., 2009a; Wen et al., 2010). The recovery of the desensitized component is an important issue in characterizing ChR variants correctly. With the exception of wild-type ChR2, the desensitized response of many of the ChRs does not recover fully after initial stimulation without "preconditioning" with light of another wavelength (Lin, 2011; Lin et al., 2009a,b; Schoenenberger et al., 2009). It should be noted that the rate of desensitization is also light intensity dependent (Lin et al., 2009a).

5. *Light sensitivity*: Different ChR variants have different light sensitivities, which influences the effectiveness of the available light-delivery devices in activating ChRs. Although ideally the light sensitivity of the ChRs should be high, they should not be activated by ambient light. In addition, increasing the light sensitivity of ChRs often has negative impacts on the kinetics of the channel (Berndt et al., 2009; Lin, 2011; Lin et al., 2009a). It should be noted that the light intensity–response curve of ChR is not linear (Ishizuka et al., 2006; Lin, 2011; Lin et al., 2009a; Wen et al., 2010) and care should be used when light intensity is used to adjust the level of depolarization or photocurrent (Berndt et al., 2011; Ishizuka et al., 2006; Zhang et al., 2008). It is also important to note that there are two methods of quantifying light sensitivity of ChR in the literature. The first one is based on the light intensity required to achieve full activation of the channel, independent of the channel desensitization, cell type or expression system (Fig. 1b; Lin, 2011; Lin et al., 2009a;

Wen et al., 2010). The second one is based on the amount of light intensity required to achieve action potential firing, which will vary with expression level (even from cell to cell), neuronal cell type, and level of desensitization, making it difficult to compare between different studies (Berndt et al., 2011; Kleinlogel et al., 2011).

6. *Spectral response*: Mutations around the retinal-binding pocket alter the spectral response of ChRs (Gunaydin et al., 2010; Lin et al., 2009a). Some ChR variants have different spectra for the fast peak and steady-state component of the photocurrents (e.g., see Fig. 1c; Lin, 2011; Lin et al., 2009a).
7. *Membrane trafficking/expression*: Not all ChRs express or traffic to the membrane well (Lin, 2011; Tsunoda and Hegemann, 2009). The ideal ChR would traffic to the membrane well with minimal intracellular aggregation, so the membrane can be depolarized more efficiently with minimal intracellular side effects (Fig. 1d). However, high levels of exogenous membrane protein expression can be toxic or have adverse effects on the membrane properties (Zimmermann et al., 2008). Careful tuning of expression is typically required to ensure sufficient depolarization without toxicity. The variation of expression levels in different cells in the same preparation would have very different effects on light-induced depolarization (see Fig. 2).

The example of the characterization of the ChR variant ChIEF is shown in Fig. 1. Recent studies have modeled several effects of these properties on using the ChRs to depolarize membrane with light (Grossman et al., 2011; Lin, 2011). It is important to note that many aspects of effects would be cell-type specific, as the membrane properties of different neurons and the level of achievable expression level can vary dramatically, especially *in vivo*. It is also important to note that the demonstration of the "maximum" rate of spiking triggered by pulsed light illustrated in the recent literatures of ChR variants should not be used as "absolute" value, but only as "relative" comparison to the variant that is directly compared, as the different experimental conditions can lead to very different results (Berndt et al., 2011; Gunaydin et al., 2010; Kleinlogel et al., 2011; Lin et al., 2009a).

Properties of ChR variants and their limitations

The following discussions of the properties and limitations of the published ChR variant are summarized in Table 2.

Wild-type ChR2 and ChR2+H134R

ChR2 is the first ChR used in experiments to excite neuronal membrane with light (Boyden et al., 2005). When stimulated with light of high intensity, ChR2 has a fast on-rate (can be fully activated <1.5ms) and channel closing rate ($\tau \sim 10$–13ms; Ishizuka et al., 2006; Lin et al., 2009a). ChR2 is maximally excited by blue light of 470nm (Ishizuka et al., 2006; Lin et al., 2009a; Nagel et al., 2003). The major shortcoming of ChR2 is the high level of desensitization even when stimulated at low light intensity at physiological pH (Ishizuka et al., 2006; Lin et al., 2009a; Nagel et al., 2003). This desensitization can lead to inconsistency in the light-induced responses with repetitive stimulation (Fig. 3). One potential solution is to use light stimulation of increased intensity to increase the photocurrent to partially compensate for the desensitization (Adesnik and Scanziani, 2010). The desensitized response can recover fully after 25s in the dark, and the recovery can be accelerated by 570nm light (Lin et al., 2009a), which coincides with maximal excitation wavelengths of VChR1 (Lin, 2011) and Halo/NpHR (Han and Boyden, 2007; Zhang et al., 2007). ChR2 traffics to the membrane well when expressed at low levels. When expressed at high levels, intracellular aggregates are typically observed (Lin, 2011).

ChR2 with the H134R mutation was the first published variant of ChR (Nagel et al., 2005). This variant has a modest reduction in desensitization

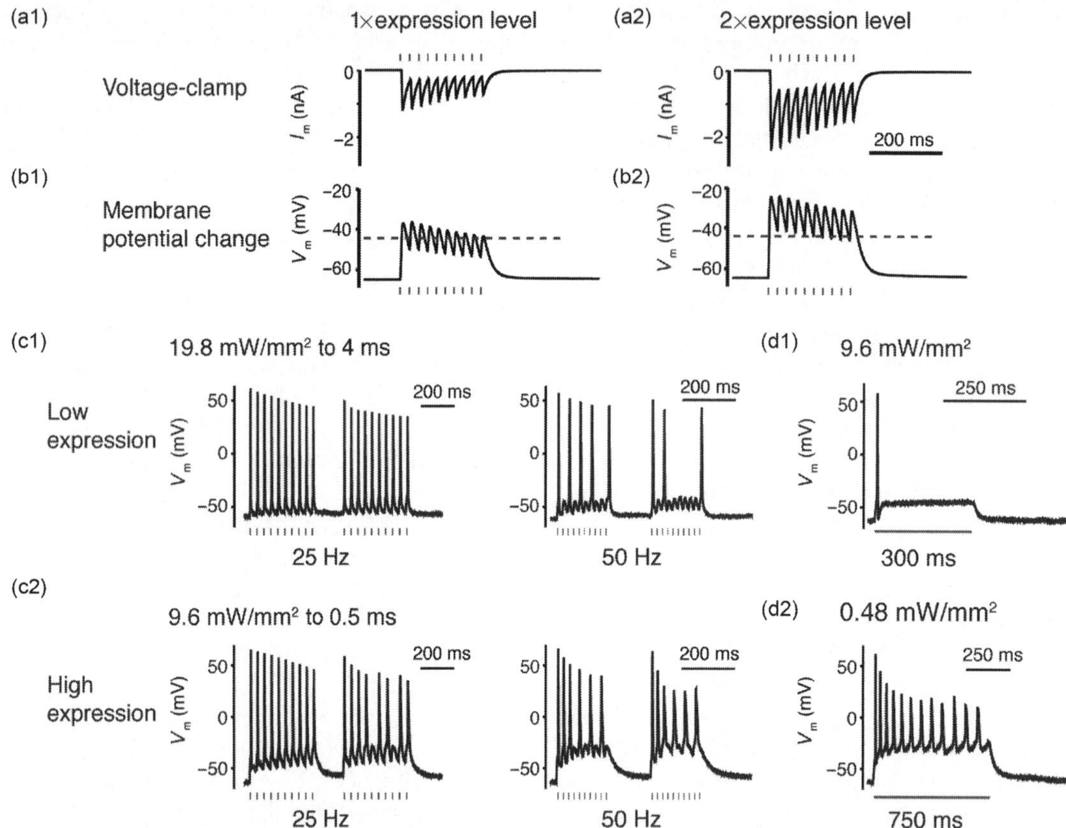

Fig. 2. The effect of expression level on the ability to trigger action potentials at high frequency with high fidelity. (a) A computer simulation of the light-induced membrane depolarization when stimulated with 10×4ms light pulses at 40Hz with a ChIEF-like ChR channel. The ChIEF-like ChR photo-responses are shown on top row (a1 and a2) and the corresponding membrane depolarization is shown below (b1 and b2). When the expression level of the ChR is increased 2× (a2 and b2), superthreshold depolarization is easily achieved by the light-induced depolarization, but there is insufficient repolarization between pulses. When the expression level is lowered (a1 and b1), superthreshold depolarization is harder to achieve consistently, but there is sufficient repolarization between pulses. Membrane parameters for the modeling: 100pF membrane capacitance, 50MΩ membrane resistance, resting potential −65mV, and hypothetical threshold of −45mV (dashed line). (c) The response of hippocampal neurons expressing low (top row) and high (bottom row) levels of ChIEF to pulsed blue light of 25 (left) and 50Hz (right). In neurons expressing low level of ChIEF (top row), strong light (19.8mW/mm^2) with longer duration (4ms) is required to achieve superthreshold depolarization, whereas in the high expresser (bottom row), weaker light (9.8mW/mm^2) with short duration (0.5ms) is sufficient to achieve superthreshold depolarization. Overexpression can easily lead to depolarization block resulting in loss of fidelity. (d) In response to long pulse of light, action potentials can be triggered easily with light of low intensity (d2), whereas light of higher intensity is required to triggered action potential in the low expresser (d1). Differential expression level is achieved with the inclusion of introns flanking ChIEF-fluorescent protein fusion in the identical vector. Blue (gray in print) bars indicate the light pulses. (For interpretation of the references to color in this figure legend, the reader is referred to the Web version of this chapter.)

(Fig. 3; decay to ~38% of the peak response), a slight increase in light sensitivity compared to ChR2, and slower kinetics compared to ChR2 (channel closing τ~18ms; Lin et al., 2009a). In many recent studies describing the use of ChR2 to excite neuronal membrane potential, ChR2 with H134R is used instead of the wild-type ChR2. There have been multiple claims of increased photocurrents or

Table 2. The features and limitations of the current published channelrhodopsin variants

Variant	Features	Limitations
ChR2, hChR2 (Boyden et al., 2005; Nagel et al., 2003)	– Original optogenetic tool – Extensive user experience and available resources – Thy-1.2 transgenic mice and rats available	– Strong desensitization
hChR2+H134R (Nagel et al., 2005)	– Early optogenetic tool – Extensive user experience and available resources	– Medium level desensitization – Slower kinetics than ChR2
VChR1 (Zhang et al., 2008)	– Wide spectral excitation from 600 to 400nm.	– Poor membrane trafficking and expression – Slow kinetics – Incomplete recovery of desensitized response
ChR2+E123T (ChETA) (Gunaydin et al., 2010)	– Fast kinetics – Small red shift in spectrum	– Reduced light sensitivity – Strong desensitization – "Smaller photocurrent" – Requirement of H134R of T159C mutation to achieve sufficient depolarization in neurons
ChD (Lin et al., 2009a)	– Fast kinetics – Consistent response to repetitive stimulation – Improved membrane trafficking	– Reduced light sensitivity – Strong desensitization to prolong light stimulation
ChEF (Lin et al., 2009a)	– Improved membrane trafficking – More light sensitive than ChR2 – Minimal desensitization with long and pulsed stimulation – Small red-shifted action spectrum	– Slower kinetics than ChR2 – Incomplete recovery of desensitized response
ChIEF (Lin et al., 2009a)	– Improved membrane trafficking – Minimal desensitization with long and pulsed stimulation	– Incomplete recovery of desensitized response – Light sensitivity lower than ChR2 (but better than ChETA and ChD)
ChRGR (Wen et al., 2010)	– Minimal desensitization – Fast kinetics – "Greater photocurrent" – Small red-shifted action spectra	– Light sensitivity unknown – Ion selectivity unknown
CatCh (Kleinlogel et al., 2011)	– Minimal desensitization to long stimulation—leading to higher sensitivity to light stimulation in neurons – Increased calcium and proton permeability	– Slower kinetic than ChR2
ChR2+T159C (TC) (Berndt et al., 2011) ChR2+ChETA+T159C (ET/TC) (Berndt et al., 2011)	– Greater photocurrent but unknown mechanism – Fast kinetics with reduced voltage dependency – Rapid kinetics in response to nanosecond light pulses – Greater photocurrent in oocytes	– Strong desensitization – Slower kinetics than ChR2 – Strong desensitization – Non-significant increase of photocurrent in neurons – Kinetics identical to ChR2 to long light pulse – Light sensitivity unknown
Step functional opsin (SFO), C128X, D156A (Bamann et al., 2010; Berndt et al., 2009)	– Sustained channel opening with short light pulse – Increased light sensitivity	– Slow kinetics – Small photocurrent – Control requires two wavelengths of light

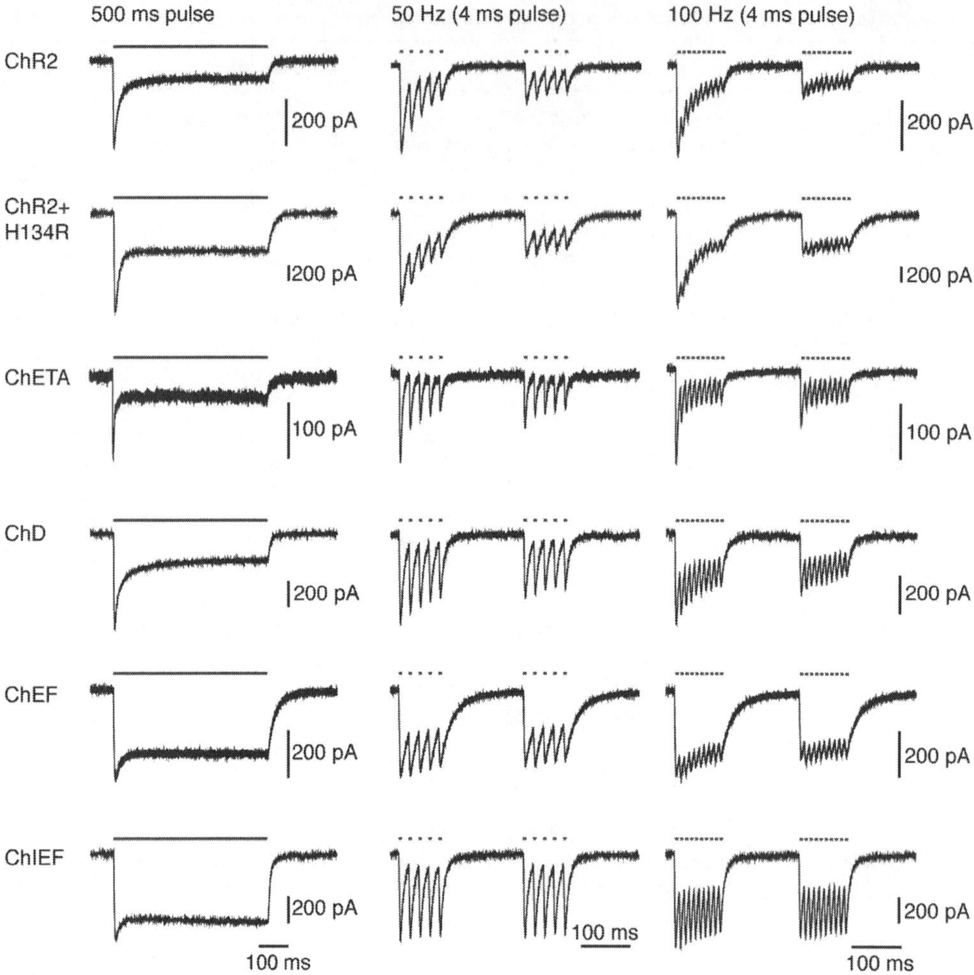

Fig. 3. The comparisons of light-induced ChR responses to long light pulses and pulsed stimulations at 50 and 100Hz. The comparisons of ChR photo-responses to 500ms 470nm light pulses (left column), 4ms 470nm light pulses at 50 (middle column) and 100Hz (right column). Each variant has different levels of desensitization and kinetics, with ChIEF and ChD outperform ChEF, ChR2, ChETA, and ChR2+H134R in all three conditions in terms of consistency and kinetics. Blue (gray in print) bars indicate the light pulses. Modified with permission from Lin et al. (2009a) and Lin (2011). (For interpretation of the references to color in this figure legend, the reader is referred to the Web version of this chapter.)

responses of this variant compared to the wild-type ChR2 (Hegemann and Moglich, 2011), but this is not associated with an increased in single-channel conductance (Feldbauer et al., 2009). The slower kinetics makes the H134R variant less temporally precise than ChR2 with high-frequency pulsed stimulation when overexpressed, with doublets of action potentials observed during a single brief stimulation pulse. Although this variant has been successfully used to depolarize neurons in multiple studies (Nagel et al., 2005; Zhang et al., 2007), the newer ChEF, ChIEF, and CatCh variants may provide more temporal precision and consistent depolarization (Kleinlogel et al., 2011; Lin et al., 2009a).

Step function opsin (SFO)/ChR2 with C128X (X=T/A/S), D156A

C128X (Berndt et al., 2009) or D156A mutations (Bamann et al., 2010) in ChR2 result in a channel that has increased light sensitivity and sustained opening after a single pulse of light stimulation. This is at the expenses of very slow kinetics (off-rate 2–100s) and reduced photocurrent of some of these variants. The first published study described these variants as "step-function opsins" and utilized these unique features to induce prolonged subthreshold level of depolarization as a "bistable" light switch (Berndt et al., 2009). However, it is difficult to control the depolarization in expressing neurons below threshold in most neuroscientific experiments, as the expression level of the protein varies greatly between cells. Short light pulses that do not fully activate the channels can achieve consistent subthreshold depolarization in most cells in culture, although it is not clear this strategy would work consistently *in vivo*. In addition, these variants have strong desensitization, further complicating the level of depolarization that can be achieved (Schoenenberger et al., 2009). To compensate for desensitization and slow kinetics, yellow light is needed to enhance recovery and speed up channel closure (Berndt et al., 2009; Schoenenberger et al., 2009). However, this increases the complexity of the instrumentation of such experiment especially for *in vivo* studies. Despite these complications, these variants may still be used in conditions where a prolonged level of depolarization is desired in cells, but the temporal precision and the level of depolarization are not critical (Schoenenberger et al., 2009).

ChR2+E123T(ChETA)/ChETA+H134R/ ChETA+T159C

ChR2 incorporating the E123T mutation (ChETA mutation) increased the channel closure rate of ChR2 (Gunaydin et al., 2010). Although this mutation was reported to have an increased steady state to peak response ratio compared to ChR2 (Gunaydin et al., 2010), this claim has not been reproduced (Fig. 3; Lin, 2011). This difference is most likely to be the result of much reduced light sensitivity of ChETA compared to ChR2, which can lead to the change of peak/steady-state ratio at non-saturating light intensities (Lin, 2011). The ChETA variant has reduced photocurrent compared to ChR2 and requires a strong promoter and the incorporation of other modifications (H134R or T159C) to achieve superthreshold depolarization (Berndt et al., 2011; Gunaydin et al., 2010). It is also important to note that the H134R or T159C mutations negate the faster kinetics of ChETA (Berndt et al., 2011). However, the reduced photocurrent can also be beneficial, as reduced photocurrent reduces the incidence of depolarization block often associated with overexpression of ChR2 under a strong promoter. This reduced photo-response may account for the higher fidelity at inducing action potentials at high frequencies stimulations (Gunaydin et al., 2010). ChETA+T159C (ET/TC) retains the kinetic features of ChETA+H134R and has slight (but statistically insignificant) increased photocurrent compared to ChR2 (Berndt et al., 2011). The kinetics of ET/TC is less voltage dependent and more consistent (Berndt et al., 2011). In response to nanosecond duration light pulses, ET/TC has channel kinetics that are 30% faster than ChR2 (in comparison, ChETA+H134R is 55% faster). It is unclear whether ChETA+H134R or ET/TC has any advantages over ChD (which has improved membrane trafficking and fast kinetics; Lin et al., 2009a), ChIEF (which has much reduced inactivation and improved membrane trafficking; Lin et al., 2009a), ChRGR (fast kinetics and reduced inactivation; Wen et al., 2010), or CatCh (which has higher efficiency at depolarizing the cell; Kleinlogel et al., 2011), as parallel comparisons in neurons have not been done in a systematic manner.

ChR2 with T159C (TC)

A recent publication with T159C in ChR2 increased the photocurrent up to 10-fold with no effects on the desensitization. The mechanism of this increase is currently unknown, but the small change in ion selectivity may partially account for the increase in photo-response (Berndt et al., 2011). The kinetics of T159C variant (off-rate~20ms) is slower than the ChR2, ChR2+H134R, ChD, ChIEF, and ChRGR. ChEF and CatCh have similar properties to this variant but reduced desensitization, and it would be interesting to compare these three variants systematically.

VChR1

VChR1 is a ChR variant from *Volvox cateri* that has a broad excitation spectrum from 610 to 410nm (Zhang et al., 2008). The peak response of this channel occurs at ~570nm and the steady-state response of this channel peaks at ~550nm (Lin, 2011; Fig. 3). In addition, VChR1 has the following limitations: (1) slow channel kinetics (channel off-rate~90ms), which can lead to temporally imprecise triggering of action potentials with pulsed light even at low frequencies (Zhang et al., 2008); (2) incomplete recovery of the desensitized response (Lin, 2011); and (3) poor expression and membrane trafficking (Lin, 2011; Tsunoda and Hegemann, 2009). Despite the initial demonstration that it is possible to stimulate two separate groups of neurons in the same preparation with VChR1 and ChR2 independently (Zhang et al., 2008), this experiment is difficult to conduct due to the strong spectral overlap with the blue/green ChRs and the varying expression levels (Lin, 2011). At its current state, the usefulness of VChR1 remains limited; however, VChR1 is the ideal template to engineer an improved red-shifted ChR. Future variants such as VCOMET (Lin, 2010) and C1V1 (Hegemann and Moglich, 2011) are likely to quickly replace the niche of this variant.

ChEF/ChIEF

ChEF and ChIEF are ChR variants engineered from the chimera of ChR1 and ChR2 (Lin et al., 2009a). Both ChEF and ChIEF are engineered for their increased steady-state phase responses without reduction of the peak photocurrent. ChEF/ChIEF variants give consistent photocurrent responses when stimulated with continuous light or repetitive light pulses at high frequencies (Figs. 2 and 3; Lin et al., 2009a). The kinetics of ChIEF is comparable to ChR2. ChIEF has slightly reduced light sensitivity relative to ChR2. ChEF has a slower kinetics and increased light sensitivity compared to ChR2 (Lin et al., 2009a). The desensitized response of ChEF/ChIEF photo-responses does not recovery fully in the dark, but the recovery can be enhanced by 570 nm light (Lin et al., 2009a). Despite the lack of complete recovery of the peak response, the increased steady-state response compensates for the desensitized responses and the reduced light sensitivity of ChIEF. ChEF and ChIEF express more highly than ChR2 in mammalian cells and traffic very efficiently to the membrane (Lin et al., 2009a; Wang et al., 2009). Although it is generally more desirable to have good membrane trafficking, it is also more likely to result in toxicity when are expressed at high level for prolonged period of expression (Karel Svoboda, personal communication). An effective approach to hinder the membrane trafficking of ChEF/ChIEF is by replacing the N-terminus with the corresponding residues from ChR2 (J.Y. Lin, unpublished observation). Overall, ChIEF is the theoretically the best ChR for conducting any ChR experiments at this time, as the response amplitudes are most consistent of any ChRs with millisecond timescale kinetics regardless of cell type, and is also the most fully characterized variant. In addition,

ChEF/ChIEF may be the best variant for use with weak cell-type-specific promoters (Fig. 2).

ChD

ChD is another variant generated by the chimeragenesis of ChR1 and ChR2 (Lin et al., 2009a; Wang et al., 2009). ChD is comparable to ChR2 with the ChETA mutation in kinetics, but with better membrane trafficking as with ChEF and ChIEF and slightly reduced desensitization at maximum activation compared to ChETA (Fig. 3; steady state ~30% of the peak response compared to ~24% of ChETA). It also produces more consistent responses with repetitive stimulation compared to ChETA without the incorporation of H134R or TC (Fig. 3; Lin, 2011). Theoretically, ChD is a good alternative to ChETA+H134R or ET/TC for high-frequency stimulation experiments although this requires experimental validation.

Channelrhodopsin green receiver (ChRGR)

ChRGR is another ChR1/ChR2 chimera-based variant that also retains ChR1-like properties and the higher conductance of ChR2 as with ChEF (Wen et al., 2010). ChRGR is reported to have a slightly red-shifted spectral peak, reduced desensitization and faster off-rate kinetics (The off-rate time constant is 0.6× of ChR2.). It is unclear whether the faster kinetics has negative impacts on the light sensitivity of this variant. The reduced desensitization and faster kinetics mean that the experimenter can control the amount of depolarization by adjusting the light intensity.

Calcium translocating channelrhodopsin (CatCh)

CatCh is based on ChR2 with the single L132C mutation which reduces the desensitization of the channel (Kleinlogel et al., 2011). Surprisingly, despite the increase (1.6×) in calcium and proton permeability compared to ChR2, the channel has identical channel conductance and reversal potential at physiological saline as ChR2. The light sensitivity of CatCh is identical to ChR2 (Kleinlogel et al., 2011), but due to its reduced desensitization, it yields greater overall photocurrent at the same light intensity compared to ChR2, a phenomenon that is commonly observed with variants that have reduced desensitization. Despite having a slower off-rate compared to ChR2, ChETA, and ChIEF, it is possible to stimulate cultured hippocampal neurons at higher frequency by utilizing the reduced desensitization of CatCh (Kleinlogel et al., 2011).

Unfortunately for the users, there have been very little attempts to systematically compare the performance of the various variants under the same testing conditions. So far, there are very few studies that characterize the variants thoroughly or compare the different variants systematically (Lin, 2011; Lin et al., 2009a). The performance of the different variants in neurons would also depend on the expression level of the protein and light intensity (Grossman et al., 2011; Lin, 2011). The "optimal" condition experiment demonstrated in many studies in neuronal cell culture may not translate well *in vivo*, with another promoter/viral system or the variable expression levels in the same preparation.

Future developments of optogenetic excitation of neurons

Illumination hardware and instrumentation

Improved LED technology is a promising way to stimulate ChR for optogenetic experiments. It already has the advantageous properties of low cost, small size, high-energy efficiency, versatility, long lifetime, and emission spectra that cover most of the visible spectrum. With ever-increasing brightness and energy efficiency, it is conceivable that LED light sources will become a good source to stimulate expressing neurons in freely moving animal. The use of lasers and arc lamps would likely continue their niche in optogenetic experiments, especially for experiments that are conducted under the microscope. The availability of off-the-shelf instrumentation would also benefit laboratories that do not have expertise in designing and constructing optical systems.

Expression system

Several aspects of the current expression system can be improved:

1. The identification of minimal size cell-type-specific promoters that can achieve sufficient levels of ChR expression. With all the approaches described in this chapter, the current limitation is our available information on cell-type-specific promoters and the level of expression that can be achieved with the promoter. Even the CaMKIIα promoter that was previously shown to be selective for excitatory neurons in the lentivirus (Dittgen et al., 2004) is nonselective in the context of AAV (Nathanson et al., 2009). The increasing availability of mouse lines with cell-type-specific expression of Cre recombinase will be beneficial, but the mouse model may not be suitable for all behavior experiments.
2. More efficient generation of transgenic animals at lower cost. New genetic techniques to generate transgenic animals quickly, efficiently, and cost-effectively would be an important step toward the use of the ChR technology to study different circuits and behaviors in species with more complex behavior. The development of techniques such as the zinc finger nuclease approach has the potential to generate transgenic animals more reliably and efficiently (Moehle et al., 2007).
3. Improvement of recombinant viral vector technology. The ability to generate virus with high infectivity and greater packaging capacity would increase the ability of the experimenter to target specific cell types. Recombinant baculovirus is potentially one of the candidates for further engineering and improvement (Kost and Condreay, 2002; Li et al., 2004).

Future developments of ChR variants

Development of new ChR variants would benefit from these improvements and developments:

1. Development of higher conductance variants. This will depolarize membrane potentials more efficiently even at low expression levels.
2. Development of ion-selective ChR. It would be desirable to achieve specific ion selectivity with high conductance for potassium, chloride, or sodium ions. This can be used to achieve more effective light-induced hyperpolarization with potassium or chloride ion-selective variants and more efficient depolarization with sodium-selective ChR.
3. Development of a ChR that is activated exclusively by red light. Despite the red shift of ChEF (490nm), ChETA (490nm), ET/TC (505nm), ChRGR (520nm), and VChR1 (530–570nm), these wavelengths have minimal effects in achieving deeper tissue penetration. Extending the spectra peak above 600nm (orange–red light) without needing overexpression would be an important development in ChR technology. A current variant of VCOMET has a spectral peak ~600nm and is a promising candidate to achieve efficient activation by red light (Lin et al., in preparation). Narrowing of the spectral response of these red variants would also be important for the stimulation of two cell groups independently or simultaneous calcium imaging.
4. Development of a reliable high-throughput assay for the activation of ChR variants. Current ChR engineering depends on the labor-intensive manual assay of variants created by individual point mutations. As with the development of

fluorescent proteins, the ability to quickly and effectively screen though the library of millions of variants generated through random or targeted mutagenesis and to select for the genetic variant with desired properties accurately would accelerate the engineering of novel variants.

Conclusions

The optogenetic excitation of neurons with ChRs holds tremendous potential for neuroscientists to explore neural circuitry underlying behavior of animals and in *ex vivo* preparations. The development and utilization of this technology will require the coordinated efforts of optical engineers, geneticists, virologists, neurophysiologists, and scientists from different disciplines. Understanding the limitation and properties of existing technology would ensure the experimenters conduct and interpret their experiments accurately and efficiently. The disclosure of limitations and resources to fellow scientists openly would be a key in the promotion of this technology.

Acknowledgements

J. Y. L. was funded by FRST postdoctoral fellowship from the Foundation of Research, Science and Technology of New Zealand. The project is supported by grants to Roger Y. Tsien from the National Institute of Health (NS027177) and Howard Hughes Medical Institute. Dr. Sharon Sann for editorial help.

Abbreviations

CatCh	calcium translocating channelrhodopsin
ChETA	chR2-E123T accelerated
ChR	channelrhodopsin
ChRGR	channelrhodopsin green receiver
LED	light-emitting diode
rAAV	recombinant adeno-associated virus
SFO	step function opsin

References

Adamantidis, A. R., Zhang, F., Aravanis, A. M., Deisseroth, K., & de Lecea, L. (2007). Neural substrates of awakening probed with optogenetic control of hypocretin neurons. *Nature, 450*, 420–424.

Adesnik, H., & Scanziani, M. (2010). Lateral competition for cortical space by layer-specific horizontal circuits. *Nature, 464*, 1155–1160.

Ajioka, I., & Nakajima, K. (2005). Birth-date-dependent segregation of the mouse cerebral cortical neurons in reaggregation cultures. *European Journal of Neuroscience, 22*, 331–342.

Andrasfalvy, B. K., Zemelman, B. V., Tang, J., & Vaziri, A. (2010). Two-photon single-cell optogenetic control of neuronal activity by sculpted light. *Proceedings of the National Academy of Sciences of the United States of America, 107*, 11981–11986.

Aravanis, A. M., Wang, L. P., Zhang, F., Meltzer, L. A., Mogri, M. Z., Schneider, M. B., et al. (2007). An optical neural interface: In vivo control of rodent motor cortex with integrated fiberoptic and optogenetic technology. *Journal of Neural Engineering, 4*, S143–S156.

Arenkiel, B. R., Peca, J., Davison, I. G., Feliciano, C., Deisseroth, K., Augustine, G. J., et al. (2007). In vivo light-induced activation of neural circuitry in transgenic mice expressing channelrhodopsin-2. *Neuron, 54*, 205–218.

Aronoff, R., Matyas, F., Mateo, C., Ciron, C., Schneider, B., & Petersen, C. C. (2010). Long-range connectivity of mouse primary somatosensory barrel cortex. *European Journal of Neuroscience, 31*, 2221–2233.

Atasoy, D., Aponte, Y., Su, H. H., & Sternson, S. M. (2008). A FLEX switch targets Channelrhodopsin-2 to multiple cell types for imaging and long-range circuit mapping. *Journal of Neuroscience, 28*, 7025–7030.

Bamann, C., Gueta, R., Kleinlogel, S., Nagel, G., & Bamberg, E. (2010). Structural guidance of the photocycle of channelrhodopsin-2 by an interhelical hydrogen bond. *Biochemistry, 49*, 267–278.

Berndt, A., Schoenenberger, P., Mattis, J., Tye, K. M., Deisseroth, K., Hegemann, P., et al. (2011). High-efficiency channelrhodopsins for fast neuronal stimulation at low light levels. *Proceedings of the National Academy of Sciences of the United States of America, 108*, 7595–7600.

Berndt, A., Yizhar, O., Gunaydin, L. A., Hegemann, P., & Deisseroth, K. (2009). Bi-stable neural state switches. *Nature Neuroscience, 12*, 229–234.

Berthold, P., Tsunoda, S. P., Ernst, O. P., Mages, W., Gradmann, D., & Hegemann, P. (2008). Channelrhodopsin-1 initiates phototaxis and photophobic responses in chlamydomonas by immediate light-induced depolarization. *The Plant Cell*, *20*, 1665–1677.

Borrell, V., Yoshimura, Y., & Callaway, E. M. (2005). Targeted gene delivery to telencephalic inhibitory neurons by directional in utero electroporation. *Journal of Neuroscience Methods*, *143*, 151–158.

Boyden, E. S., Zhang, F., Bamberg, E., Nagel, G., & Deisseroth, K. (2005). Millisecond-timescale, genetically targeted optical control of neural activity. *Nature Neuroscience*, *8*, 1263–1268.

Cardin, J. A., Carlen, M., Meletis, K., Knoblich, U., Zhang, F., Deisseroth, K., et al. (2009). Driving fast-spiking cells induces gamma rhythm and controls sensory responses. *Nature*, *459*, 663–667.

Chater, T. E., Henley, J. M., Brown, J. T., & Randall, A. D. (2010). Voltage- and temperature-dependent gating of heterologously expressed channelrhodopsin-2. *Journal of Neuroscience Methods*, *193*, 7–13.

Dittgen, T., Nimmerjahn, A., Komai, S., Licznerski, P., Waters, J., Margrie, T. W., et al. (2004). Lentivirus-based genetic manipulations of cortical neurons and their optical and electrophysiological monitoring in vivo. *Proceedings of the National Academy of Sciences of the United States of America*, *101*, 18206–18211.

Drew, P. J., Shih, A. Y., Driscoll, J. D., Knutsen, P. M., Blinder, P., Davalos, D., et al. (2010). Chronic optical access through a polished and reinforced thinned skull. *Nature Methods*, *7*, 981–984.

Feldbauer, K., Zimmermann, D., Pintschovius, V., Spitz, J., Bamann, C., & Bamberg, E. (2009). Channelrhodopsin-2 is a leaky proton pump. *Proceedings of the National Academy of Sciences of the United States of America*, *106*, 12317–12322.

Feng, G., Mellor, R. H., Bernstein, M., Keller-Peck, C., Nguyen, Q. T., Wallace, M., et al. (2000). Imaging neuronal subsets in transgenic mice expressing multiple spectral variants of GFP. *Neuron*, *28*, 41–51.

Franich, N. R., Fitzsimons, H. L., Fong, D. M., Klugmann, M., During, M. J., & Young, D. (2008). AAV vector-mediated RNAi of mutant huntingtin expression is neuroprotective in a novel genetic rat model of Huntington's disease. *Molecular Therapy*, *16*, 947–956.

Ghosh, A., & Duan, D. (2007). Expanding adeno-associated viral vector capacity: A tale of two vectors. *Biotechnology and Genetic Engineering Reviews*, *24*, 165–177.

Ghosh, A., Yue, Y., Lai, Y., & Duan, D. (2008). A hybrid vector system expands adeno-associated viral vector packaging capacity in a transgene-independent manner. *Molecular Therapy*, *16*, 124–130.

Grossman, N., Nikolic, K., Toumazou, C., & Degenaar, P. (2011). Modeling study of the light stimulation of a neuron cell with channelrhodopsin-2 mutants. *IEEE Transactions on Biomedical Engineering*, *58*, 1742–1751.

Gunaydin, L. A., Yizhar, O., Berndt, A., Sohal, V. S., Deisseroth, K., & Hegemann, P. (2010). Ultrafast optogenetic control. *Nature Neuroscience*, *13*, 387–392.

Han, X., & Boyden, E. S. (2007). Multiple-color optical activation, silencing, and desynchronization of neural activity, with single-spike temporal resolution. *PLoS One*, *2*, e299.

Hegemann, P., & Moglich, A. (2011). Channelrhodopsin engineering and exploration of new optogenetic tools. *Nature Methods*, *8*, 39–42.

Huber, D., Petreanu, L., Ghitani, N., Ranade, S., Hromadka, T., Mainen, Z., et al. (2008). Sparse optical microstimulation in barrel cortex drives learned behaviour in freely moving mice. *Nature*, *451*, 61–64.

Ishizuka, T., Kakuda, M., Araki, R., & Yawo, H. (2006). Kinetic evaluation of photosensitivity in genetically engineered neurons expressing green algae light-gated channels. *Neuroscience Research*, *54*, 85–94.

Kato, S., Kobayashi, K., Inoue, K., Kuramochi, M., Okada, T., Yaginuma, H., et al. (2011). A lentiviral strategy for highly efficient retrograde gene transfer by pseudotyping with fusion envelope glycoprotein. *Human Gene Therapy*, *22*, 197–206.

Kleinlogel, S., Feldbauer, K., Dempski, R. E., Fotis, H., Wood, P. G., Bamann, C., et al. (2011). Ultra light-sensitive and fast neuronal activation with the Ca(2+)-permeable channelrhodopsin CatCh. *Nature Neuroscience*, *14*, 513–518.

Kost, T. A., & Condreay, J. P. (2002). Recombinant baculoviruses as mammalian cell gene-delivery vectors. *Trends in Biotechnology*, *20*, 173–180.

Kumar, M., Keller, B., Makalou, N., & Sutton, R. E. (2001). Systematic determination of the packaging limit of lentiviral vectors. *Human Gene Therapy*, *12*, 1893–1905.

Lanyi, J. K., & Luecke, H. (2001). Bacteriorhodopsin. *Current Opinion in Structural Biology*, *11*, 415–419.

Leifer, A. M., Fang-Yen, C., Gershow, M., Alkema, M. J., & Samuel, A. D. (2011). Optogenetic manipulation of neural activity in freely moving Caenorhabditis elegans. *Nature Methods*, *8*, 147–152.

Levskaya, A., Weiner, O. D., Lim, W. A., & Voigt, C. A. (2009). Spatiotemporal control of cell signalling using a light-switchable protein interaction. *Nature*, *461*, 997–1001.

Li, Y., Wang, X., Guo, H., & Wang, S. (2004). Axonal transport of recombinant baculovirus vectors. *Molecular Therapy*, *10*, 1121–1129.

Lin, J. Y. (2010). Choosing the right channelrhodopsin variant for your neuroscientific experiment. In: *Society for Neuroscience Meeting Planner.* San Diego, CA.

Lin, J. Y. (2011). User's guide to channelrhodopsin variants: Features, limitations and future developments. *Experimental Physiology*, *96*, 19–25.

Lin, J. Y., Lin, M. Z., Steinbach, P., & Tsien, R. Y. (2009a). Characterization of engineered channelrhodopsin variants with improved properties and kinetics. *Biophysical Journal*, *96*, 1803–1814.

Lin, J. Y., Yang, J., & Tsien, R. Y. (2009b). Development and characterization of channelrhodopsin variants for improved optical control of neuronal excitability. In *2009 Neuroscience meeting planner.* Program No. 485.9. Chicago: Society for Neuroscience.

Lin, J. Y., Knutsen, P. M., Muller, A., Kleinfeld, D., & Tsien, R.Y. Novel channelrhodopsin variant for non-intrusive deep-tissue optogenetic excitation with red light. (In preparation).

Moehle, E. A., Rock, J. M., Lee, Y. L., Jouvenot, Y., DeKelver, R. C., Gregory, P. D., et al. (2007). Targeted gene addition into a specified location in the human genome using designed zinc finger nucleases. *Proceedings of the National Academy of Sciences of the United States of America*, *104*, 3055–3060.

Nagel, G., Brauner, M., Liewald, J. F., Adeishvili, N., Bamberg, E., & Gottschalk, A. (2005). Light activation of channelrhodopsin-2 in excitable cells of Caenorhabditis elegans triggers rapid behavioral responses. *Current Biology*, *15*, 2279–2284.

Nagel, G., Ollig, D., Fuhrmann, M., Kateriya, S., Musti, A. M., Bamberg, E., et al. (2002). Channelrhodopsin-1: A light-gated proton channel in green algae. *Science*, *296*, 2395–2398.

Nagel, G., Szellas, T., Huhn, W., Kateriya, S., Adeishvili, N., Berthold, P., et al. (2003). Channelrhodopsin-2, a directly light-gated cation-selective membrane channel. *Proceedings of the National Academy of Sciences of the United States of America*, *100*, 13940–13945.

Nathanson, J. L., Yanagawa, Y., Obata, K., & Callaway, E. M. (2009). Preferential labeling of inhibitory and excitatory cortical neurons by endogenous tropism of adeno-associated virus and lentivirus vectors. *Neuroscience*, *161*, 441–450.

Papagiakoumou, E., Anselmi, F., Begue, A., de Sars, V., Gluckstad, J., Isacoff, E. Y., et al. (2010). Scanless two-photon excitation of channelrhodopsin-2. *Nature Methods*, *7*, 848–854.

Petreanu, L., Mao, T., Sternson, S. M., & Svoboda, K. (2009). The subcellular organization of neocortical excitatory connections. *Nature*, *457*, 1142–1145.

Rickgauer, J. P., & Tank, D. W. (2009). Two-photon excitation of channelrhodopsin-2 at saturation. *Proceedings of the National Academy of Sciences of the United States of America*, *106*, 15025–15030.

Schoenenberger, P., Gerosa, D., & Oertner, T. G. (2009). Temporal control of immediate early gene induction by light. *PLoS One*, *4*, e8185.

Schoonheim, P. J., Arrenberg, A. B., Del Bene, F., & Baier, H. (2010). Optogenetic localization and genetic perturbation of saccade-generating neurons in zebrafish. *Journal of Neuroscience*, *30*, 7111–7120.

Schroll, C., Riemensperger, T., Bucher, D., Ehmer, J., Voller, T., Erbguth, K., et al. (2006). Light-induced activation of distinct modulatory neurons triggers appetitive or aversive learning in Drosophila larvae. *Current Biology*, *16*, 1741–1747.

Tabata, H., & Nakajima, K. (2008). Labeling embryonic mouse central nervous system cells by in utero electroporation. *Development, Growth & Differentiation*, *50*, 507–511.

Tomita, H., Sugano, E., Fukazawa, Y., Isago, H., Sugiyama, Y., Hiroi, T., et al. (2009). Visual properties of transgenic rats harboring the channelrhodopsin-2 gene regulated by the thy-1.2 promoter. *PLoS One*, *4*, e7679.

Tsunoda, S. P., & Hegemann, P. (2009). Glu 87 of channelrhodopsin-1 causes pH-dependent color tuning and fast photocurrent inactivation. *Photochemistry and Photobiology*, *85*, 564–569.

Volgraf, M., Gorostiza, P., Numano, R., Kramer, R. H., Isacoff, E. Y., & Trauner, D. (2006). Allosteric control of an ionotropic glutamate receptor with an optical switch. *Nature Chemical Biology*, *2*, 47–52.

Wang, H., Sugiyama, Y., Hikima, T., Sugano, E., Tomita, H., Takahashi, T., et al. (2009). Molecular determinants differentiating photocurrent properties of two channelrhodopsins from Chlamydomonas. *Journal of Biological Chemistry*, *284*, 5685–5696.

Wen, L., Wang, H., Tanimoto, S., Egawa, R., Matsuzaka, Y., Mushiake, H., et al. (2010). Opto-current-clamp actuation of cortical neurons using a strategically designed channelrhodopsin. *PLoS One*, *5*, e12893.

Wong, L. F., Azzouz, M., Walmsley, L. E., Askham, Z., Wilkes, F. J., Mitrophanous, K. A., et al. (2004). Transduction patterns of pseudotyped lentiviral vectors in the nervous system. *Molecular Therapy*, *9*, 101–111.

Zemelman, B. V., Nesnas, N., Lee, G. A., & Miesenbock, G. (2003). Photochemical gating of heterologous ion channels: Remote control over genetically designated populations of neurons. *Proceedings of the National Academy of Sciences of the United States of America*, *100*, 1352–1357.

Zhang, Y. P., & Oertner, T. G. (2007). Optical induction of synaptic plasticity using a light-sensitive channel. *Nature Methods*, *4*, 139–141.

Zhang, F., Prigge, M., Beyriere, F., Tsunoda, S. P., Mattis, J., Yizhar, O., et al. (2008). Red-shifted optogenetic excitation: A tool for fast neural control derived from Volvox carteri. *Nature Neuroscience*, *11*, 631–633.

Zhang, F., Wang, L. P., Brauner, M., Liewald, J. F., Kay, K., Watzke, N., et al. (2007). Multimodal fast optical interrogation of neural circuitry. *Nature*, *446*, 633–639.

Zimmermann, D., Zhou, A., Kiesel, M., Feldbauer, K., Terpitz, U., Haase, W., et al. (2008). Effects on capacitance by overexpression of membrane proteins. *Biochemical and Biophysical Research Communications*, *369*, 1022–1026.

T.Knöpfel and E.Boyden (Eds.)
Progress in Brain Research, Vol. 196
ISSN: 0079-6123

CHAPTER 3

Genetically encoded molecular tools for light-driven silencing of targeted neurons

Brian Y. Chow[†], Xue Han[‡] and Edward S. Boyden[§,¶,||,*]

[†] *Department of Bioengineering, University of Pennsylvania, Philadelphia, PA, USA*
[‡] *Department of Biomedical Engineering, Boston University, Boston, MA, USA*
[§] *MIT Media Lab, MIT, Cambridge, MA, USA*
[¶] *Department of Biological Engineering, MIT, Cambridge, MA, USA*
[||] *Department of Brain and Cognitive Sciences, MIT, Cambridge, MA, USA*

Abstract: The ability to silence, in a temporally precise fashion, the electrical activity of specific neurons embedded within intact brain tissue, is important for understanding the role that those neurons play in behaviors, brain disorders, and neural computations. "Optogenetic" silencers, genetically encoded molecules that, when expressed in targeted cells within neural networks, enable their electrical activity to be quieted in response to pulses of light, are enabling these kinds of causal circuit analyses studies. Two major classes of optogenetic silencer are in broad use in species ranging from worm to monkey: light-driven inward chloride pumps, or halorhodopsins, and light-driven outward proton pumps, such as archaerhodopsins and fungal light-driven proton pumps. Both classes of molecule, when expressed in neurons via viral or other transgenic means, enable the targeted neurons to be hyperpolarized by light. We here review the current status of these sets of molecules, and discuss how they are being discovered and engineered. We also discuss their expression properties, ionic properties, spectral characteristics, and kinetics. Such tools may not only find many uses in the quieting of electrical activity for basic science studies but may also, in the future, find clinical uses for their ability to safely and transiently shut down cellular electrical activity in a precise fashion.

Keywords: optogenetics; opsins; neural silencing; halorhodopsin; archaerhodopsin; channelrhodopsin; control; cell types; neural circuits; causality.

Introduction

The ability to silence, in a temporally precise fashion, the electrical activity of specific neurons embedded within intact brain tissue, is important

*Corresponding author.
Tel.: +617 324-3085
E-mail: eboyden3@gmail.com

DOI: 10.1016/B978-0-444-59426-6.00003-3

for understanding the role that those neurons play in behaviors, brain disorders, and neural computations. Over the past several years, we and others have discovered that a large number of light-driven ion pumps naturally occurring in archaea, fungi, and other species, can be genetically expressed in neurons, enabling them to be hyperpolarized in response to light. These molecules, microbial (type I) opsins, are seven-transmembrane proteins, which translocate specific ions from one side of the membrane in which they are expressed, to the other, in response to light. For example, halorhodopsins are light-driven inward chloride pumps, which, when expressed in neurons, enable them to be hyperpolarized by orange light (Fig. 1ai). Archaerhodopsins are light-driven outward proton pumps which, when expressed in neurons, also result in hyperpolarization (Fig. 1bi). These proteins are monolithic, and encoded for by relatively small genes, under a kilobase long, small enough that genetic delivery to specific cells (e.g., via viruses, transfection, electroporation, transgenesis, or other methods) is achievable using many conventional methods in animals commonly used in neuroscience, from *Caenorhabditis elegans* to nonhuman primate (Adamantidis et al., 2007; Atasoy et al., 2008; Bi et al., 2006; Chan et al., 2010; Chhatwal et al., 2007; Dittgen et al., 2004; Douglass et al., 2008; Han et al., 2009a,b; Huber et al., 2008; Ishizuka et al., 2006; Kuhlman and Huang, 2008; Lagali et al., 2008; Li et al., 2005; Nagel et al., 2005; Petreanu et al., 2007; Schroll et al., 2006; Tan et al., 2008; Toni et al., 2008; Wang et al., 2007; Zhang et al., 2006). Further, thanks to the proliferation of optical devices driven by microscopy innovation as well as from other fields (e.g., telecommunications and computing), many devices exist or are easily engineered for the delivery of light to specific brain regions, even deep within the mammalian brain (conveyed via implanted optical fibers or waveguides; e.g., a typical 200-μm diameter fiber with end-of-tip irradiance of 100–200 mW/mm^2 can illuminate a volume of a few cubic millimeters), and even in a wirelessly controlled fashion (Aravanis et al., 2007; Gradinaru et al., 2007; Wentz et al., 2011; Zorzos et al., 2010). *In vitro*, conventional microscope fluorescence illuminators, such as xenon lamps, mercury lamps, confocal or two-photon lasers, and LEDs, equipped with fast shutters, deflectors, or controllers (Boyden et al., 2005; Campagnola et al., 2008; Farah et al., 2007; Guo et al., 2009; Petreanu et al., 2007, 2009; Rickgauer and Tank, 2008), suffice to drive these molecules, the majority of which operate under illumination with incident light in the range of 0.1–10 mW/mm^2 at typical levels of expression in typical cells.

The speed of operation of these molecules is high, as they respond to light within milliseconds, and shut off rapidly after cessation of light delivery. For operation, these proteins require the cofactor all-*trans*-retinal as their essential chromophore, which acts to capture light. This chemical, being a natural beta carotene derivative, occurs in many organisms (such as mice, rats, and primates) at high enough background levels so that external chemical supplementation is not needed; for species such as *C. elegans* and *Drosophila*, whose background levels are lower, feeding of the all-*trans*-retinal to the organism suffices to enable supplementation. Because of this ease of use, adoption of these molecules for use in neuroscience has been rapid.

Overview of light-driven ion pumps

The first fully genetically encoded optogenetic tool to be utilized in neuroscience was the light-gated cation channel channelrhodopsin-2 (ChR2) from the green alga *Chlamydomonas reinhardtii*. This light-gated nonspecific cation channel, when expressed in neurons, enable the neurons to be depolarized with blue light (Boyden et al., 2005; Nagel et al., 2003). The channelrhodopsins, and the origins of optogenetic tools, are reviewed elsewhere in this volume (Chapter 2), as well as in other earlier references (Boyden, 2011; Chow et al., 2011a). Like halorhodopsins and light-driven

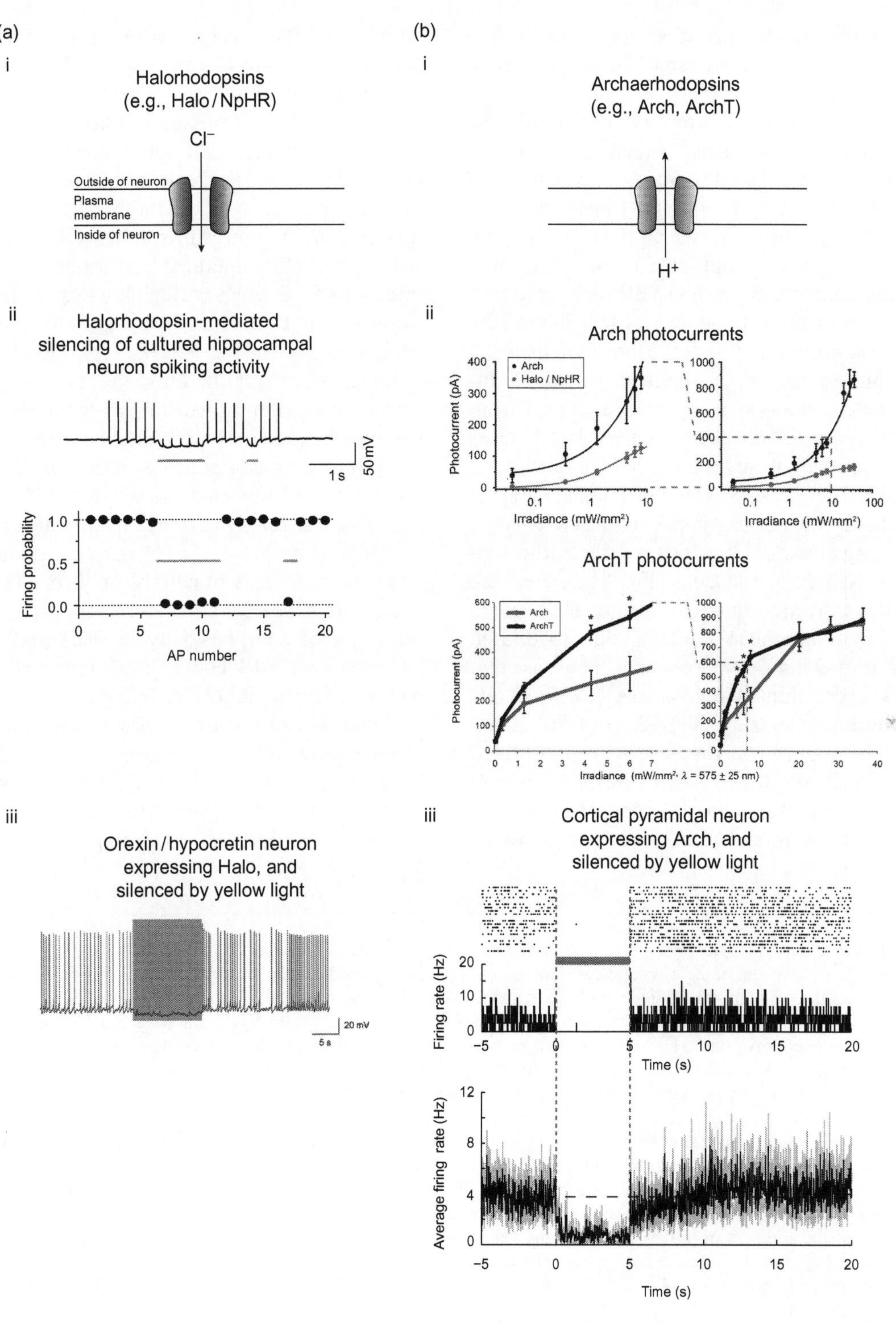
(a)
i
Halorhodopsins
(e.g., Halo / NpHR)
Cl⁻
Outside of neuron
Plasma membrane
Inside of neuron
ii
Halorhodopsin-mediated silencing of cultured hippocampal neuron spiking activity
1 s
50 mV
Firing probability
AP number
iii
Orexin / hypocretin neuron expressing Halo, and silenced by yellow light
20 mV
5 s
(b)
i
Archaerhodopsins
(e.g., Arch, ArchT)
H⁺
ii
Arch photocurrents
Arch
Halo / NpHR
Photocurrent (pA)
Irradiance (mW/mm²)
ArchT photocurrents
Arch
ArchT
Irradiance (mW/mm², λ = 575 ± 25 nm)
iii
Cortical pyramidal neuron expressing Arch, and silenced by yellow light
Firing rate (Hz)
Time (s)
Average firing rate (Hz)
Time (s)

proton pumps, channelrhodopsins are all-*trans*-retinal binding seven-transmembrane proteins, and so there are some similarities in structure and function across these different opsin classes. Unlike channelrhodopsins, which open up a channel pore when gated by light, halorhodopsins and light-driven proton pumps translocate one ion per photon absorbed, which may in principle limit light sensitivity, but also means that they can pump ions against a concentration gradient, causing significant voltage swings less limited by the reversal potential of specific ions in neurons.

Halorhodopsins, as exemplified by the light-driven inward chloride pump halorhodopsin from the species *Natronomonas pharaonis* (Halo/NpHR; Lanyi et al., 1990), were found in 2007 to function in mammalian neurons, where they mediate hyperpolarizing currents of 40–100pA (Han and Boyden, 2007; Zhang et al., 2007), which is sufficient to support modest neural silencing (Fig. 1aii). Because of the low currents that the original Halo could produce in mammalian neurons, due to limited protein trafficking in mammalian neurons that hampers expression at high levels (Chow et al., 2010; Gradinaru et al., 2008; Zhao et al., 2008), usage of the original Halo has largely been confined to use in invertebrates such as *C. elegans*, where it appears to function sufficiently in neurons and other excitable cells, to modulate behavior (Zhang et al., 2007). In 2011, however, a group reported for the first time the use of the original Halo to mediate behavioral changes in living mammals (Tsunematsu et al., 2011), expressing Halo in the orexin/hypocretin neurons of the mouse brain via the generation of a transgenic mouse line. The orexin/hypocretin neurons of this mouse, when illuminated, undergo reductions in spiking (Fig. 1aiii), and when such illumination is performed in the awake mouse, the animals enter slow-wave sleep, thus showing that brief silencing of this pathway is sufficient to induce sleep. Light-driven proton pumps, as exemplified by the archaeal protein archaerhodopsin-3 (Arch/aR-3) from *Halorubrum sodomense* (Fig. 1bi), were found in 2010 to function well in mammalian neurons in the service of light-driven neural silencing (Chow et al., 2010). Arch, and its close relative ArchT, from *Halorubrum* sp. *TP009* (Han et al., 2011), can mediate photocurrents of 900pA in cultured neurons (Fig. 1bii), and can mediate complete light-driven neural silencing of neural activity in mice and primates (Chow et al., 2010; Han et al., 2011), in response to yellow or green light (Fig. 1biii).

Light-driven inward chloride pumps and light-driven outward proton pumps are found in species of archaea (Ihara et al., 1999; Klare et al., 2008; Lanyi, 2004, 1986; Mukohata et al., 1999), bacteria (Antón et al., 2005; Balashov et al., 2005;

Fig. 1. Two major classes of microbial opsin for light-driven hyperpolarization of neurons. (a) Halorhodopsins, light-driven inward chloride pumps. (i) Diagram of the physiological response of halorhodopsins when expressed in the plasma membranes of neurons and exposed to light. (ii) Light-driven spike blockade, demonstrated for a representative cultured hippocampal neuron (*top*), as well as a population of cultured hippocampal neurons (*bottom*). Neurons expressed Halo/NpHR, and received 20 pulses of somatic current injection (~300pA, 4ms, 5Hz), accompanied by two periods of yellow light delivery (yellow bars). Adapted from Han and Boyden (2007). (iii) A neuron in a transgenic mouse expressing Halo in orexin/hypocretin neurons, being whole-cell patch clamp recorded during yellow light delivery, and exhibiting neural silencing. Adapted from Tsunematsu et al. (2011). (b) Archaerhodopsins, light-driven outward proton pumps. (i) Diagram of the physiological response of archaerhodopsins when expressed in the plasma membranes of neurons and exposed to light. (ii) Photocurrents of Arch (*top*) and ArchT (*bottom*), measured as a function of 575±25nm light irradiance, in patch-clamped cultured neurons, for low (*left*) and high (*right*) light powers. Adapted from Chow et al. (2010) and Han et al. (2011). (iii) Neural activity in a representative neuron recorded in awake mouse brain (*top*, *middle*), as well as a population of neurons in awake mouse brain (*bottom*). Shown is neural activity before, during, and after 5s of yellow light illumination, displayed both as a spike raster plot (*top*) and as a histogram of instantaneous firing rate averaged across trials (*middle*, *bottom*; bin size, 20ms). Adapted from Chow et al. (2010). (For interpretation of the references to color in this figure legend, the reader is referred to the Web version of this chapter.)

Beja et al., 2001, 2000; Friedrich et al., 2002; Kelemen et al., 2003; Kim et al., 2008), fungi (Brown, 2004; Waschuk et al., 2005), and algae (Tsunoda et al., 2006). Because halorhodopsins, bacteriorhodopsins, and archaerhodopsins have been crystallized and characterized by spectroscopy, mutagenesis, and physiology, much is known about their structure–function relationships (Enami et al., 2006; Essen, 2002; Kolbe et al., 2000; Luecke et al., 1999; Yoshimura and Kouyama, 2008). To date, these light-driven ion pumps have been found chiefly by analyzing genomes for sequences that resemble those of light-activated ion pumps and channels, followed by assessment of these gene products in heterologous expression systems (chiefly cell lines, neurons, and intact mouse brain) (Chow et al., 2011b). Interestingly, a great many light-activated ion pumps from species all over the tree of life—even the very first one ever characterized, the *Halobium salinarum* bacteriorhodopsin—were able to mediate neural hyperpolarizations when assessed in cultured mouse neurons (Chow et al., 2010).

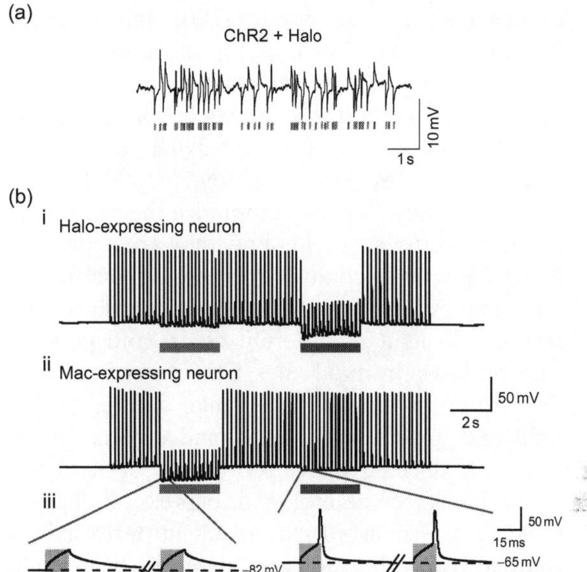

Fig. 2. Silencing of neurons with specific colors of light. (a) Hyperpolarization and depolarization events induced in a representative cultured hippocampal neuron expressing both ChR2 and Halo, by a Poisson train (mean interpulse interval λ=100ms) of alternating pulses of yellow and blue light (10 ms duration), denoted by yellow and blue bars, respectively. Adapted from Han and Boyden (2007). (b) Multicolor silencing of two neural populations, enabled by blue- and red-light drivable ion pumps of different genomic classes. Action potentials evoked by current injection into patch-clamped cultured neurons transfected with the archaeal opsin Halo (i) were selectively silenced by the red light but not by blue light, and vice versa in neurons expressing the fungal opsin Mac (ii). Gray boxes in the inset (iii) indicate periods of patch clamp current injection. Adapted from Chow et al. (2010). (For interpretation of the references to color in this figure legend, the reader is referred to the Web version of this chapter.)

Properties of opsins and how they affect performance

Each of these opsins has a different characteristic action spectrum, which describes the set of colors that optimally drive the opsin to function. Although chloride pumps are chiefly driven by yellow light and possess largely stereotyped action spectra, proton pumps exist in many different kingdoms of life that are driven by many different colors of light. For example, the fungal light-driven proton pump from *Leptosphaeria maculans* (Mac) is drivable by blue-green light, and thus, alongside the earlier molecule Halo, which can be driven by yellow-red light, enables two-color silencing of different sets of neurons (Fig. 2b; Chow et al., 2010). That is, a Mac-expressing neuron will be quieted by blue light, but not red light; in contrast, a Halo-expressing neuron will be quieted by red light, and not blue light. ChR2, which is blue light driven, can also be used alongside silencers such as Halo which are yellow or red light driven, even in the same cell, enabling depolarization and hyperpolarization of the doubly targeted cell by blue and yellow light, respectively (Fig. 2a). The delivery of multiple opsins to the same cell can be facilitate by the use of 2A linker peptides to combine the genes for multiple opsins into a single open reading frame (Han and Boyden, 2007; Han et al., 2009b), useful for bidirectionally assessing the causal role that a given set of neurons plays in a single animal, or for perturbing complex properties of neural dynamics such as neural synchrony (Han and Boyden, 2007).

In this review, we will summarize the properties of these opsins, surveying how they are expressed in cells, how they conduct ions across membranes, how they cycle in response to light, and how they respond to light of different colors and powers. The optogenetic molecules here described were isolated from archaea and fungi, species whose lipid and ionic compositions, and external environment, differ greatly from those of neurons. Accordingly, the heterologous expression of these molecules in neurons may result in performance characteristics that are very different from what would be expected from their expression in their native cells (Chow et al., 2011b). Indeed, this means that the performance of opsins must be assessed directly in the cell types of interest, as it can be difficult to predict how these high-speed molecules will function in a specific milieu, from their properties as characterized in another one.

Some general guidelines and considerations apply, however. First, to insure efficient translation, it is useful to codon-optimize molecules for the target species. Second, to boost membrane expression, it is often helpful to append extra protein sequences to opsins to improve their protein folding, membrane trafficking and localization, and even subcellular compartmentalization (Chow et al., 2010; Gradinaru et al., 2008; Greenberg et al., 2011; Zhao et al., 2008). Third, although many groups use these molecules to change the voltage of neurons, it is also useful to explicitly consider the ions translocated by these pumps; for example, they can be used to change the levels of specific ions in cells or subcellular compartments, in a temporally precise way. Fourth, the photocycle kinetics can be used to determine which molecule might be best for a given scientific application; for example, various light-driven ion pumps may enter different states that may render them nonfunctional for periods of time (Fig. 4). It is critical to remember that these light-driven ion pumps are not simple on–off switches; they undergo a series of structural rearrangements, with many intermediate states between light reception and final restoration of the molecule to the initial state. Over the past several decades, a myriad of structure–function studies have been performed on these molecules (Bamberg et al., 1993; Blanck and Oesterhelt, 1987; Braiman et al., 1987; Brown et al., 1996; Essen, 2002; Gilles-Gonzalez et al., 1991; Hegemann et al., 1985; Henderson and Schertler, 1990; Kolbe et al., 2000; Lanyi, 2004, 1986; Lanyi et al., 1990; Luecke et al., 1999; Marinetti et al., 1989; Marti et al., 1991; Mogi et al., 1987, 1988, 1989a,b; Rudiger and Oesterhelt, 1997; Subramaniam et al., 1992; Tittor et al., 1997, 1995; Varo et al., 1995a,b), linking specific amino acids with specific kinetic and spectral properties of the molecules. Fifth, the light-sensitivity of a molecule can determine how well it will function in the brain; improving light sensitivity can enable larger volumes of brain to be controlled than possible with less light-sensitive molecules (e.g., Fig. 1bii, *bottom*). In the following sections, we explore these properties, for light-driven chloride pumps and light-driven proton pumps.

Protein expression and photocurrent magnitude of light-driven ion pumps

The *N. pharaonis* halorhodopsin Halo/NpHR, the first microbial opsin to be used for optical neural silencing (Fig. 1a), has a reversal potential of approximately −400mV (Seki et al., 2007), and although it comes from an archaeon that lives in very high salinity environments, the optimal chloride concentration for its operation is actually very close to that found in neurons (in contrast to that of other halorhodopsins such as the *H. salinarum* halorhodopsin, which demands far higher chloride concentrations) (Okuno et al., 1999). This may be one of the reasons why it works in some neurons in some species (Fig. 1aii and aiii). However, at high expression levels, Halo forms intracellular aggregates (Gradinaru et al., 2008; Zhao et al., 2008), an issue that can be ameliorated by the appending of trafficking sequences from the Kir2.1 channel (Gradinaru et al., 2008, 2010; Zhao et al., 2008), which reduces aggregation and increases photocurrent manifold over that of the baseline molecule. Other signal sequences, such as a prolactin localization sequence, also increase halorhodopsin photocurrent (Chow et al., 2011a); it is important to note, however, that different sequences may do different things in different cell types in different species.

The *H. sodomense* archaerhodopsin Arch (Fig. 1b), possesses strong photocurrents that exceed the original *N. pharaonis* halorhodopsin currents by an order of magnitude, both at low and at high light power (Fig. 1bii, *top*) (Chow et al., 2010). Arch can mediate nearly complete silencing of neural activity in awake behaving mice (Fig. 1biii). Arch does not require trafficking sequences to achieve these high levels of performance, although they can help to boost expression and/or current yet further (Chow et al., 2010). Other opsins that are similar in amino acid sequence to Arch have been described in genomic and gene sequence databases; all of these Arch relatives, found in other species of the *Halorubrum* genus, also express well in neurons (Han et al., 2011). One of these Arch relatives, the archaerhodopsin from *Halorubrum* sp. *TP009*, presents photocurrents that are about 3.5× more light sensitive than those of Arch (Fig. 1bii, *bottom*), supporting the silencing of broad brain regions, and neurons in the cortex of the awake behaving macaque (Han et al., 2011). Light-driven proton pumps are powerful at neural silencing, a perhaps surprising fact given the relative scarcity of protons relative to other major charge carriers in the brain (such as Na^+, K^+, or Cl^-); the fast kinetics of archaerhodopsins (Lukashev et al., 1994; Ming et al., 2006), and their good trafficking in mammalian cells, among other attributes, may support the high performance of archaerhodopsins. These light-driven outward proton pumps, interestingly, do not change cellular pH to a greater extent than do opsins such as ChR2 (whose proton conductance is 10^6 times its sodium conductance). Specifically, illuminating Arch with relatively bright light when expressed in cultured neurons resulted in alkalinizations of neurons by 0.1–0.15 pH units, a change in pH of smaller magnitude than that seen with ChR2, perhaps due to the high proton conductance of channelrhodopsins (Berthold et al., 2008; Lin et al., 2009; Nagel et al., 2003). From an end-user standpoint, illumination of Arch- and ArchT-expressing neurons, in awake mice and macaques, for periods of many minutes, did not alter spike waveform or spike frequency when comparing before versus after silencing.

Kinetics of light-driven ion pumps

From a kinetic standpoint, all of the halorhodopsins we have examined to date inactivate by roughly 30% every 15s under typical illumination conditions (1–10mW/mm^2, of yellow, e.g., 593nm, light; see Fig. 3a and b), consistent with the photocycle topology shown in Fig. 4a (Chow et al., 2010; Han and Boyden, 2007). Further, recovery in the dark of halorhodopsins from this inactivated state is slow, taking tens of minutes for full recovery (Bamberg et al., 1993; Han and Boyden, 2007; Hegemann et al., 1985). Although this rundown may somewhat compromise performance over very long illumination periods, it is possible to drive the molecule out of this inactive state by using UV or blue light (Bamberg et al., 1993; Han and Boyden, 2007; Fig. 3aii and b). The entire photocycle is summarized in Fig. 4a, with the limiting time constants noted in the diagram, and with the states denoted by the light absorption properties of the state, as is conventional. The outer circle of the photocycle describes the dominant mode of operation, in which, after photon absorption, a chloride ion is released into the cytoplasm, and then a second

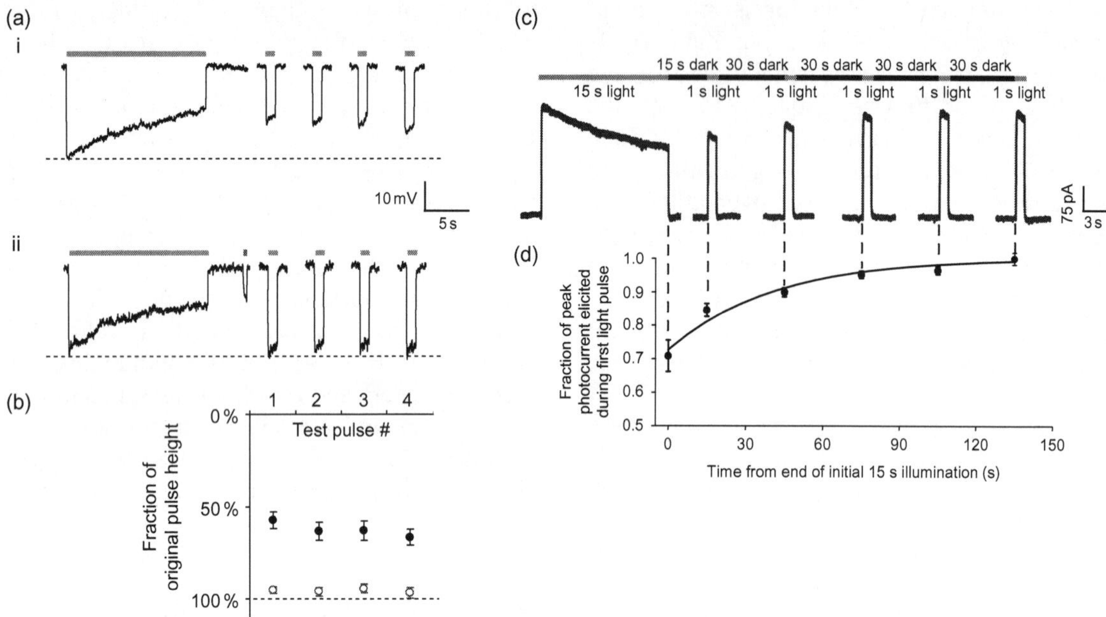

Fig. 3. Kinetic properties of the two major optical neural silencer classes. (a) (i) Halo-mediated hyperpolarizations in a representative current-clamped hippocampal neuron during 15s of continuous yellow light, followed by four 1-s test pulses of yellow light (one every 30s, starting 10s after the end of the first 15-s period of yellow light). (ii) Halo-mediated hyperpolarizations for the same cell exhibited in (i), but when Halo function is facilitated by a 400-ms pulse of blue light in between the 15-s period of yellow light and the first 1-s test pulse. Adapted from Han and Boyden (2007). (b) Population data for blue-light facilitation of Halo recovery. Plotted are the hyperpolarizations elicited by the four 1-s test pulses of yellow light, normalized to the peak hyperpolarization induced by the original 15-s yellow light pulse (mean±std. err.). Black dots represent experiments when no blue light pulse was delivered (as in ai). Open blue dots represent experiments when 400ms of blue light was delivered to facilitate recovery (as in aii). Adapted from Han and Boyden (2007). (c) Raw current trace of a neuron lentivirally infected with Arch, illuminated by a 15-s light pulse (575±25nm, irradiance 7.8mW/mm^2), followed by 1s test pulses delivered starting 15, 45, 75, 105, and 135s after the end of the 15s light pulse. Adapted from Chow et al. (2010). (d) Population data of averaged Arch photocurrents sampled at the times indicated by the vertical dotted lines that extend into (c). Adapted from Chow et al. (2010). (For interpretation of the references to color in this figure legend, the reader is referred to the Web version of this chapter.)

chloride ion is taken up from the extracellular space, thus fulfilling the chloride pumping action. In the center of the photocycle diagram, however, is a proton-pumping series of states that involve the long-lasting inactive state described above, here denoted HR410. This HR410 state is the inactivated state that can be reprimed back to the initial state using blue light.

In contrast to the kinetic properties of halorhodopsins, archaerhodopsins such as Arch and ArchT spontaneously recover in the dark, even after extensive illumination (Fig. 3c and d). The time constant of recovery is in the range of tens of seconds, in contrast to that of halorhodopsins, which is in the range of tens of minutes. Although archaerhodopsin photocycles are still relatively uncharacterized compared to those of earlier-discovered opsins, some insight into proton pump operation can be derived from consideration of the *H. salinarum* bacteriorhodopsin photocycle (Fig. 4b). The topology of the photocycle shares some similarities with that of

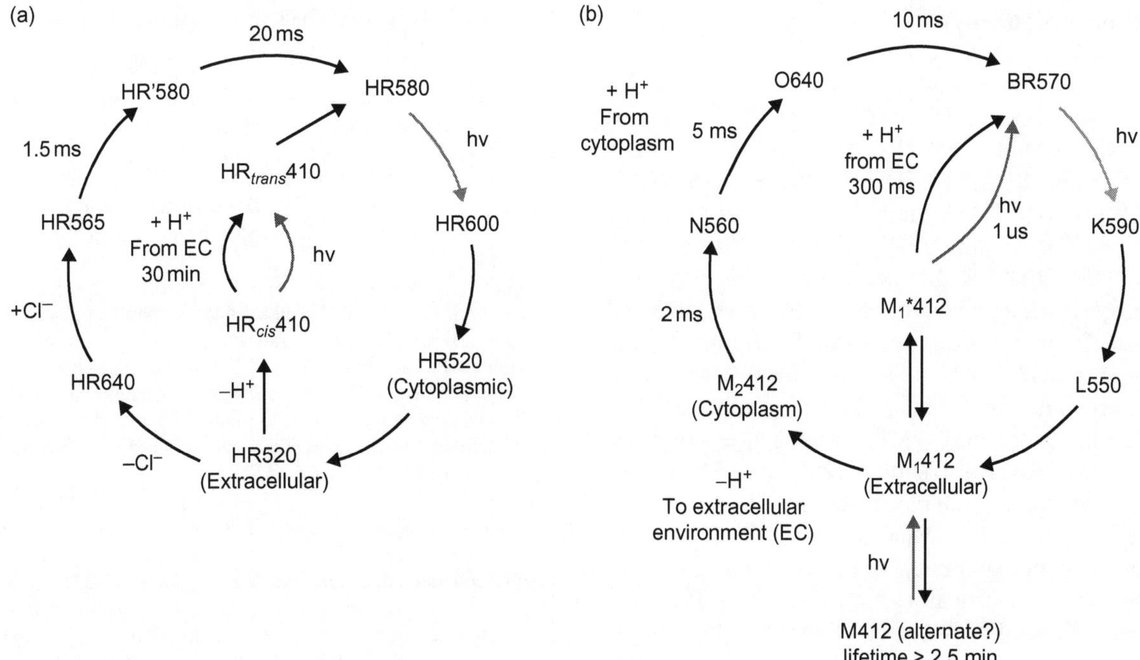

Fig. 4. Photocycle models for the two major optical neural silencer classes. (a) The halorhodopsin photocycle at high-power continuous illumination on the timescale of a typical photocycle (i.e., conditions used for neural silencing, greater than few milliseconds). The HR410 intermediate is the origin of long-lived inactivation in neural silencing (e.g., Bamberg et al., 1993; Han and Boyden, 2007). (b) The photocycle of the *H. salinarum* bacteriorhodopsin. The photocycle has been simplified to reflect the dominant photocycle at large continuous illumination on the timescale of a typical photocycle (i.e., conditions used for neural silencing, greater than few milliseconds). The M412 alternate intermediate is the origin of long-lived inactivation with bacteriorhodopsin. In contrast, Arch spontaneously quickly recovers from this state in the dark. (For color version of this figure, the reader is referred to the Web version of this chapter.)

halorhodopsins, although many of the timescales are significantly different.

Toward the future

Improvements in optogenetic silencers are rapidly transpiring, driven by the burgeoning amount of microbial genome being sequenced as well as the influx of bioengineering approaches such as mutagenesis and protein engineering. It is likely that the kinetic and amplitude properties of these molecules will continue to improve, given that only a small part of genomic space has been explored, and only a small number of mutations and protein changes (e.g., the appending of Kir2.1 trafficking sequences) have been explored. From a scientific standpoint, the optical silencers have been broadly applied in the service of assessing the causal role that specific neurons play in the behaviors and neural computations implemented by brain circuits. Now that these molecules have been safely and efficaciously used to modulate neural activity in awake macaque (Diester et al., 2011; Han et al., 2009a), it is also intriguing to ponder whether new neural control prosthetics will arise, capable of silencing overactive neurons or deleting pathological neural activity patterns, in human patients, in the future.

Acknowledgments

E. S. B. acknowledges funding by the NIH Director's New Innovator Award (DP2OD002002) as well as NIH Grants 1R01NS075421, 1R01DA029639, 1RC1MH088182, 1RC2DE020919, 1R01NS067199, and 1R43NS070453; the NSF CAREER award as well as NSF Grants EFRI 0835878, DMS 0848804, and DMS 1042134; Benesse Foundation, Jerry and Marge Burnett, Department of Defense CDMRP Post-Traumatic Stress Disorder Program, Google, Harvard/MIT Joint Grants Program in Basic Neuroscience, Human Frontiers Science Program, MIT Alumni Class Funds, MIT Intelligence Initiative, MIT McGovern Institute and the McGovern Institute Neurotechnology Award Program, MIT Media Lab, MIT Mind-Machine Project, MIT Neurotechnology Fund, NARSAD, Paul Allen Distinguished Investigator Award, Alfred P. Sloan Foundation, SFN Research Award for Innovation in Neuroscience, and the Wallace H. Coulter Foundation.

The authors declare no competing financial interests.

References

Adamantidis, A. R., Zhang, F., Aravanis, A. M., Deisseroth, K., & de Lecea, L. (2007). Neural substrates of awakening probed with optogenetic control of hypocretin neurons. *Nature*, *450*(7168), 420–424.

Antón, J., Peña, A., Valens, M., Santos, F., Glöckner, F. O., Bauer, M., et al. (2005). Salinibacter ruber: Genomics and biogeography. In N. Gunde-Cimerman, A. Plemenitas & A. Oren (Eds.), *Adaptation to life in high salt concentrations in archaea, bacteria and eukarya* (pp. 257–266). Dordrecht: Kluwer Academic Publishers.

Aravanis, A. M., Wang, L. P., Zhang, F., Meltzer, L. A., Mogri, M. Z., Schneider, M. B., et al. (2007). An optical neural interface: In vivo control of rodent motor cortex with integrated fiberoptic and optogenetic technology. *Journal of Neural Engineering*, *4*(3), S143–S156.

Atasoy, D., Aponte, Y., Su, H. H., & Sternson, S. M. (2008). A FLEX switch targets Channelrhodopsin-2 to multiple cell types for imaging and long-range circuit mapping. *Journal of Neuroscience*, *28*(28), 7025–7030.

Balashov, S. P., Imasheva, E. S., Boichenko, V. A., Anton, J., Wang, J. M., & Lanyi, J. K. (2005). Xanthorhodopsin: A proton pump with a light-harvesting carotenoid antenna. *Science*, *309*(5743), 2061–2064.

Bamberg, E., Tittor, J., & Oesterhelt, D. (1993). Light-driven proton or chloride pumping by halorhodopsin. *Proceedings of the National Academy of Sciences of the United States of America*, *90*(2), 639–643.

Beja, O., Aravind, L., Koonin, E. V., Suzuki, M. T., Hadd, A., Nguyen, L. P., et al. (2000). Bacterial rhodopsin: Evidence for a new type of phototrophy in the sea. *Science*, *289* (5486), 1902–1906.

Beja, O., Spudich, E. N., Spudich, J. L., Leclerc, M., & DeLong, E. F. (2001). Proteorhodopsin phototrophy in the ocean. *Nature*, *411*(6839), 786–789.

Berthold, P., Tsunoda, S. P., Ernst, O. P., Mages, W., Gradmann, D., & Hegemann, P. (2008). Channelrhodopsin-1 initiates phototaxis and photophobic responses in chlamydomonas by immediate light-induced depolarization. *The Plant Cell*, *20*(6), 1665–1677.

Bi, A., Cui, J., Ma, Y. P., Olshevskaya, E., Pu, M., Dizhoor, A. M., et al. (2006). Ectopic expression of a microbial-type rhodopsin restores visual responses in mice with photoreceptor degeneration. *Neuron*, *50*(1), 23–33.

Blanck, A., & Oesterhelt, D. (1987). The halo-opsin gene. II. Sequence, primary structure of halorhodopsin and comparison with bacteriorhodopsin. *EMBO Journal*, *6*(1), 265–273.

Boyden, E. S. (2011). A history of optogenetics: The development of tools for controlling brain circuits with light. *F1000 Biology Reports*, *3*, 11.

Boyden, E. S., Zhang, F., Bamberg, E., Nagel, G., & Deisseroth, K. (2005). Millisecond-timescale, genetically targeted optical control of neural activity. *Nature Neuroscience*, *8*(9), 1263–1268.

Braiman, M. S., Stern, L. J., Chao, B. H., & Khorana, H. G. (1987). Structure-function studies on bacteriorhodopsin. IV. Purification and renaturation of bacterio-opsin polypeptide expressed in *Escherichia coli*. *Journal of Biological Chemistry*, *262*(19), 9271–9276.

Brown, L. S. (2004). Fungal rhodopsins and opsin-related proteins: Eukaryotic homologues of bacteriorhodopsin with unknown functions. *Photochemical and Photobiological Sciences*, *3*(6), 555–565.

Brown, L. S., Needleman, R., & Lanyi, J. K. (1996). Interaction of proton and chloride transfer pathways in recombinant bacteriorhodopsin with chloride transport activity: Implications for the chloride translocation mechanism. *Biochemistry*, *35*(50), 16048–16054.

Campagnola, L., Wang, H., & Zylka, M. J. (2008). Fiber-coupled light-emitting diode for localized photostimulation of neurons expressing channelrhodopsin-2. *Journal of Neuroscience Methods*, *169*(1), 27–33.

Chan, S. C., Bernstein, J., & Boyden, E. (2010). Scalable fluidic injector arrays for viral targeting of intact 3-D brain circuits. *Journal of Visualized Experiments*, 35. pii: 1489. doi: 10.3791/1489.

Chhatwal, J. P., Hammack, S. E., Jasnow, A. M., Rainnie, D. G., & Ressler, K. J. (2007). Identification of cell-type-specific promoters within the brain using lentiviral vectors. *Gene Therapy, 14*(7), 575–583.

Chow, B. Y., Chuong, A. S., Klapoetke, N. C., & Boyden, E. S. (2011a). Synthetic physiology strategies for adapting tools from nature for genetically targeted control of fast biological processes. *Methods in Enzymology, 497*, 425–443.

Chow, B. Y., Han, X., Bernstein, J. G., Monahan, P. E., & Boyden, E. S. (2011b). Light-activated ion pumps and channels for temporally precise optical control of activity in genetically targeted neurons. In J. J. Chambers & R. H. Kramer (Eds.), *Photosensitive molecules for controlling biological function* (pp. 99–132). New York: Humana Press.

Chow, B. Y., Han, X., Dobry, A. S., Qian, X., Chuong, A. S., Li, M., et al. (2010). High-performance genetically targetable optical neural silencing by light-driven proton pumps. *Nature, 463*(7277), 98–102.

Diester, I., Kaufman, M. T., Mogri, M., Pashaie, R., Goo, W., Yizhar, O., et al. (2011). An optogenetic toolbox designed for primates. *Nature Neuroscience, 14*(3), 387–397.

Dittgen, T., Nimmerjahn, A., Komai, S., Licznerski, P., Waters, J., Margrie, T. W., et al. (2004). Lentivirus-based genetic manipulations of cortical neurons and their optical and electrophysiological monitoring in vivo. *Proceedings of the National Academy of Sciences of the United States of America, 101*(52), 18206–18211.

Douglass, A. D., Kraves, S., Deisseroth, K., Schier, A. F., & Engert, F. (2008). Escape behavior elicited by single, channelrhodopsin-2-evoked spikes in zebrafish somatosensory neurons. *Current Biology, 18*(15), 1133–1137.

Enami, N., Yoshimura, K., Murakami, M., Okumura, H., Ihara, K., & Kouyama, T. (2006). Crystal structures of archaerhodopsin-1 and -2: Common structural motif in archaeal light-driven proton pumps. *Journal of Molecular Biology, 358*(3), 675–685.

Essen, L. O. (2002). Halorhodopsin: Light-driven ion pumping made simple? *Current Opinion in Structural Biology, 12*(4), 516–522.

Farah, N., Reutsky, I., & Shoham, S. (2007). Patterned optical activation of retinal ganglion cells. *Conference Proceedings: Annual International Conference of the IEEE Engineering in Medicine and Biology Society, 2007*, 6369–6371.

Friedrich, T., Geibel, S., Kalmbach, R., Chizhov, I., Ataka, K., Heberle, J., et al. (2002). Proteorhodopsin is a light-driven proton pump with variable vectoriality. *Journal of Molecular Biology, 321*(5), 821–838.

Gilles-Gonzalez, M. A., Engelman, D. M., & Khorana, H. G. (1991). Structure-function studies of bacteriorhodopsin XV. Effects of deletions in loops B-C and E-F on bacteriorhodopsin chromophore and structure. *Journal of Biological Chemistry, 266*(13), 8545–8550.

Gradinaru, V., Thompson, K. R., & Deisseroth, K. (2008). eNpHR: A Natronomonas halorhodopsin enhanced for optogenetic applications. *Brain Cell Biology, 36*(1–4), 129–139.

Gradinaru, V., Thompson, K. R., Zhang, F., Mogri, M., Kay, K., Schneider, M. B., et al. (2007). Targeting and readout strategies for fast optical neural control in vitro and in vivo. *Journal of Neuroscience, 27*(52), 14231–14238.

Gradinaru, V., Zhang, F., Ramakrishnan, C., Mattis, J., Prakash, R., Diester, I., et al. (2010). Molecular and cellular approaches for diversifying and extending optogenetics. *Cell, 141*(1), 154–165.

Greenberg, K. P., Pham, A., & Werblin, F. S. (2011). Differential targeting of optical neuromodulators to ganglion cell soma and dendrites allows dynamic control of center-surround antagonism. *Neuron, 69*(4), 713–720.

Guo, Z. V., Hart, A. C., & Ramanathan, S. (2009). Optical interrogation of neural circuits in *Caenorhabditis elegans*. *Nature Methods, 6*(12), 891–896.

Han, X., & Boyden, E. S. (2007). Multiple-color optical activation, silencing, and desynchronization of neural activity, with single-spike temporal resolution. *PLoS One, 2*(3), e299.

Han, X., Chow, B. Y., Zhou, H., Klapoetke, N., Chuong, A., Rajimehr, R., et al. (2011). A high-light sensitivity optical neural silencer: Development and application to optogenetic control of non-human primate cortex. *Frontiers in Systems Neuroscience, 5*, 18.

Han, X., Qian, X., Bernstein, J. G., Zhou, H. H., Franzesi, G. T., Stern, P., et al. (2009a). Millisecond-timescale optical control of neural dynamics in the nonhuman primate brain. *Neuron, 62*(2), 191–198.

Han, X., Qian, X., Stern, P., Chuong, A. S., & Boyden, E. S. (2009b). Informational lesions: Optical perturbation of spike timing and neural synchrony via microbial opsin gene fusions. *Frontiers in Molecular Neuroscience*, doi:10.3389/neuro.02.012.2009.

Hegemann, P., Oesterhelt, D., & Steiner, M. (1985). The photocycle of the chloride pump halorhodopsin. I: Azide-catalyzed deprotonation of the chromophore is a side reaction of photocycle intermediates inactivating the pump. *EMBO Journal, 4*(9), 2347–2350.

Henderson, R., & Schertler, G. F. (1990). The structure of bacteriorhodopsin and its relevance to the visual opsins and other seven-helix G-protein coupled receptors. *Philosophical Transactions of the Royal Society of London. Series B, Biological Sciences, 326*(1236), 379–389.

Huber, D., Petreanu, L., Ghitani, N., Ranade, S., Hromadka, T., Mainen, Z., et al. (2008). Sparse optical microstimulation in barrel cortex drives learned behaviour in freely moving mice. *Nature, 451*(7174), 61–64.

Ihara, K., Umemura, T., Katagiri, I., Kitajima-Ihara, T., Sugiyama, Y., Kimura, Y., et al. (1999). Evolution of the archaeal rhodopsins: Evolution rate changes by gene duplication and functional differentiation. *Journal of Molecular Biology, 285*(1), 163–174.

Ishizuka, T., Kakuda, M., Araki, R., & Yawo, H. (2006). Kinetic evaluation of photosensitivity in genetically engineered neurons expressing green algae light-gated channels. *Neuroscience Research*, *54*(2), 85–94.

Kelemen, B. R., Du, M., & Jensen, R. B. (2003). Proteorhodopsin in living color: Diversity of spectral properties within living bacterial cells. *Biochimica et Biophysica Acta*, *1618*(1), 25–32.

Kim, S. Y., Waschuk, S. A., Brown, L. S., & Jung, K. H. (2008). Screening and characterization of proteorhodopsin color-tuning mutations in Escherichia coli with endogenous retinal synthesis. *Biochimica et Biophysica Acta*, *1777*(6), 504–513.

Klare, J. P., Chizhov, I., & Engelhard, M. (2008). Microbial rhodopsins: Scaffolds for ion pumps, channels, and sensors. *Results and Problems in Cell Differentiation*, *45*, 73–122.

Kolbe, M., Besir, H., Essen, L. O., & Oesterhelt, D. (2000). Structure of the light-driven chloride pump halorhodopsin at 1.8 A resolution. *Science*, *288*(5470), 1390–1396.

Kuhlman, S. J., & Huang, Z. J. (2008). High-resolution labeling and functional manipulation of specific neuron types in mouse brain by Cre-activated viral gene expression. *PLoS One*, *3*(4), e2005.

Lagali, P. S., Balya, D., Awatramani, G. B., Munch, T. A., Kim, D. S., Busskamp, V., et al. (2008). Light-activated channels targeted to ON bipolar cells restore visual function in retinal degeneration. *Nature Neuroscience*, *11*(6), 667–675.

Lanyi, J. K. (1986). Halorhodopsin: A light-driven chloride ion pump. *Annual Review of Biophysics and Biophysical Chemistry*, *15*(1), 11–28.

Lanyi, J. K. (2004). Bacteriorhodopsin. *Annual Review of Physiology*, *66*(1), 665–688.

Lanyi, J. K., Duschl, A., Hatfield, G. W., May, K., & Oesterhelt, D. (1990). The primary structure of a halorhodopsin from Natronobacterium pharaonis. Structural, functional and evolutionary implications for bacterial rhodopsins and halorhodopsins. *Journal of Biological Chemistry*, *265*(3), 1253–1260.

Li, X., Gutierrez, D. V., Hanson, M. G., Han, J., Mark, M. D., Chiel, H., et al. (2005). Fast noninvasive activation and inhibition of neural and network activity by vertebrate rhodopsin and green algae channelrhodopsin. *Proceedings of the National Academy of Sciences of the United States of America*, *102*(49), 17816–17821.

Lin, J. Y., Lin, M. Z., Steinbach, P., & Tsien, R. Y. (2009). Characterization of engineered channelrhodopsin variants with improved properties and kinetics. *Biophysical Journal*, *96*(5), 1803–1814.

Luecke, H., Schobert, B., Richter, H. T., Cartailler, J. P., & Lanyi, J. K. (1999). Structure of bacteriorhodopsin at 1.55 A resolution. *Journal of Molecular Biology*, *291*(4), 899–911.

Lukashev, E. P., Govindjee, R., Kono, M., Ebrey, T. G., Sugiyama, Y., & Mukohata, Y. (1994). pH dependence of the absorption spectra and photochemical transformations of the archaerhodopsins. *Photochemistry and Photobiology*, *60*(1), 69–75.

Marinetti, T., Subramaniam, S., Mogi, T., Marti, T., & Khorana, H. G. (1989). Replacement of aspartic residues 85, 96, 115, or 212 affects the quantum yield and kinetics of proton release and uptake by bacteriorhodopsin. *Proceedings of the National Academy of Sciences of the United States of America*, *86*(2), 529–533.

Marti, T., Otto, H., Mogi, T., Rosselet, S. J., Heyn, M. P., & Khorana, H. G. (1991). Bacteriorhodopsin mutants containing single substitutions of serine or threonine residues are all active in proton translocation. *Journal of Biological Chemistry*, *266*(11), 6919–6927.

Ming, M., Lu, M., Balashov, S. P., Ebrey, T. G., Li, Q., & Ding, J. (2006). pH dependence of light-driven proton pumping by an archaerhodopsin from Tibet: Comparison with bacteriorhodopsin. *Biophysical Journal*, *90*(9), 3322–3332.

Mogi, T., Marti, T., & Khorana, H. G. (1989a). Structure-function studies on bacteriorhodopsin. IX. Substitutions of tryptophan residues affect protein-retinal interactions in bacteriorhodopsin. *Journal of Biological Chemistry*, *264* (24), 14197–14201.

Mogi, T., Stern, L. J., Chao, B. H., & Khorana, H. G. (1989b). Structure-function studies on bacteriorhodopsin. VIII. Substitutions of the membrane-embedded prolines 50, 91, and 186: The effects are determined by the substituting amino acids. *Journal of Biological Chemistry*, *264*(24), 14192–14196.

Mogi, T., Stern, L. J., Hackett, N. R., & Khorana, H. G. (1987). Bacteriorhodopsin mutants containing single tyrosine to phenylalanine substitutions are all active in proton translocation. *Proceedings of the National Academy of Sciences of the United States of America*, *84*(16), 5595–5599.

Mogi, T., Stern, L. J., Marti, T., Chao, B. H., & Khorana, H. G. (1988). Aspartic acid substitutions affect proton translocation by bacteriorhodopsin. *Proceedings of the National Academy of Sciences of the United States of America*, *85*(12), 4148–4152.

Mukohata, Y., Ihara, K., Tamura, T., & Sugiyama, Y. (1999). Halobacterial rhodopsins. *Journal of Biochemistry*, *125*(4), 649–657.

Nagel, G., Brauner, M., Liewald, J. F., Adeishvili, N., Bamberg, E., & Gottschalk, A. (2005). Light activation of channelrhodopsin-2 in excitable cells of Caenorhabditis elegans triggers rapid behavioral responses. *Current Biology*, *15*(24), 2279–2284.

Nagel, G., Szellas, T., Huhn, W., Kateriya, S., Adeishvili, N., Berthold, P., et al. (2003). Channelrhodopsin-2, a directly light-gated cation-selective membrane channel. *Proceedings of the National Academy of Sciences of the United States of America*, *100*(24), 13940–13945. (PMCID: 283525).

Okuno, D., Asaumi, M., & Muneyuki, E. (1999). Chloride concentration dependency of the electrogenic activity of halorhodopsin. *Biochemistry*, *38*(17), 5422–5429.

Petreanu, L., Huber, D., Sobczyk, A., & Svoboda, K. (2007). Channelrhodopsin-2-assisted circuit mapping of long-range callosal projections. *Nature Neuroscience*, *10*(5), 663–668.

Petreanu, L., Mao, T., Sternson, S. M., & Svoboda, K. (2009). The subcellular organization of neocortical excitatory connections. *Nature*, *457*(7233), 1142–1145.

Rickgauer, J. P., & Tank, D. W. (Eds.), (2008). Optimizing two-photon activation of channelrhodopsin-2 for stimulation at cellular resolution. *Neuroscience*. Washington, DC: Society for Neuroscience.

Rudiger, M., & Oesterhelt, D. (1997). Specific arginine and threonine residues control anion binding and transport in the light-driven chloride pump halorhodopsin. *EMBO Journal, 16*(13), 3813–3821.

Schroll, C., Riemensperger, T., Bucher, D., Ehmer, J., Voller, T., Erbguth, K., et al. (2006). Light-induced activation of distinct modulatory neurons triggers appetitive or aversive learning in Drosophila larvae. *Current Biology, 16* (17), 1741–1747.

Seki, A., Miyauchi, S., Hayashi, S., Kikukawa, T., Kubo, M., Demura, M., et al. (2007). Heterologous expression of Pharaonis halorhodopsin in Xenopus laevis oocytes and electrophysiological characterization of its light-driven Cl^- pump activity. *Biophysical Journal, 92*(7), 2559–2569.

Subramaniam, S., Greenhalgh, D. A., & Khorana, H. G. (1992). Aspartic acid 85 in bacteriorhodopsin functions both as proton acceptor and negative counterion to the Schiff base. *Journal of Biological Chemistry, 267*(36), 25730–25733.

Tan, W., Janczewski, W. A., Yang, P., Shao, X. M., Callaway, E. M., & Feldman, J. L. (2008). Silencing pre-Botzinger complex somatostatin-expressing neurons induces persistent apnea in awake rat. *Nature Neuroscience, 11*(5), 538–540.

Tittor, J., Haupts, U., Haupts, C., Oesterhelt, D., Becker, A., & Bamberg, E. (1997). Chloride and proton transport in bacteriorhodopsin mutant D85T: Different modes of ion translocation in a retinal protein. *Journal of Molecular Biology, 271*(3), 405–416.

Tittor, J., Oesterhelt, D., & Bamberg, E. (1995). Bacteriorhodopsin mutants D85N, D85T and D85, 96N as proton pumps. *Biophysical Chemistry, 56*(1–2), 153–157.

Toni, N., Laplagne, D. A., Zhao, C., Lombardi, G., Ribak, C. E., Gage, F. H., et al. (2008). Neurons born in the adult dentate gyrus form functional synapses with target cells. *Nature Neuroscience, 11*(8), 901–907.

Tsunematsu, T., Kilduff, T. S., Boyden, E. S., Takahashi, S., Tominaga, M., & Yamanaka, A. (2011). Acute optogenetic silencing of orexin/hypocretin neurons induces slow-wave sleep in mice. *Journal of Neuroscience, 31*(29), 10529–10539.

Tsunoda, S. P., Ewers, D., Gazzarrini, S., Moroni, A., Gradmann, D., & Hegemann, P. (2006). H+-pumping rhodopsin from the marine alga acetabularia. *Biophysical Journal, 91*(4), 1471–1479.

Varo, G., Brown, L. S., Sasaki, J., Kandori, H., Maeda, A., Needleman, R., et al. (1995a). Light-driven chloride ion transport by halorhodopsin from Natronobacterium pharaonis. 1. The photochemical cycle. *Biochemistry, 34* (44), 14490–14499.

Varo, G., Needleman, R., & Lanyi, J. K. (1995b). Light-driven chloride ion transport by halorhodopsin from Natronobacterium pharaonis. 2. Chloride release and uptake, protein conformation change, and thermodynamics. *Biochemistry, 34*(44), 14500–14507.

Wang, H., Peca, J., Matsuzaki, M., Matsuzaki, K., Noguchi, J., Qiu, L., et al. (2007). High-speed mapping of synaptic connectivity using photostimulation in Channelrhodopsin-2 transgenic mice. *Proceedings of the National Academy of Sciences of the United States of America, 104*(19), 8143–8148.

Waschuk, S. A., Bezerra, A. G., Shi, L., & Brown, L. S. (2005). Leptosphaeria rhodopsin: Bacteriorhodopsin-like proton pump from a eukaryote. *Proceedings of the National Academy of Sciences of the United States of America, 102*(19), 6879–6883.

Wentz, C. T., Bernstein, J. G., Monahan, P., Guerra, A., Rodriguez, A., & Boyden, E. S. (2011). A wirelessly powered and controlled device for optical neural control of freely-behaving animals. *Journal of Neural Engineering, 8* (4), 046021.

Yoshimura, K., & Kouyama, T. (2008). Structural role of bacterioruberin in the trimeric structure of archaerhodopsin-2. *Journal of Molecular Biology, 375*(5), 1267–1281.

Zhang, F., Wang, L. P., Boyden, E. S., & Deisseroth, K. (2006). Channelrhodopsin-2 and optical control of excitable cells. *Nature Methods, 3*(10), 785–792.

Zhang, F., Wang, L. P., Brauner, M., Liewald, J. F., Kay, K., Watzke, N., et al. (2007). Multimodal fast optical interrogation of neural circuitry. *Nature, 446*(7136), 633–639.

Zhao, S., Cunha, C., Zhang, F., Liu, Q., Gloss, B., Deisseroth, K., et al. (2008). Improved expression of halorhodopsin for light-induced silencing of neuronal activity. *Brain Cell Biology, 36*(1–4), 141–154.

Zorzos, A. N., Boyden, E. S., & Fonstad, C. G. (2010). Multi-waveguide implantable probe for light delivery to sets of distributed brain targets. *Optics Letters, 35*(24), 4133–4135.

T. Knöpfel and E. Boyden (Eds.)
Progress in Brain Research, Vol. 196
ISSN: 0079-6123

CHAPTER 4

Genetically encoded probes for optical imaging of brain electrical activity

Amélie Perron, Walther Akemann, Hiroki Mutoh and Thomas Knöpfel*

RIKEN Brain Science Institute, Hirosawa, Wako City, Saitama, Japan

Abstract: The combination of optical imaging methods with targeted expression of protein-based fluorescent probes constitutes a powerful approach for functional analysis of selected cell populations within intact neuronal circuitries. Herein, we lay out the conceptual motivation for optogenetic recording of brain electrical activity using genetically encoded voltage-sensitive fluorescent proteins (VSFPs), describe how the current generation of VSFPs has evolved, and demonstrate how VSFPs report membrane voltage signals in isolated cells, brain slices, and living animals. We conclude with a critical appraisal of VSFPs for voltage recording and highlight promising applications of this emerging methodology for bridging cellular and intact systems biology.

Keywords: genetically encoded voltage sensors; fluorescence; optical imaging; neuronal activity.

Introduction

Brain function relies on coordinated signaling between neurons organized into functional microcircuits, which in turn, assemble into larger scale systems. Electrical events play an essential role for signal propagation and information processing at all levels of the structural and functional organization of the brain. These events include unitary currents through ion channels (molecular level), synaptic potentials (subcellular level), action potentials (cellular level), local field potentials (microcircuit level), and electroencephalographic voltage transients (system level). For the past several decades, a solid characterization of the electrophysiological properties of ion channels and individual neurons has been gained from microelectrode recordings of mammalian neurons. Likewise, the functional significance of electrical signals mediating information transfer between pairs or small groups of neurons is also well understood. But, how the activity of large populations of neurons is orchestrated and assembled into coding patterns remains poorly understood. Concepts for neural population coding are perhaps best established in the case of rhythmic activities observed in local field potential and electroencephalographic recordings

*Corresponding author.
E-mail: tknopfel@brain.riken.jp; tknopfel@knopfel-lab.net

DOI: 10.1016/B978-0-444-59426-6.00004-5

(e.g., see Buzsáki, 2006; Traub and Whittington, 2010) but a comprehensive understanding of population coding at the intact system or behavioral level is still elusive. Until very recently, the principal approach for investigating neuronal population responses relied on the use of single or multiple electrodes. However, further progress in our understanding of the dynamics of complex circuits such as cortical columns and the interaction of such circuits at the mesoscopic level requires an experimental approach that allows the monitoring of electrical activity from very large numbers of neurons simultaneously in a cell type-specific manner. In the following, we will demonstrate that genetically encoded voltage probes have the potential to achieve this goal.

The idea to use optical imaging to map brain activities goes back several decades ago with the report of activity-dependent changes of endogenous fluorophores and absorption or scattering of light through neural tissue (Cohen et al., 1968; reviewed in Villringer, 1997). The main advantage of optical imaging is, in principle, its minimal invasiveness in comparison to recording techniques based on needle electrodes because even the finest microelectrode will mechanically lesion or disintegrate brain tissue. In addition, optical methods can provide a superior spatial resolution since photons interact at the molecular level while electrodes are larger than molecules. The use of exogenously applied dyes for reporting neuronal activities has been pioneered by Cohen and coworkers who demonstrated changes in dye absorption during action potentials in giant axons from squid followed by the first multisite optical recordings of network activities by means of voltage-sensitive dyes (Grinvald et al., 1977; Ross et al., 1974). While absorption dyes were initially the focus of development, fluorescent dyes became dominant afterwards as they provide signals with higher contrast due to lower background. They also allow for epi-illumination and, hence, application to non-transparent preparations like living rodents. Despite its enormous potential, classical voltage imaging has a small signal-to-noise ratio (SNR) and is also hindered by significant cell toxicity. It is largely for these reasons that voltage-sensitive dyes have lately been superseded by calcium indicators for the imaging of network activities.

Although well performing calcium and voltage dyes have been developed during the past 20 years, genetically encoded probes are highly sought after as they allow the monitoring of neuronal populations based on their identity or connectivity rather than their location. Since protein-based sensors are encoded in DNA, they can be expressed in practically any cell type and have the advantage of staining only the cell population determined by the promoter used to drive the expression. This is a major advance over classical system neurophysiological approaches relying on the detection of action potentials generated by sensory input, motor output, or internal processes (e.g., expectations, decision making) with limited consideration of whether the recorded neurons are excitatory or inhibitory. A second advantage of genetically encoded probes is the possibility to generate transgenic animals with stable labeling of cells, resulting in preparations ready for acute and chronic *in vitro* or *in vivo* imaging without the need for time consuming and invasive staining procedures. In addition to expediency, genetic labeling is more reliable than conventional organic dyes due to fewer variations between individual cells within or between preparations. Finally, genetically encoded probes facilitate observations over long periods of time (weeks and months) as required for developmental and regeneration studies in contrast to other staining procedures.

Genetically encoded calcium indicators have already been proven to successfully report neural activity in zebrafish, worms, flies, and mice with intensity changes larger than those of voltage-sensitive dyes (Horikawa et al., 2010; Tian et al., 2009). So, why are genetically encoded voltage probes so eagerly needed? One important difference between calcium and voltage imaging of network activities is that voltage signals report subthreshold activities while calcium imaging mainly captures action potential activities in neural network analysis. Although action potentials represent the main output for postsynaptic cells, the time structure of action potential events is the result of a neuronal computation that is reflected in the time course of subthreshold electrical activity in dendritic and somatic membranes. Moreover, subthreshold activity resolves information about the spiking activities of the ensemble of neurons that are synaptically interconnected while most of this information is lost in the action potential firing information of a single neuron. Indeed, voltage imaging at the population level usually favors the recording of subthreshold postsynaptic signals from dendrites; since, they constitute a significant fraction of labeled membranes. Another important difference between calcium and voltage imaging is that voltage signals are much faster than calcium responses. Because calcium enters the cell during action potential firing, calcium imaging can be used as an indirect measure of voltage transients. However, calcium imaging usually provides sufficient single-sweep sensitivity only for action potential spiking activities (Cossart et al., 2005; Muri and Knöpfel, 1994) but not for subthreshold synaptic communication.

Evolution of VSFPs

Over the past 15 years, genetically encoded optical voltage reporter proteins have considerably evolved (for reviews, see Knöpfel et al., 2006; Mutoh et al., 2011; Perron et al., 2009a). Initial design concepts exploited voltage-dependent conformational changes associated with voltage-gated ion channels (FlaSh: Siegel and Isacoff, 1997; SPARC: Ataka and Pieribone, 2002) or their isolated voltage sensor domain (VSFP1: Sakai et al., 2001), resulting in modulation of the reporter protein fluorescence intensity. Although these first generation voltage probes were shown to optically report changes in membrane potential, their application in mammalian systems is considerably limited by their poor membrane targeting in transfected cells (Baker et al., 2007). More recently, largely improved second generation of voltage-sensitive proteins (VSFP2s) were developed by replacing the actuator protein in our VSFP1 (Sakai et al., 2001) with the voltage-sensing domain of the non ion channel-forming protein *Ciona intestinalis* voltage sensor-containing phosphatase (Ci-VSP), leading to sensors with increased targeting to the plasma membrane and reliable responsiveness to membrane potential signaling in isolated cells, brain slices, and living mice (Fig. 1a; Akemann et al., 2010; Dimitrov et al., 2007; Lundby et al., 2008; Mutoh et al., 2009, 2011; Tsutsui et al., 2008). We also generated a series of monochromatic fluorescent probes (termed VSFP3s; Fig. 1b) with spectral variants spanning the cyan to far-red region of the visible light spectrum (Lundby et al., 2008; Perron et al., 2009b). Whereas VSFP2s were based on a FRET reporting mechanism for protein conformational rearrangements, VSFP3s respond to voltage changes with a modulation in the fluorescence intensity of a single fluorescent protein through a FRET-independent mechanism that is only partially understood. VSFP3s offer the advantages of a broad coverage of the color spectrum with fast overall kinetics, but with signal amplitudes smaller than VSFP2s. Despite these limitations in signal size, VSFP3s were the first probes to report spontaneous electrical activity in neuronal cultures (Perron et al., 2009b). We have also explored a voltage-reporting design based on circularly permuted fluorescent proteins, resulting in proof-of-principle voltage-sensitive probes, albeit with small fluorescence changes (Fig. 1c; Gautam et al., 2009). Finally, we have been exploring additional design approaches. Most notable are new VSFP variants that we named “Butterflies” in which the voltage sensor domain is sandwiched between two fluorescent proteins (Fig. 1d). The rationale behind this design was to combine the advantageous properties of VSFP2s (large dynamic range) with the fast kinetics of VSFP3s.

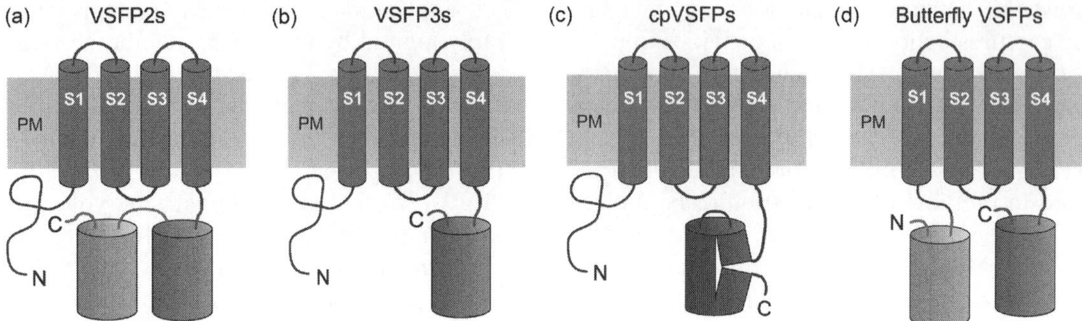

Fig. 1. Schematic design of VSFP probes derived from a combination of a voltage sensor domain and fluorescent proteins. (a) VSFP2s are FRET-based voltage sensors comprising an acceptor (e.g., Citrine) and donor (e.g., Cerulean) fluorophore fused to the fourth transmembrane segment (S4) of the voltage-sensing domain of Ci-VSP. (b) VSFP3s involve a single FP. (c) cpVSFPs are based on circularly permuted fluorescent proteins (e.g., mKate). (d) In Butterfly VSFPs, the voltage-sensing domain is sandwiched between a fluorescent protein FRET pair. PM: plasma membrane. (For color version of this figure, the reader is referred to the Web version of this chapter.)

The domain structures of presently available VSFP variants are illustrated in Figure 1.

It is important to stress that many fusion proteins involving FPs are mistargeted or form intracellular aggregates (Mutoh et al., 2011). Additional fluorescence originating from these aggregates adds photon noise and decreases the signal amplitude after normalization to baseline fluorescence values ($\Delta F/F_0$). Aggregate formation is particularly troublesome in sensors that are derived from fluorescent proteins isolated from the Anthozoa class as these proteins have a high tendency to form various kinds of deposits in mammalian brain cells even when expressed in the absence of a fusion partner (Hirrlinger et al., 2005). To some extent, this behavior may reflect their origin from tetrameric proteins but it might also indicate retention within intracellular compartments (e.g., endoplasmic reticulum, ER, or Golgi apparatus) or some yet unknown affinity for intracellular organelles. It is therefore not surprising that VSFP imaging of neuronal responses in mouse brain preparations has revealed signal amplitudes that are much smaller than those obtained in the more simple Xenopus model system (Tsutsui et al., 2008; Villalba-Galea et al., 2009) or in selected regions devoid of background fluorescence (Mutoh et al., 2009). As the effectiveness of membrane potential indicators largely depends on the protein fraction that is correctly targeted to the plasma membrane, we have applied combinational molecular trafficking strategies (Gradinaru et al., 2010; Zhao et al., 2008) to increase the sensitivity of our biosensors. Such modifications include the introduction of Golgi trafficking signals and additional ER export motifs to favor transport along the secretory pathway to the cell surface (unpublished observations).

State of the art VSFPs

The FRET-based VSFP2.3 (Akemann et al., 2010; Lundby et al., 2008; Mutoh et al., 2009) is not only the first sensor that allowed voltage imaging in living mice but also remains one of the best characterized and validated VSFP variants reported so far. As for all VSFP variants tested in our laboratory, we initially characterized VSFP2.3 in PC12 cells. This model system is convenient and provides reliable data, a feature that is essential when comparing different probes. Figure 2 illustrates a data set obtained in this system, wherein PC12 cells were voltage clamped

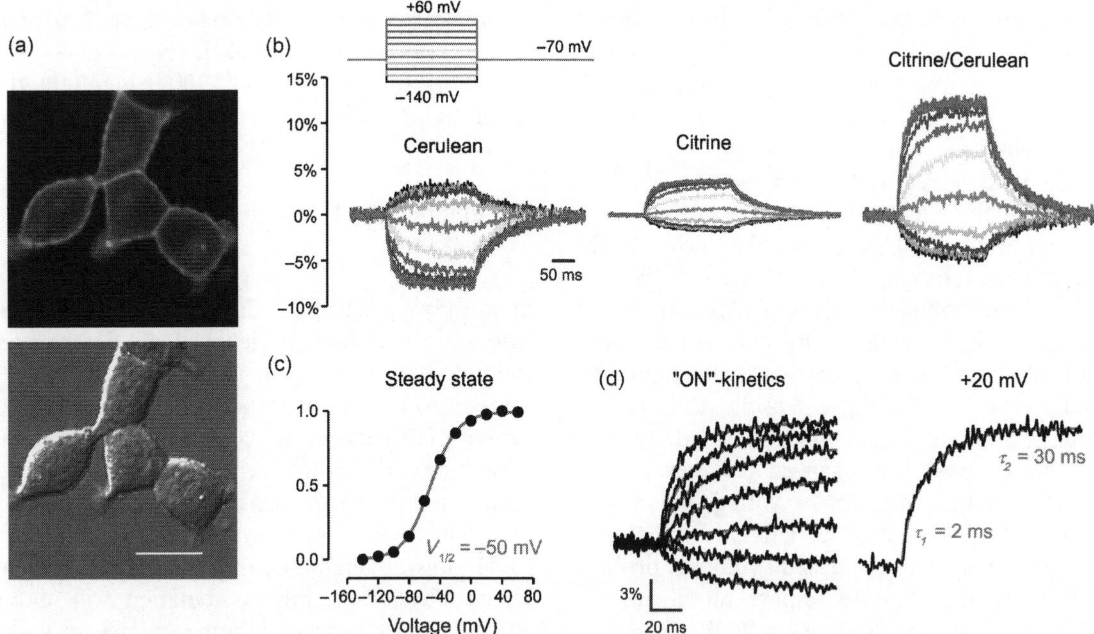

Fig. 2. Expression and functional properties of VSFP2.3 in PC12 cells. (a) Confocal and transmission images of PC12 cells transiently expressing VSFP2.3. (b) Changes in VSFP2.3 fluorescence evoked by a family of 200ms voltage steps (−140 to +60 mV; 20mV increments) from a holding potential of −70mV. Cerulean and Citrine emission were simultaneously recorded with fast photodiodes and averaged over five trials in a single cell. (c) Fluorescence–voltage relationship of VSFP2.3 fitted with a single Boltzmann equation. (d) Voltage response kinetics of VSFP2.3. *Right panel*, activation kinetics fitted with two exponentials. The left panel illustrates the fluorescence response to +20mV with fast and slow components. (For color version of this figure, the reader is referred to the Web version of this chapter.)

and fluorescence signals to voltage steps were simultaneously recorded in the donor and acceptor spectral bands. The fluorescence–voltage relationship is reminiscent of the activation curve of ion channels with a $V_{1/2}$ value around −50mV. The optical response to membrane depolarization is initially dominated by a rather fast ON time component of 2ms accompanied by a slow time constant in the order of 30ms at +20mV.

In addition to a large dynamic range, the tandem fluorescent protein design provides bidirectional donor and acceptor signals that allow the separation of hemodynamic and movement-related signals from membrane voltage signals during *in vivo* recordings (Akemann et al., 2010). Most importantly, VSFP2.3 exhibits efficient targeting to neuronal plasma membranes as opposed to variants derived from anthozoan proteins (Mutoh et al., 2011; Tsutsui et al., 2008), which makes it the sensor of choice for voltage imaging experiments in mammalian brain slices and living mice. The best red-shifted VSFP variants are based on the Butterfly design illustrated in Figure 1d.

Space for future improvement of VSFPs

An important concern raised during early phases of VSFP development relates to the voltage-reporting speed of the probes, which is slow compared to fast neuronal action potentials. Originally, the argument was that a change in membrane potential cannot exert sufficient force to translocate a bulky fluorescent protein with suitably fast kinetics. While initial prototypic VSFP2s essentially reported a low-pass filtered component of fast voltage transients, newer probes such as VSFP2.42 and VSFP3.1 are reaching voltage-reporting responses with time constants in the millisecond time range (1–4 ms; Akemann et al., 2010; Perron et al., 2009b). Such speed is very close to the limit set out by the kinetics of the voltage-dependent structural rearrangement of the Ci-VSP voltage sensor domain as indicated by sensing current measurements (Lundby et al., 2008). These sensing currents are analogous to gating currents generated by the movement of positively charged residues within the voltage sensor domain of voltage-gated ion channels during voltage-dependent structural rearrangements. Ci-VSP-based voltage probes with faster kinetics would require modifications of the voltage sensor domain itself. We envision achieving this goal by developing chimeric voltage sensor proteins that will include structural components from fast voltage-gated ion channels. To exploit the enhanced time resolution of faster reporters the light sampling intervals need to be shortened accordingly, but then instrumental limitations or limitations of SNR may become a restraining factor in the end.

Future efforts might also lead to probes that are not only faster in responding to voltage changes but that also exhibit larger voltage sensitivity than currently available VSFP2s. This goal may be achieved by using conformation-sensitive circularly permuted fluorescent proteins (Baird et al., 1999). Circular permutation rearranges the ends of the fluorescent proteins as to place a sensor protein domain in close proximity to the chromophore environment, which can ultimately lead to direct changes in the fluorescence signal upon structural conformational changes initiated by the sensor domain. Although circularly permuted FP-based VSFP variants (Gautam et al., 2009) are not yet outperforming the FRET-based VSFP2s, work on genetically encoded calcium sensors based on green or yellow circularly permuted fluorescent proteins (Akemann et al., 2001; Baird et al., 1999) such as GCaMP (Nakai et al., 2001) and Pericam (Nagai et al., 2001) highlighted the potential of this approach.

Combining VSFP readouts with optogenetic control tools

Naturally occurring light-activated Channelrhodopsin-2 (ChR2; Nagel et al., 2003) or chloride pumping halorhodopsin (Kolbe et al., 2000) has proven to be powerful tools for controlling neural activity (Boyden et al., 2005; Zhang et al., 2006). Combining these light-driven actuators with light-emitting reporter proteins such as VSFPs will open the door to an all-optogenetic approach to electrophysiology. In analogy to conventional intracellular electrophysiology, this combination will allow to control and record membrane voltage in a single preparation simply by using light. In contrast to conventional intracellular electrophysiology, optogenetic electrophysiology will reveal connections in large numbers of genetically defined groups of neurons. The combination of VSFPs and opsins requires careful consideration of the spectral properties of the two optogenetic components. A particularly critical factor is an adequate separation of the opsin activation spectrum from the VSFP excitation spectrum such that VSFP imaging does not interfere with opsin activation. For compatibility with ChR2 and derivates with similar activation spectra, VSFPs with excitation in the near infrared range (>600 nm) are likely to be needed to permit multimodal stimulation and voltage imaging. Bright fluorescent reporter proteins emitting beyond the far-red range will be required to advance further in that direction.

Alternative genetic approaches for optical voltage report

In addition to fully genetically encoded voltage sensors, alternative approaches using genetics in combination with classical organic chromophores were also explored (Chanda et al., 2005; DiFranco et al., 2007; Sjulson and Miesenbock, 2008). Chanda et al. (2005) introduced the hybrid voltage sensor (hVOS) design comprising a genetically encoded fluorescent reporter protein (GFP) and a voltage-sensitive absorbance dye, dipicrylamine (DPA), wherein DPA acts as a mobile charge and FRET acceptor within the membrane electric field to quench a membrane-bound GFP in a voltage-dependent manner. While hVOS exhibits the largest fluorescence signals among all genetically encoded voltage sensors reported to date, the required DPA concentrations (>3 μM) increase the membrane capacitance (Fernandez et al., 1983; Zimmermann et al., 2008) to a level that might affect neural activity (Sjulson and Miesenbock, 2008). An improved probe for hVOS imaging was subsequently shown to resolve single action potentials in cultured hippocampal neurons but with much smaller fractional changes in fluorescence (Wang et al., 2010). At the present, the hVOS has not been proven to successfully resolve brain electrical activity *in vivo*. Indeed, the main concern of this approach is related to well-known issues associated with organic dyes such as limited access to certain brain areas and toxicity. A second semi-genetic concept relies on the activation of a precursor voltage-sensitive dye by a genetically encoded enzyme (Hinner et al., 2006; Ng and Fromherz, 2011). To explore this principle for cell-selective staining, Hinner et al. (2006) synthesized an amphiphilic hemicyanine dye that loses its water solubility after cleavage of phosphate groups. Cell specific staining was achieved by genetically expressing a phosphatase on the extracellular side of the plasma membrane that removes the hydrophilic phosphate group, resulting in a lipophilic product that integrates and specifically stains the membrane of phosphatase-expressing cells. While this approach could potentially be applied to a variety of classical lipophilic voltage-sensitive dyes, its applicability in intact neuronal tissues remains unproven.

Instrumental challenges and sensitivity of VSFPs

High spatiotemporal resolution optical voltage imaging has been an active field of research and development over the past 20 years but is merely mastered by few neurophysiology laboratories worldwide. As microscopes designed for functional optical imaging are now already available in most neuroscience departments, we will mainly comment on some specific aspects related to signal detection using VSFP imaging. In earlier days, sufficient temporal resolution was obtained only with photodiode array detectors but this approach has been surpassed by readout noise-optimized charge-coupled device (CCD) and complementary metal–oxide–semiconductor (CMOS)-based cameras that can operate at up to 10 kHz frame rate. Single photodiodes however remain a simple and inexpensive solution for development and initial characterization of VSFPs in individual voltage-clamped cells (Akemann et al., 2010; Dimitrov et al., 2007).

In addition to a large $\Delta F/F_0$ value, reliable voltage-sensitive probes should also have a low level of noise associated with the optical signal to faithfully report electrical activity under experimental conditions suitable for neurophysiological studies. This is particularly important for the imaging of fast voltage transients requiring minimal frame rates of 100–1000 Hz as shorter exposure times yield fewer photons and higher baseline-normalized noise levels. We like to remind that photon shot noise represents the fluctuation of photons associated with the detected signal and is determined by the stochastic nature of photon emission following a Poisson distribution function (noise $= \sqrt{}$number of photons). Therefore, the SNR increases proportionally with the square root of the fluorescence intensity. A SNR greater than 5 for a signal of 1% $\Delta F/F_0$ would require a noise amplitude minimized to 0.2% of baseline fluorescence, which necessitates the detection of about 140,000 photons per sampling time interval (e.g., 10 ms if the rate is 100 Hz). This requirement therefore constrains the choice of light detector considerably. Indeed, a suitable CCD or CMOS must not only allow for fast imaging but also needs to be able to sample large numbers of photons (specified as "well depth"). Under the imaging conditions discussed herein, the charge readout noise of the detector, although critical, is a frequently overrated specification parameter. Readout noise increases with the square root of the readout speed, and can be greater than 100 electrons per pixel in high-speed CCDs (Pawley, 2006). It is important, however, to realize from the above calculations that the detector noise is already overwhelmed by the shot noise associated with the minimal number of detected photons required to resolve, for instance, 1% change in fluorescence intensity. The need to collect large numbers of photons from membranes of interest then establishes a scenario in which the optical components and experimental conditions have to be optimized. This entails suitably high expression levels of the probe (without interfering with proper membrane targeting due to overexpression), sufficient output power from the excitation light source (with minimal photobleaching of the probe), low fluorescence background, optimal light path for the collection of emitted photons, and a high quantum yield of the photodetector (for details, see Akemann et al., 2009a). Another important but frequently underrated consideration is the selection of high-quality optical filters and beam splitters carefully optimized for the spectral response properties of each of the VSFP probes (Mutoh et al., 2009).

Gene transfer challenge

One of the most important features of genetically encodable sensors is the possibility to specifically target different neuronal populations or locations in the nervous system. Several gene delivery approaches have been explored so far and these methods will be discussed in more detail in Chapter 9 of this issue. In this section, we will mainly review some aspects particularly relevant for use of VSFPs in transgenic mouse models since *in utero* electroporation and DNA transfer techniques have already been covered elsewhere (Mutoh et al., 2011). Expression of fluorescent proteins in genetically specified populations of neural cells has been achieved using several regulatory gene sequences such as promoters for glutamic acid decarboxylase (Tamamaki et al., 2003), odorant receptors (Potter et al., 2001), and K^+ channel Kv3.1 (Metzger et al., 2002) to name a few. In particular, a promoter that specifically drives synapto-pHluorin expression to olfactory sensory neurons has been used for imaging neurotransmitter release in glomeruli of the olfactory bulb *in vivo* (Bozza et al., 2004). The releasing events triggered by single action potentials produced only small synapto-pHluorin signals but this work clearly exemplifies the potential of genetic targeting of fluorescent probes. The regulatory sequences of Kv3.1 were also used to generate transgenic mice expressing the fluorescent protein Ca^{2+} sensor GCaMP2 (Akemann et al., 2009b; Díez-Gar et al., 2007; Díez-García et al., 2005). Physiological experimentation using these GCaMP2 mice has provided several important insights about achievable spatiotemporal resolution of functional imaging based on GFP-derived indicator proteins. However, the generation of transgenic mice that exhibit sufficiently high VSFP expression levels in selected populations of neurons with minimal adverse effects on cell function or viability has been particularly troublesome. Most importantly, the formation of intracellular clusters resulting from mistargeting and/or aggregation (see section "Evolution of VSFPs" for further details) should be minimized not only to optimize SNR but also to preserve the functional integrity of neuronal circuitries. Since synthesis, maturation, and trafficking of membrane proteins is a relatively slow process, expression of VSFPs in intact tissue (brain slices or *in situ*) requires gene transfer *in vivo*. In principle, gene delivery could be archived via viruses carrying cell type-specific promoters but most, if not all, promoters tested to date for this purpose are either not specific enough or too weak to drive sufficient expression levels of transgene construct. In most recent studies that employed optogenetic stimulation techniques, viruses have been used to target the expression of light-activated channels to multiple cell types using a strong promoter in a loxP/Cre recombinase-dependent system under a FLEX configuration (Atasoy et al., 2008; Tsai et al., 2009). In this approach, Cre recombinase is used to invert a reverse coding sequence with flanking antiparallel loxP-type recombination sites. The inversion of the sequence then turns on the transcription of the gene of interest. The advantage of this approach is that expression of Cre recombinase can be driven by a weak but highly cell type-specific promoter while a strong but ubiquitous promoter ensures sufficient high expression levels of the recombinant protein. Viral gene delivery has, however, several limitations including low transgene capacity and spatially inhomogeneous transduction efficacy (typically concentrated around the injection site). This limitation can be overcome by the generation of mice that carry VSFP coding sequences under a strong promoter in a Cre-dependent configuration. Such transgenic mice can be crossed with available Cre-driver mouse lines to target the expression of the transgene to specific cells or tissues without any size limitation.

Marker genes define cell populations that may be subdivided on the basis of functional criteria such as recruitment under distinct tasks. Therefore, future work should aim at the development of a genetic approach for targeting groups of neurons according to their spatiotemporal activity patterns (rather than their genetic identity). Such advance would allow the imaging of coactive neurons or neural assemblies that carry information to guide behavior in complex mammalian circuits. This could be achieved by using activity-dependent regulatory sequences such as immediate early genes Arc, Egr1, and Fos (Peron and Svoboda, 2011; Shepherd and Bear, 2011).

VSFP imaging in brain slices

Classical voltage-sensitive dyes have successfully reported voltage responses from large neuronal populations in acute brain slice preparations. Most of these experiments were limited to conditions where most (if not all) cells were synchronously recruited with the aim to image the spread of excitation following electrical stimulation, or to analyze spontaneous rhythmic activity (Kimura et al., 1999; Tsau et al., 1998). The boundaries of these studies were those described in the section "Introduction", namely the difficulty to attribute fluorescence output signals to a defined cell population due to averaging over heterologous cell groups and a modest SNR resulting from large amounts of background florescence from cells that are not the focus of interest. These caveats can be overcome by injecting a voltage-sensitive dye directly into a single cell but derived conclusions will remain at the single-cell level and will not be suitable for population signaling and, therefore, inapplicable to the analysis of large number of neurons.

As genetically encoded proteins can be targeted to specific cell populations, VSFP voltage imaging is a powerful tool for electrophysiological studies in brain slices. Such preparations have the advantage of easy manipulability and low complexity as compared to intact animals while basic circuit properties are preserved. Hence, Akemann et al. (2010) reported that VSFP2 imaging in acute cortical brain slices can resolve synaptic potentials from single neurons or large populations of selected neurons in single trials (Fig. 3). Action potentials were also clearly resolved against slower voltage transients during synaptic stimulation and steady-state current injection in averaged traces (Akemann et al., 2010). VSFP expression levels in mice electroporated *in utero* thus allow the imaging of brain electrical activity with sufficient SNR without interfering with action potential generation. In the context of voltage imaging, the absence of confounding signal components from hemodynamic processes in brain slices preparations is also a valuable advantage. In particular, the absence of movement artifacts from heartbeat and respiration usually observed in living animals

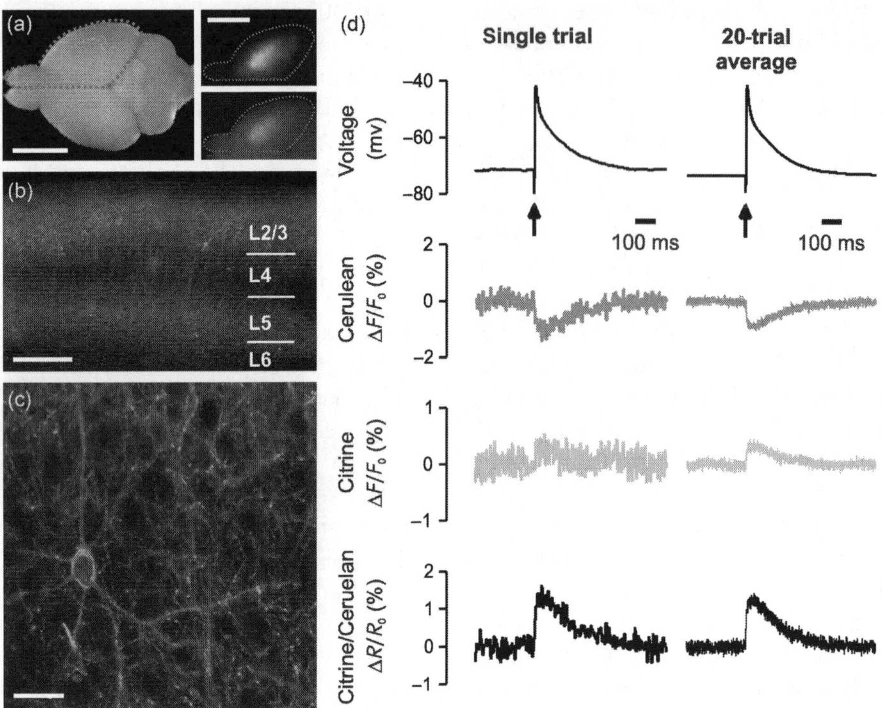

Fig. 3. Expression and voltage responses of VSFP2.3 in mouse cerebral cortex. (a) Epi-illumination image of the brain of a mouse electroporated *in utero* with VSFP2.3 showing the expression area with a green oval (left). Citrine and Cerulean fluorescence images with the contour of the right hemisphere and olfactory bulb outlined in red (right). Confocal fluorescence images of VSFP2.3 in brain slices at low (b) and high (c) magnification. Cortical layers L2/3 to L6 are labeled. (d) Subthreshold postsynaptic voltage responses evoked by a single-shock electrical field stimulation. Single traces (left) and 20-trial averages (right) are shown. Scale bars: 5mm in (a), 250μm in (b), and 25μm in (c). Adapted from Akemann et al. (2010). (For interpretation of the references to color in this figure legend, the reader is referred to the Web version of this chapter.)

along with more favorable optical conditions facilitate the use of monochromatic VSFP3s and, more importantly, their different color combination for multiple and simultaneous labeling of different cell types.

VSFP imaging in living animals

The main motivation for the development of VSFP imaging was the need of an optical method for reporting interaction dynamics within and between large neuronal populations with high spatiotemporal resolution in intact, behaving animals. As mentioned above, *in vivo* imaging is inherently challenging due to additional noise sources such as mechanical movements caused by heartbeat and breathing rhythms, hemodynamic volume fluctuations, changes in hemoglobin oxygenation, and activity-dependent tissue autofluorescence. To address these issues, VSFPs can be imaged through a thinned skull reinforced with a cover glass (Drew et al., 2010), as shown in Figure 4. This approach maintains the intracranial pressure and stabilizes the brain, thereby minimizing motion-induced imaging artifacts considerably. In addition, capturing of voltage transients through a thinned but otherwise intact skull is much less invasive than the craniotomy usually needed for organic dye imaging and

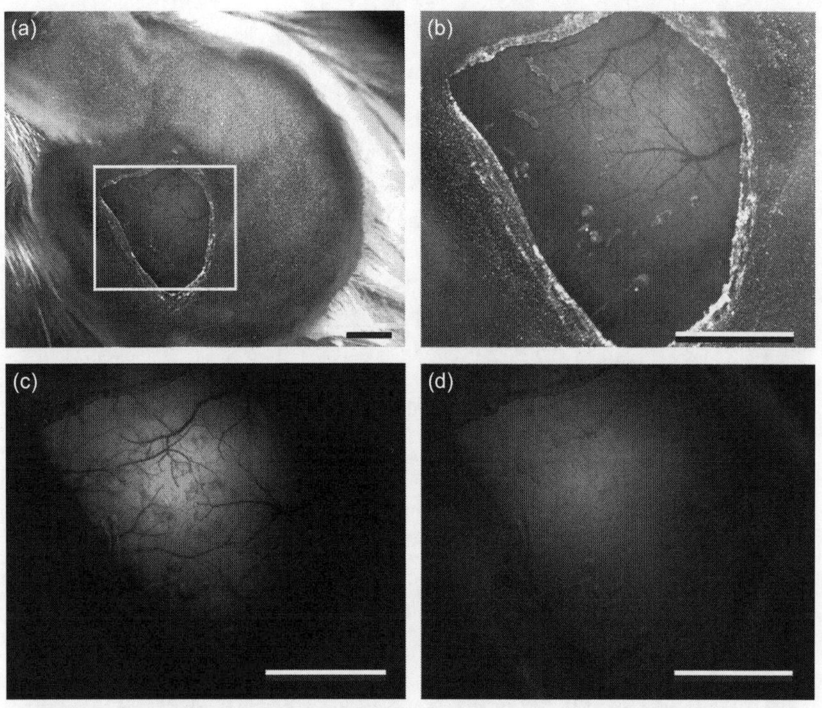

Fig. 4. VSFP imaging through a thinned-skull cranial window in a living mouse. (a) Wide field view of a mouse head after the embedment of a cover glass to the thinned bone. The surgery was performed in anesthetized animals. The boxed area outlining the brain window is shown at a higher magnification in (b). (c and d) Fluorescence images of a yellow–red VSFP variant captured through the glass window. Scale bars: 2mm. (For interpretation of the references to color in this figure legend, the reader is referred to the Web version of this chapter.)

conventional electrophysiology. Moreover, VSFP2s have the advantage of providing a ratiometric signal (a decrease in the donor fluorescence readout and an increase in the fluorescence intensity of the acceptor) that facilitates both signal identification and separation from movements and intrinsic signals as the ratio of donor and acceptor fluorescence excludes voltage-independent intensity fluctuations of the same polarity. This feature was demonstrated recently by Akemann et al. (2010) in an experiment paradigm, wherein mechanical stimulus was applied sequentially to different whiskers in anesthetized mice (Fig. 5). VSFP2 imaging enabled the construction of maps of cortical representation of different whiskers in the same mouse with receptive areas mainly overlapping with the cortical barrel field representation (Fig. 6).

An alternative strategy to entirely eliminate hemodynamic and heartbeat-related signals for *in vivo* VSFP recordings builds on VSFP variants derived from far-red shifted fluorescent proteins that lack spectral overlap with hemoglobin oxygenation states. These variants will further expand the toolbox of VSFP probes designed for *in vivo* experimentation.

So far, *in vivo* VSFP imaging has only been established for superficial brain structures at the circuit (mesoscopic) level. Single-cell resolution imaging of deeper brain structures is however problematic with conventional epifluorescence approaches due to light scattering which decreases the number of photons available for

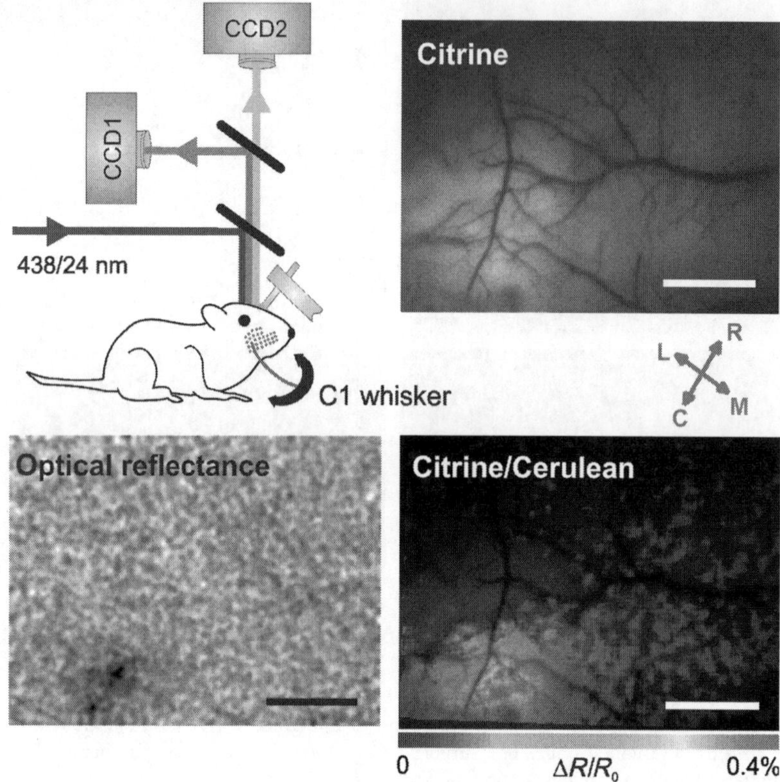

Fig. 5. VSFP2.3 imaging in the somatosensory cortex *in vivo*. *Upper left*, schematic diagram of the experimental configuration for imaging donor and acceptor fluorescence by two synchronized CCD cameras in living animals. *Upper right*, example of an image of Citrine fluorescence signal captured with CCD2. *Lower left*, optical reflectance signal (intrinsic imaging technique) evoked by repetitive C1 whisker stimulation. *Lower right*, ratiometric ($\Delta R/R_0$) VSFP2.3 signal to a single whisker deflection at the time of the maximal response. Arrows indicate the rostral (R), caudal (C), lateral (L), and medial (M) directions. Scale bars: 1mm. Adapted from Akemann et al. (2010). (For color version of this figure, the reader is referred to the Web version of this chapter.)

detection and therefore limits the depth of observation. As such, voltage imaging in deep brain tissue adds new technical challenges (nonlinear optics, light patterning) that will need to be addressed in a near future.

Perspective

VSFP imaging has already succeeded important benchmark experiments including the demonstration of its practicability in resolving responses to sensory stimuli in living mice, yet this emerging methodology still requires additional evolutional developments that will need to be undertaken. We predict that VSFP variants will have an imminent impact on system physiology in addressing two overlapping fundamental questions. First, VSFP imaging will be especially valuable in experiments designed for unraveling the contribution of specific cell types in the dynamics of local neuronal networks in the cerebral cortex. A specific topic relevant to this context is the interplay between inhibition and excitation during the high frequency oscillations classically observed by local field potential recordings (Adesnik and

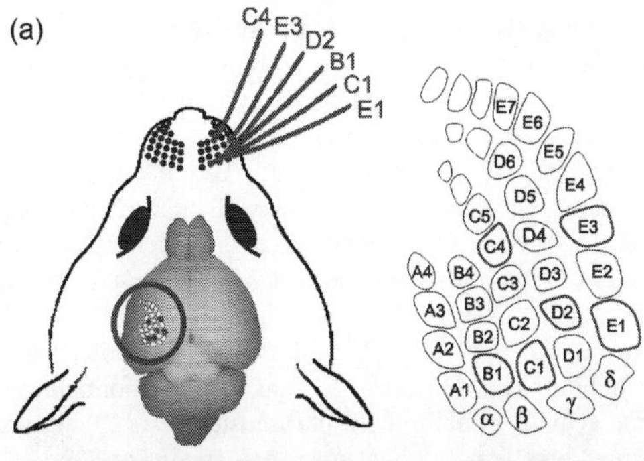

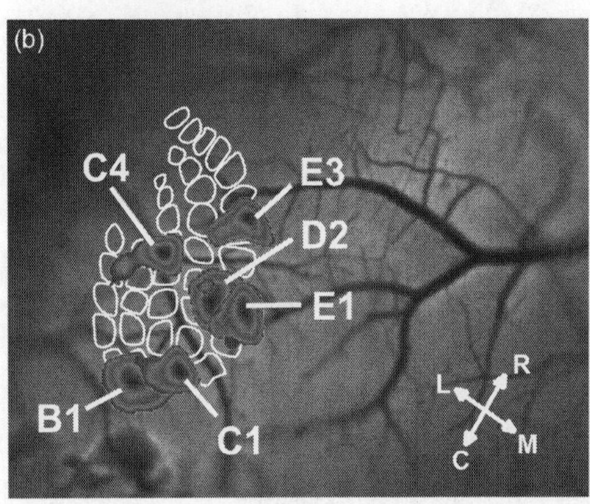

Fig. 6. Construction of receptive field maps with VSFP2.3 imaging in living mice. (a) *Left panel*, selection of six whiskers (C4, E3, D2, E1, B1, and C1) for sequential stimulation and receptive field imaging. *Right panel*, identification of selected whiskers in a schematic representation of a cortical barrel field in the mouse somatosensory cortex. (c) Composite image showing an overlay of the responsive areas from each of the six stimulated whiskers from a single VSFP2.3 mouse. Responsive areas are shown in color. Arrows indicate the rostral (R), caudal (C), lateral (L), and medial (M) directions. Scale bar: 1mm. Data are from Akemann et al. (2010). (For interpretation of the references to color in this figure legend, the reader is referred to the Web version of this chapter.)

Scanziani, 2010; Buzsáki, 2006; Traub and Whittington, 2010). Secondly, VSFP imaging will be a powerful method to unfold interaction dynamics across distant cortical areas during perception, cognition, and motor behavior (Ferezou et al., 2007; Huang et al., 2010; Matyas et al., 2010; Wu et al., 2008). Both types of studies will certainly profit from the new experimental opportunities provided by the VSFPs in performing stable recordings over periods of weeks and probably months without the need to resupply the probe as in the case of organic dyes. Finally, it should be noted that single-cell *in vivo* imaging with VSFPs has not been established yet. We expect that emerging optical methods such as two-photon excitation fluorescence microscopy with temporal focusing and light-sculpturing techniques will be instrumental to achieve this goal.

References

Adesnik, H., & Scanziani, M. (2010). Lateral competition for cortical space by layer-specific horizontal circuits. *Nature*, *464*, 1155–1160.

Akemann, W., Lundby, A., Mutoh, H., & Knöpfel, T. (2009a). Effect of voltage sensitive fluorescent proteins on neuronal excitability. *Biophysical Journal*, *96*, 3959–3976.

Akemann, W., Middleton, S. J., & Knöpfel, T. (2009b). Optical Imaging as a link between cellular neurophysiology and circuit modeling. *Frontiers in Cellular Neuroscience*, *3*, 5.

Akemann, W., Mutoh, H., Perron, A., Rossier, J., & Knöpfel, T. (2010). Imaging brain electric signals with genetically targeted voltage-sensitive fluorescent proteins. *Nature Methods*, *7*, 643–649.

Akemann, W., Raj, C. D., & Knöpfel, T. (2001). Functional characterization of permuted enhanced green fluorescent proteins comprising varying linker peptides. *Photochemistry and Photobiology*, *74*, 356–363.

Ataka, K., & Pieribone, V. A. (2002). A genetically targetable fluorescent probe of channel gating with rapid kinetics. *Biophysical Journal*, *82*, 509–516.

Atasoy, D., Aponte, Y., Su, H. H., & Sternson, S. M. (2008). A FLEX switch targets Channelrhodopsin-2 to multiple cell types for imaging and long-range circuit mapping. *Journal of Neuroscience*, *28*, 7025–7030.

Baird, G. S., Zacharias, D. A., & Tsien, R. Y. (1999). Circular permutation and receptor insertion within green fluorescent proteins. *Proceedings of the National Academy of Sciences of the United States of America*, *96*, 11241–11246.

Baker, B. J., Lee, H., Pieribone, V. A., Cohen, L. B., Isacoff, E. Y., Knöpfel, T., et al. (2007). Three fluorescent protein voltage sensors exhibit low plasma membrane expression in mammalian cells. *Journal of Neuroscience Methods*, *161*, 32–38.

Boyden, E. S., Zhang, F., Bamberg, E., Nagel, G., & Deisseroth, K. (2005). Millisecond-timescale, genetically targeted optical control of neural activity. *Nature Neuroscience*, *8*, 1263–1268.

Bozza, T., McGann, J. P., Mombaerts, P., & Wachowiak, M. (2004). In vivo imaging of neuronal activity by targeted expression of a genetically encoded probe in the mouse. *Neuron*, *42*, 9–21.

Buzsáki, G. (2006). *Rhythms of the brain*. Oxford: Oxford University Press.

Chanda, B., Blunck, R., Faria, L. C., Schweizer, F. E., Mody, I., & Bezanilla, F. (2005). A hybrid approach to measuring electrical activity in genetically specified neurons. *Nature Neuroscience*, *8*, 1619–1626.

Cohen, L. B., Keynes, R. D., & Hille, B. (1968). Light scattering and birefringence changes during nerve activity. *Nature*, *218*, 438–441.

Cossart, R., Ikegaya, Y., & Yuste, R. (2005). Calcium imaging of cortical networks dynamics. *Cell Calcium*, *37*, 451–457.

Díez-García, J., Akemann, W., & Knöpfel, T. (2007). In vivo calcium imaging from genetically specified target cells in mouse cerebellum. *NeuroImage*, *34*, 859–869.

Díez-García, J., Matsushita, S., Mutoh, H., Nakai, J., Ohkura, M., Yokoyama, J., et al. (2005). Activation of cerebellar parallel fibers monitored in transgenic mice expressing a fluorescent Ca^{2+} indicator protein. *European Journal of Neuroscience*, *22*, 627–635.

DiFranco, M., Capote, J., Quinonez, M., & Vergara, J. L. (2007). Voltage-dependent dynamic FRET signals from the transverse tubules in mammalian skeletal muscle fibers. *Journal of General Physiology*, *130*, 581–600.

Dimitrov, D., He, Y., Mutoh, H., Baker, B. J., Cohen, L., Akemann, W., et al. (2007). Engineering and characterization of an enhanced fluorescent protein voltage sensor. *PLoS One*, *2*, e440.

Drew, P. J., Shih, A. Y., Driscoll, J. D., Knutsen, P. M., Blinder, P., Davalos, D., et al. (2010). Chronic optical access through a polished and reinforced thinned skull. *Nature Methods*, *7*, 981–984.

Ferezou, I., Haiss, F., Gentet, L. J., Aronoff, R., Weber, B., & Petersen, C. C. H. (2007). Spatiotemporal dynamics of cortical sensorimotor integration in behaving mice. *Neuron*, *56*, 907–923.

Fernandez, J. M., Taylor, R. E., & Bezanilla, F. (1983). Induced capacitance in the squid giant axon. Lipophilic ion displacement currents. *Journal of General Physiology*, *82*, 331–346.

Gautam, S. G., Perron, A., Mutoh, H., & Knöpfel, T. (2009). Exploration of fluorescent protein voltage probes based on circularly permuted fluorescent proteins. *Frontiers in Neuroengineering*, *2*, 14.

Gradinaru, V., Zhang, F., Ramakrishnan, C., Mattis, J., Prakash, R., Diester, I., et al. (2010). Molecular and cellular approaches for diversifying and extending optogenetics. *Cell*, *141*, 154–165.

Grinvald, A., Salzberg, B. M., & Cohen, L. B. (1977). Simultaneous recording from several neurones in an invertebrate central nervous system. *Nature*, *268*, 140–142.

Hinner, M. J., Hubener, G., & Fromherz, P. (2006). Genetic targeting of individual cells with a voltage-sensitive dye through enzymatic activation of membrane binding. *ChemBioChem*, *7*, 495–505.

Hirrlinger, P. G., Scheller, A., Braun, C., Quintela-Schneider, M., Fuss, B., Hirrlinger, J., et al. (2005). Expression of reef coral fluorescent proteins in the central nervous system of transgenic mice. *Molecular and Cellular Neuroscience*, *30*, 291–303.

Horikawa, K., Yamada, Y., Matsuda, T., Kobayashi, K., Hashimoto, M., Matsu-ura, T., et al. (2010). Spontaneous network activity visualized by ultrasensitive $Ca(2^+)$ indicators, yellow Cameleon-Nano. *Nature Methods*, *7*, 729–732.

Huang, X., Xu, W., Liang, J., Takagaki, K., Gao, X., & Wu, J. Y. (2010). Spiral wave dynamics in neocortex. *Neuron*, *68*, 978–990.

Kimura, F., Fukuda, M., & Tsumoto, T. (1999). Acetylcholine suppresses the spread of excitation in the visual cortex revealed by optical recording: Possible differential effect depending on the source of input. *European Journal of Neuroscience*, *11*, 3597–3609.

Knöpfel, T., Diez-Garcia, J., & Akemann, W. (2006). Optical probing of neuronal circuit dynamics: Genetically encoded versus classical fluorescent sensors. *Trends in Neurosciences*, *29*, 160–166.

Kolbe, M., Besir, H., Essen, L. O., & Oesterhelt, D. (2000). Structure of the light-driven chloride pump halorhodopsin at 1.8 A resolution. *Science*, *288*, 1390–1396.

Lundby, A., Mutoh, H., Dimitrov, D., Akemann, W., & Knöpfel, T. (2008). Engineering of a genetically encodable fluorescent voltage sensor exploiting fast Ci-VSP voltage-sensing movements. *PLoS One*, *3*, e2514.

Matyas, F., Sreenivasan, V., Marbach, F., Wacongne, C., Barsy, B., Mateo, C., et al. (2010). Motor control by sensory cortex. *Science*, *330*, 1240–1243.

Metzger, F., Repunte-Canonigo, V., Matsushita, S., Akemann, W., Diez-Garcia, J., Ho, C. S., et al. (2002). Transgenic mice expressing a pH and Cl^- sensing yellow-fluorescent protein under the control of a potassium channel promoter. *European Journal of Neuroscience*, *15*, 40–50.

Muri, R., & Knöpfel, T. (1994). Activity induced elevations of intracellular calcium concentration in neurons of the deep cerebellar nuclei. *Journal of Neurophysiology*, *71*, 420–428.

Mutoh, H., Perron, A., Akemann, W., Iwamoto, Y., & Knöpfel, T. (2011). Optogenetic monitoring of membrane potentials. *Experimental Physiology*, *96*, 13–18.

Mutoh, H., Perron, A., Dimitrov, D., Iwamoto, Y., Akemann, W., Chudakov, D. M., et al. (2009). Spectrally-resolved response properties of the three most advanced FRET based fluorescent protein voltage probes. *PLoS One*, *4*, e4555.

Nagai, T., Sawano, A., Park, E. S., & Miyawaki, A. (2001). Circularly permuted green fluorescent proteins engineered to sense Ca^{2+}. *Proceedings of the National Academy of Sciences of the United States of America, 98*, 3197–3202.

Nagel, G., Szellas, T., Huhn, W., Kateriya, S., Adeishvili, N., Berthold, P., et al. (2003). Channelrhodopsin-2, a directly light-gated cation-selective membrane channel. *Proceedings of the National Academy of Sciences of the United States of America, 100*, 13940–13945.

Nakai, J., Ohkura, M., & Imoto, K. (2001). A high signal-to-noise Ca(2+) probe composed of a single green fluorescent protein. *Nature Biotechnology, 19*, 137–141.

Ng, D. N., & Fromherz, P. (2011). Genetic targeting of a voltage-sensitive dye by enzymatic activation of phosphonooxymethyl-ammonium derivative. *ACS Chemical Biology, 6*, 444–451.

Pawley, J. B. (2006). *Handbook of biological confocal microscopy*. New York, NY: Springer.

Peron, S., & Svoboda, K. (2011). From cudgel to scalpel: Toward precise neural control with optogenetics. *Nature Methods, 8*, 30–34.

Perron, A., Mutoh, H., Akemann, W., Gautam, S. G., Dimitrov, D., Iwamoto, Y., et al. (2009a). Second and third generation voltage-sensitive fluorescent proteins for monitoring membrane potential. *Frontiers in Molecular Neuroscience, 2*, 5.

Perron, A., Mutoh, H., Launey, T., & Knöpfel, T. (2009b). Red-shifted voltage-sensitive fluorescent proteins. *Chemistry and Biology, 16*, 1268–1277.

Potter, S. M., Zheng, C., Koos, D. S., Feinstein, P., Fraser, S. E., & Mombaerts, P. (2001). Structure and emergence of specific olfactory glomeruli in the mouse. *Journal of Neuroscience, 21*, 9713–9723.

Ross, W. N., Salzberg, B. M., Cohen, L. B., & Davila, H. V. (1974). A large change in dye absorption during the action potential. *Biophysical Journal, 14*, 983–986.

Sakai, R., Repunte-Canonigo, V., Raj, C. D., & Knöpfel, T. (2001). Design and characterization of a DNA-encoded, voltage-sensitive fluorescent protein. *European Journal of Neuroscience, 13*, 2314–2318.

Shepherd, J. D., & Bear, M. F. (2011). New views of Arc, a master regulator of synaptic plasticity. *Nature Neuroscience, 14*, 279–284.

Siegel, M. S., & Isacoff, E. Y. (1997). A genetically encoded optical probe of membrane voltage. *Neuron, 19*, 735–741.

Sjulson, L., & Miesenbock, G. (2008). Rational optimization and imaging in vivo of a genetically encoded optical voltage reporter. *Journal of Neuroscience, 28*, 5582–5593.

Tamamaki, N., Yanagawa, Y., Tomioka, R., Miyazaki, J., Obata, K., & Kaneko, T. (2003). Green fluorescent protein expression and colocalization with calretinin, parvalbumin, and somatostatin in the GAD67-GFP knock-in mouse. *The Journal of Comparative Neurology, 467*, 60–79.

Tian, L., Hires, S. A., Mao, T., Huber, D., Chiappe, M. E., Chalasani, S. H., et al. (2009). Imaging neural activity in worms, flies and mice with improved GCaMP calcium indicators. *Nature Methods, 6*, 875–881.

Traub, R. D., & Whittington, M. A. (2010). *Cortical oscillations in health and disease*. Oxford: Oxford University Press.

Tsai, H. C., Zhang, F., Adamantidis, A., Stuber, G. D., Bonci, A., de Lecea, L., et al. (2009). Phasic firing in dopaminergic neurons is sufficient for behavioral conditioning. *Science, 324*, 1080–1084.

Tsau, Y., Guan, L., & Wu, J. Y. (1998). Initiation of spontaneous epileptiform activity in the neocortical slice. *Journal of Neurophysiology, 80*, 978–982.

Tsutsui, H., Karasawa, S., Okamura, Y., & Miyawaki, A. (2008). Improving membrane voltage measurements using FRET with new fluorescent proteins. *Nature Methods, 5*, 683–685.

Villalba-Galea, C. A., Sandtner, W., Dimitrov, D., Mutoh, H., Knöpfel, T., & Bezanilla, F. (2009). Charge movement of a voltage-sensitive fluorescent protein. *Biophysical Journal, 96*, L19–L21.

Villringer, A. (1997). Functional neuroimaging. Optical approaches. *Advances in Experimental Medicine and Biology, 413*, 1–18.

Wang, D., Zhang, Z., Chanda, B., & Jackson, M. B. (2010). Improved probes for hybrid voltage sensor imaging. *Biophysical Journal, 99*, 2355–2365.

Wu, J. Y., Huang, X., & Zhang, C. (2008). Propagating waves of activity in the neocortex: What they are, what they do. *The Neuroscientist, 14*, 487–502.

Zhang, F., Wang, L. P., Boyden, E. S., & Deisseroth, K. (2006). Channelrhodopsin-2 and optical control of excitable cells. *Nature Methods, 3*, 785–792.

Zhao, S., Cunha, C., Zhang, F., Liu, Q., Gloss, B., Deisseroth, K., et al. (2008). Improved expression of halorhodopsin for light-induced silencing of neuronal activity. *Brain Cell Biology, 36*, 141–154.

Zimmermann, D., Kiesel, M., Terpitz, U., Zhou, A., Reuss, R., Kraus, J., et al. (2008). A combined patch-clamp and electrorotation study of the voltage- and frequency-dependent membrane capacitance caused by structurally dissimilar lipophilic anions. *Journal of Membrane Biology, 221*, 107–121.

T. Knöpfel and E. Boyden (Eds.)
Progress in Brain Research, Vol. 196
ISSN: 0079-6123

CHAPTER 5

Neural activity imaging with genetically encoded calcium indicators

Lin Tian*[,1], Jasper Akerboom, Eric R. Schreiter and Loren L. Looger[†]

Howard Hughes Medical Institute, Janelia Farm Research Campus, Ashburn, VA, USA

Abstract: Genetically encoded calcium indicators (GECIs), together with modern microscopy, allow repeated activity measurement, in real time and with cellular resolution, of defined cellular populations. Recent efforts in protein engineering have yielded several high-quality GECIs that facilitate new applications in neuroscience. Here, we summarize recent progress in GECI design, optimization, and characterization, and provide guidelines for selecting the appropriate GECI for a given biological application. We focus on the unique challenges associated with imaging in behaving animals.

Keywords: genetically encoded calcium indicators; protein calcium sensors; protein engineering; GCaMP3; calcium imaging; neural activity imaging.

Introduction

One of the primary challenges of neuroscience is to link complex neural phenomena to the structure and function of their composite neural circuits. Addressing this problem requires a thorough understanding of patterns of neural activity and the ability to relate this to physiological processes, behavior, and disease states. An essential step toward this goal is the simultaneous recording of neural activity in large, defined populations, ideally in intact circuitry.

Traditional electrophysiological approaches provide excellent sensitivity and temporal resolution (Scanziani and Hausser, 2009) but are limited in the number of cells that can be recorded simultaneously. More importantly, assigning this activity to specific cells is quite difficult (and impossible for more than a few cells at a time), limiting the ability to create high-resolution circuit maps. Modern fluorescence imaging techniques (Svoboda and Yasuda, 2006), combined with high-quality fluorescent indicators (both small molecules and proteins), can potentially overcome these limitations. There is

*Corresponding author.
Tel.: +571-209-4155; Fax: +571-209-4905
E-mail: lintian@ucdavis.edu
†Co-corresponding author.
E-mail: loogerl@janelia.hhmi.org

[1]Current address: Department of Biochemistry and Molecular Medicine, School of Medicine, University of California, Davis, 95817

DOI: 10.1016/B978-0-444-59426-6.00005-7

a rapidly growing toolkit of reagents that transduce changes in neural state (e.g., membrane potential or calcium ion flux, [Ca^{2+}], following action potentials (APs) or synaptic input), to fluorescence (or in some instances, luminescence, etc.) observables (for reviews, see Miyawaki, 2011; Palmer et al., 2011). Protein sensors are genetically encoded and can thus be used to label large populations of defined cell types and/or subcellular compartments (Borghuis et al., 2011; Dreosti et al., 2009; Mittmann et al., 2011; Shigetomi et al., 2010a,b), unlike small molecule dyes, whose delivery and targeting can be problematic (Hendel et al., 2008; Shigetomi et al., 2010a, b). In principle, genetically encoded sensors allow long-term measurements of activity *in vivo*, simultaneously across a neural population. Since the creation of the first generation of genetically encoded calcium indicators (GECIs), a decade ago (Miyawaki et al., 1997; Romoser et al., 1997), their performance has been iteratively optimized for applications in neurophysiology (as well as in other excitable cells such as cardiomyocytes (Tallini et al., 2006); recently, GECIs have also been used to monitor Ca^{2+} transients in nonexcitable cell types, such as astrocytes (Gourine et al., 2010; Shigetomi et al., 2010a). The culmination of these efforts has led to activity measurements of defined neuronal populations in awake, behaving animals (Chiappe et al., 2010; Dombeck et al., 2010; Lütcke et al., 2010; Muto et al., 2011; O'Connor et al., 2010; Seelig et al., 2010; Tian et al., 2009).

In neurons, APs generate small calcium transients that can occur over a wide range of frequencies. To enable quantitative measurements of neural activity, one, in principle, desires a sensor with fast rise and decay kinetics, broad dynamic range, and calcium affinity appropriate for the cells in question (Hires et al., 2008). In addition, the basal brightness of the sensor should be high enough to permit identification of positive cells and improve the signal-to-noise ratio (SNR) of baseline measurements. Improved brightness also facilitates imaging with lower excitation power, important for minimizing indicator photobleaching and illumination phototoxicity.

In this chapter, we highlight recent progress in GECI design, optimization, and testing protocol standardization. We provide guidelines for selecting the appropriate GECI for a given biological application and discuss the remaining hurdles to perfect chronic, robust neural activity imaging *in vivo*.

GECI development and neuroscience applications

The general paradigm of GECI design is to fuse a calcium-binding domain (e.g., Calmodulin, CaM, or Troponin C, TnC) to one or two fluorescent proteins (FPs). In single-FP GECIs, the fluorescence intensity of a circularly permuted or split FP is modulated by calcium binding-dependent changes in the chromophore environment (Baird et al., 1999; Nagai et al., 2001; Nakai et al., 2001). In two-FP GECIs, calcium binding allosterically modulates the relative donor–acceptor emission spectra through distance- and orientation-dependent changes in Förster resonance energy transfer (FRET) (Miyawaki et al., 1997; Heim and Griesbeck, 2004; Palmer et al., 2006). In many cases, a conformational actuator, such as the Ca^{2+}/CaM-binding peptide of myosin light chain kinase (the "M13" peptide), is included to enhance the allosteric regulation of the chromophore environment (for single-FP GECIs) or FRET (for two-FP GECIs). The molecular architecture of the major GECI classes is shown in Fig. 1.

In 1997, the first GECIs were developed by the groups of Romoser (1997) and Miyawaki (1997). FIP-CB_{SM} developed by Romoser et al., employed an avian smooth muscle myosin light chain kinase (smMLCK) M13 peptide (interestingly, a conservative point mutation, glutamine to asparagine, was inadvertently introduced), sandwiched between blue- and red-shifted versions of GFP ("BGFP" and "RGFP," respectively), and relied on endogenous CaM for modulating the FRET between the two. Cameleon developed by Miyawaki et al., included its own CaM as well as a

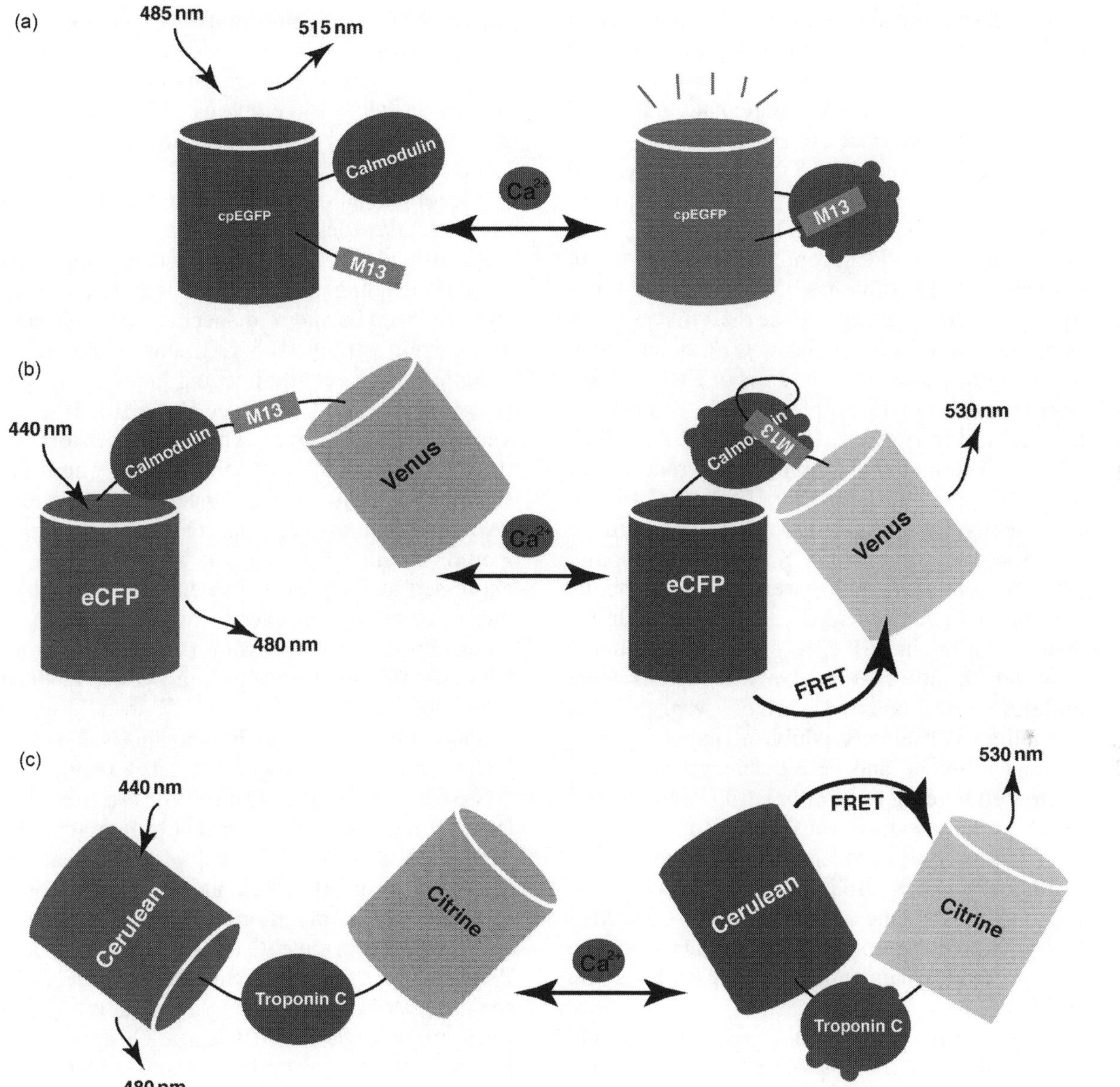

Fig. 1. Schematic representation of the three major GECI classes. GECIs are based either on florescence intensity changes of circularly permuted single FPs (a) or changes in Förster resonance energy transfer (FRET) efficiency between two FPs (b, c). (a) Schematic of the GCaMP-type sensing mechanism. Upon calcium binding, conformational changes in the CaM/M13 complex induce fluorescence changes in the circularly permuted GFP. (b) The Cameleon family of FRET-based GECIs. A calcium-dependent increase in FRET between a CFP and YFP FRET pair is coupled to the binding of calmodulin to the M13 peptide. (c) Troponin C-based FRET GECIs. Binding of calcium to troponin C induces conformational changes and an increase in FRET between CFP and YFP. (For a color version of this figure, the reader is referred to the Web version of this chapter.)

mammalian skeletal/cardiac muscle myosin light chain kinase (skMLCK) M13 peptide, came in blue/green and cyan/yellow ("Yellow Cameleon" or "YC") combinations, and had improved fluorescence response relative to FIP-CB_{SM} (Miyawaki et al., 1997). Intriguingly, all Cameleon sequences appear to contain the inadvertent mutation Asp64Tyr, which is a Ca^{2+}-chelating residue in the second EF hand. It is not known what effect this has on their binding affinity, kinetics, or specificity. Since their first publication, the Cameleon family of GECIs has been incrementally improved in terms of FRET signal change, calcium affinity, pH stability, and folding efficiency (for a review, see Palmer et al., 2011). Of note, rational and computational redesign of the CaM/M13 interface diversified the family of Cameleons (D1 (Palmer et al., 2004) was designed by inspection from skMLCK/CaM; D2, D3, and D4 (Palmer et al., 2006) were computationally designed from avian smMLCK/CaM; the Gln-to-Asn mutation in FIP-CB_{SM} was not included; however, a different inadvertent conservative mutation, arginine to lysine, was). These diversified Cameleons exhibit a range of Ca^{2+}-binding affinities and FRET changes and show decreased binding to CaM *in vitro* (Palmer et al., 2006). Whether this translates into decreased interference with CaM signaling pathways *in situ* is not known. It is also impossible to deconvolve the effects of the computational CaM/M13 bump/hole designs of D2, D3, and D4 from the change from mammalian skMLCK to avian smMLCK.

Several of these Cameleon variants have found wide utility in neuroscience. The interface-redesigned sensor D3cpVenus (D3cpV) has been used to detect single APs in organotypic mouse brain slice and *in vivo* in layer 2/3 somatosensory cortical neurons (Wallace et al., 2008). Recently, another Cameleon variant, YC3.60 (Nagai et al., 2004), has been reported to allow *in vivo* detection of single APs with a sensitivity comparable to D3cpV, but with faster kinetics and minimal saturation up to 10 APs (Lütcke et al., 2010). Recently, a very high-affinity variant of YC3.60 (YC-Nano) was shown to detect subtle Ca^{2+} transients associated with spontaneous motor activity in zebrafish embryos (Horikawa et al., 2010), as well as to differentiate resting $[Ca^{2+}]$ levels in different cell types.

The other major two-FP GECI family utilizes the skeletal/cardiac muscle Ca^{2+}-binding protein TnC, instead of CaM. CaM is a ubiquitous intracellular signaling molecule, whereas TnC has historically been considered specific to muscle cells. The current variant, TN-XXL, has been used for chronic *in vivo* activity imaging in mouse and fly (Heim et al., 2007; Reiff et al., 2010). It is not known to what extent TnC-based sensors are truly "orthogonal" to endogenous neuronal signaling mechanisms, since other studies (Aubin-Horth et al., 2005; Fine et al., 1975; Roisen et al., 1983) suggest that TnC may be expressed at high levels in neurons, particularly during development (Schevzov et al., 1997). In mouse brain, overexpression produced a phenotype similar to GCaMP3 and D3cpV (Tian et al., 2009).

Single-FP GECIs include pericam (Nagai et al., 2001), camgaroo (Baird et al., 1999), and GCaMPs. The GCaMP scaffold consists of a circularly permuted green fluorescent protein (cpGFP) with CaM and the M13 peptide (the identical smMLCK as in FIP-CB_{SM}, including the inadvertent Gln-to-Asn mutation) linked to its C- and N-termini, respectively (Nakai et al., 2001). Pericam has a similar architecture to GCaMPs; camgaroo consists of CaM inserted into a yellow FP. Among single-FP GECIs, the GCaMP family has been iteratively optimized and achieved the broadest usage across multiple model organisms. A summary of the historical development of the GCaMP family is shown in Fig. 2.

The first major improvement made to the GCaMP scaffold was the incorporation of GFP thermal stability mutations, which led to the development of GCaMP1.6 (Ohkura et al., 2005). Subsequent random mutagenesis produced the brighter variant GCaMP2 (Diez-Garcia

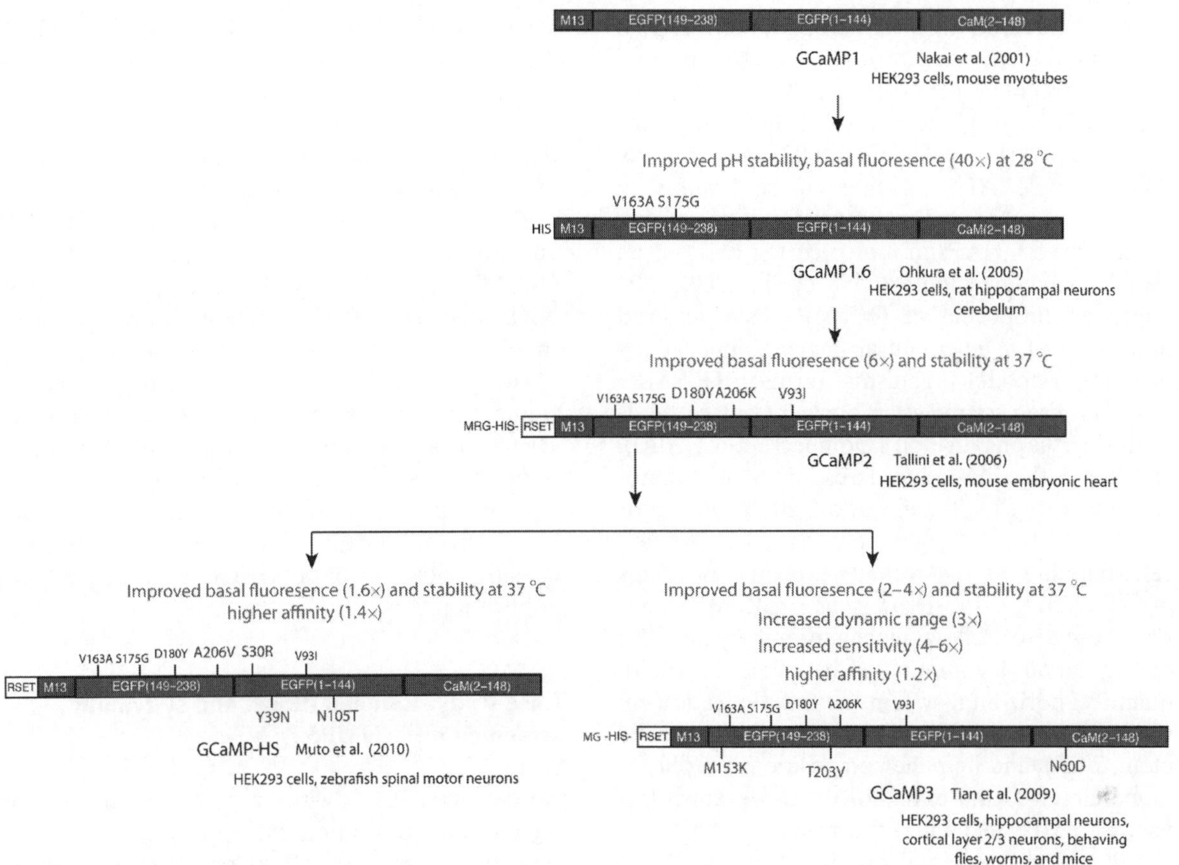

Fig. 2. Summary of the historical development of the GCaMP family. GCaMP family has been iteratively optimized and achieved the broadest usage across multiple model organisms. (For color version of this figure, the reader is referred to the Web version of this chapter.)

et al., 2005; Tallini et al., 2006). GCaMP2 has been used for *in vivo* imaging across multiple model organisms (Chalasani et al., 2007; Diez-Garcia et al., 2005; Hasan et al., 2004, 2008). GCaMP2 has been further optimized by at least three different groups, resulting in GCaMP-HS (Muto et al., 2011) and GCaMP4.1 (Shindo et al., 2010), Case12 and Case16 (Souslova et al., 2007), and GCaMP3 (Tian et al., 2009). A direct comparison of these GCaMPs in terms of their brightness and signal change in neurons has not been performed. GCaMP-HS includes a subset of the "superfolder GFP" mutations (Pedelacq et al., 2006), apparently resulting in increased stability and signal change in zebrafish spinal motor neurons (Muto et al., 2011). Imaging in *Xenopus* embryos has been reported with "GCaMP4.1" (Shindo et al., 2010), although the mutations and/or improvements have not been published. Case12 and Case16 encode a mutation Thr203Phe (near the chromophore); Case16 has an additional Thr145Ser mutation in the second linker (connecting cpGFP to CaM). Both Case sensors showed improvement in *in vitro* response compared to GCaMP2. Case12 has been used to report calcium transients in astrocytes *in vivo* (Gourine et al., 2010), although no *in situ* responses of either Case12 or Case16 have been reported in neurons.

We engineered GCaMP3 from GCaMP2 by a combination of structure-guided design and library screening (Tian et al., 2009). GCaMP3 has increased brightness, Ca^{2+} affinity, and protein stability relative to GCaMP2, and robustly detects single APs in acute mouse brain slice (Tian et al., 2009). This translates into reliable detection of ~3 APs in motor cortex (M1) of an awake, behaving mouse (Tian et al., 2009). The improved properties of GCaMP3 have allowed new types of *in vivo* neural imaging applications in multiple model organisms. In mice, GCaMP3 has been used to image hippocampal place fields during virtual navigation (Dombeck et al., 2010), whisker deflection-induced responses in somatosensory cortex (Mittmann et al., 2011; O'Connor et al., 2010), and light responses of targeted cell populations in retinal explant (Borghuis et al., 2011). In *Drosophila*, GCaMP3 has been used to follow mechanosensory neurons during larval locomotion (Cheng et al., 2010), quantify horizontal-system neural activation in walking adults (Chiappe et al., 2010; Seelig et al., 2010), and map the composition of a pheromone circuit (Ruta et al., 2010). In zebrafish larvae, GCaMP3 facilitated the mapping of a tectal circuit underlying visual discrimination of object size (Del Bene et al., 2010).

Most important, GCaMP3 has made it possible to repeatedly image large neuronal populations over months in mouse M1 (Tian et al., 2009), which has facilitated investigation of learning-mediated changes in neural ensemble representations (D. Huber and K. Svoboda, personal communication). Long-term expression has evinced no apparent behavioral phenotype in model organisms, unlike other GECIs in *Caenorhabditis elegans* (Tian et al., 2009). Extremely high levels of sensor (e.g., at the site of viral injection or *in utero* electroporation) can, however, give rise to an optical correlate for the GCaMP3-induced "cytomorbid" cellular phenotype, in which cells become brightly fluorescent in both cytosol and nucleus, and less excitable (Borghuis et al., 2011; Tian et al., 2009). Similar results were observed with other GECIs such as D3cpV and TN-XXL (Tian et al., 2009). The mechanistic details of this cytotoxicity are not clear; further study will reveal to what extent Ca^{2+} buffering and CaM signaling pathway disruption are implicated, and whether next-generation indicators may alleviate these concerns.

The current set of state-of-the-art GECIs encompasses the single-wavelength indicator, GCaMP3, and the FRET sensors, D3cpV, TN-XXL, and YC3.60. We systematically compared the first three of these in acute mouse brain slice (Tian et al., 2009), with a number of parameters summarized in Table 1. The exact selection of GECI for a given application depends on a number of factors, regarding both indicator properties and experimental constraints. In the next sections, we discuss the factors that need to be taken into consideration in both GECI engineering and application.

Case study: Rational design and systematic screening of GCaMP3

To develop GECIs with properties optimized for a particular application, the appropriate combination of intrinsic GECI parameters should be matched to the extrinsic factors of the system studied. Intrinsic GECI parameters include sensor affinity, kinetics, dynamic range, brightness, expression level, fluorescence properties, and independence from endogenous interference (for recent reviews, see Hires et al., 2008; Tian et al., 2011). Extrinsic parameters include the size, speed, time-course, and frequency of calcium transients, and basal $[Ca^{2+}]$ of the target cell type. Here, using GCaMP3 as an example, we discuss the principles we followed and lessons we learned in improving the GCaMP protein for *in vivo* neurophysiology. We performed structure-guided and library-based protein engineering of the parent scaffold GCaMP2 to optimize each of the following parameters: basal fluorescence intensity, protein stability, dynamic range, and calcium affinity/kinetics.

Table 1. Properties of state-of-the-art GECIs in mouse

In vitro (acute mouse brain slice)							
	$\Delta R/R$, $\Delta F/F$ (%)[a]		Kinetics[a] rise time decay time		Photostability (% fluorescence after 30-min imaging)		
Indicators	1 AP	40 AP	(ms)	(ms)		*In vivo*	References
D3cpV	5±3	40±20	110±30	9500±3400	59 CFP, 84 YFP	Single-spike detection (5% $\Delta R/R$)	Tian et al. (2009), Wallace et al. (2008)
TN-XXL	4±2	50±10	80±20	1600±600	36 CFP, 70 YFP	N/A	Tian et al. (2009)
GCaMP3	14±3	500±200	100±30	650±200	109 GFP	2–3 AP detection (12–20% $\Delta F/F$), linear up to 20 APs	Tian et al. (2009)
YC3.60	6±0.3	N/A	N/A	800±60	N/A	Single-spike detection (2% $\Delta R/R$), minimal saturation to 10 AP	Lütcke et al. (2010)

[a]Mean±s.d.

The crystal structures of GCaMP2 in both the Ca^{2+}-free and Ca^{2+}-saturated states provide insights into the molecular mechanism of calcium sensing (Akerboom et al., 2009; Wang et al., 2008) and form the basis for rational library design. In the Ca^{2+}-free (apo) structure, much of CaM and the M13 peptide are disordered, suggesting a large amount of flexibility between the M13, cpGFP, and CaM sensor components. In the Ca^{2+}-saturated state, GCaMP2 crystallizes as both a monomer and a dimer, but size-exclusion chromatography experiments suggest that the monomeric form predominates at intracellular concentrations of GCaMP (~10μM) (Akerboom et al., 2009; Hires et al., 2008). The monomer structure shows considerable domain movement and conformational differences compared to the apo-state of GCaMP; CaM is tightly wrapped around M13 and packs against the GFP barrel opening created by circular permutation (Akerboom et al., 2009). This induces a reorganization of the GFP chromophore chemical environment, resulting in a decrease in solvent access and a downward shift of the tyrosyl hydroxyl pK_a that moves the equilibrium toward the deprotonated, bright state at physiological pH. Other factors such as alteration of dynamics influencing excited state lifetime or extinction coefficient are also possible but are not directly addressed by the available biophysical data. An important result of the crystal structures was to delineate the interface between CaM and cpGFP such that these positions could be targeted for mutagenic screening. To increase the dynamic range of GCaMP2, we selected a group of amino acids in close proximity to the chromophore in both the apo and the Ca^{2+}-bound (sat) states for mutagenesis. To increase protein stability at physiological temperatures, we tuned the expression and turnover of GCaMP2 by changing the N-terminally encoded proteasomal degradation sequence; we also mutated positions known to regulate GFP thermodynamic stability. Finally, we made libraries of the Ca^{2+}-binding EF hands and the CaM/M13 binding surface to screen for mutants with higher Ca^{2+} affinity.

Our initial screens were performed in *Escherichia coli* lysates. However, we found that correlation between sensor performance in bacterial lysate or purified protein, and that in slice, was relatively low; the GCaMP2-LIA mutant is an example of this (Tian et al., 2009). As each system or cell type has a unique calcium signaling toolkit and may handle GECIs and Ca^{2+} differently, we first sought to establish a standardized screening paradigm balancing throughput with downstream predictability. For developmental efficiency, such an initial testing system should allow screening for brightness, calcium sensitivity, dynamic range, and kinetics with decent throughput. It should also capture as many aspects of intact preparations (e.g., calcium buffering and cytotoxicity) as possible, allowing better prediction of GECI performance in more challenging environments. The method we settled on uses agonized endogenous G-protein-coupled receptors in cultured mammalian cells to create synthetic Ca^{2+} transients lasting tens of seconds (Tian et al., 2009). The biggest disadvantage of this method is that the use of alternative cell types may obscure the dynamic aspects of Ca^{2+} signaling unique to neuronal morphology and physiology. As an alternative, testing GECI performance in dissociated neuronal cultures coupled to field potential stimuli may provide a better platform (Janelia Farm GECI Project, unpublished data). Even cultured neurons, however, may differ dramatically from an *in vivo* setting (e.g., disrupted network connectivity), and more intact preparations are invariably required to validate candidate hits.

Following an initial high-throughput screen, a small number of high-quality sensors can be further characterized in cortical pyramidal neurons in acute slice with simultaneous two-photon imaging and electrophysiology (Mao et al., 2008; Tian et al., 2009). Imaging GECI responses to evoked back-propagating APs is the most reliable way to assess relevant sensor kinetics and signal change, and captures many aspects of *in vivo* functional imaging (e.g., possible long-term expression and developmental effects), while allowing a moderate throughput (together with *in utero* electroporation). To optimize GCaMP signal, the laser is typically tuned to 910nm, although the interval 900–1000nm all works well for excitation. The typical imaging configuration is line-scan mode (500Hz) across the apical dendrite, 20–50μm from the base of the neuron (Tian et al., 2009). We triggered defined numbers of APs at specific intervals with intermittent short depolarizing pulses (Hendel et al., 2008). Alternatively, neurons can be depolarized by continuous current injection from a patch pipette (Pologruto et al., 2004).

Finally, the performance of the best candidates needs to be confirmed in an *in vivo* preparation, with such complicating factors as sample access, imaging depth, motion artifacts, and hemodynamics. To do this, we expressed a few of the best GCaMPs postnatally in neurons of M1 by adeno-associated virus (AAV)-mediated gene transfer. We observed uniform, bright labeling of layer 2/3 pyramidal cells, of cell bodies and neurites alike. We performed simultaneous electrophysiological recordings and two-photon imaging via a loose seal cell-attached configuration. We observed robust fluorescent transients following spontaneous, sensory-related, or AP-evoked calcium transients (Tian et al., 2009). Responses were imaged for several months, with no apparent decline in sensor performance or neuronal morphology. Of course, each system will have unique considerations; comparison of new GECIs across a set of model organisms (e.g., *C. elegans*, *Drosophila*, mouse, zebrafish) will best give a sense of an indicator's strengths and weaknesses.

In addition to ensuring that a GECI faithfully reports neural activity, it is equally important to evaluate the potential endogenous interference in cells caused by GECI expression. Every GECI by definition will sequester Ca^{2+}. Expressing high levels of indicator to achieve good SNR may magnify such calcium buffering. GECIs composed of endogenous Ca^{2+}-binding proteins run the further risk of activating cellular signal transduction pathways. These two (at least) mechanisms of endogenous interference may lead to perturbations in cell physiology, synaptic and circuit activity, and ultimately behavior. A variety of biochemical and physiological approaches, such as immunohistochemistry and patch clamping, can be used to evaluate neuronal morphology and electrophysiological parameters (e.g., R_m, C_m, E_m) in GECI-positive and -negative cells. At the network level, laser-scanning photostimulation circuit mapping (Petreanu et al., 2007, 2009) can test for changes in synaptic properties and connectivity patterns (Tian et al., 2009). In the end, a robust behavioral paradigm is required to test for possible perturbations to the system and to the model organism as a whole.

Choosing the best GECI for a given application

For end users, choosing the most appropriate GECI at the start of a project is paramount, since experimental optimization requires significant investments in time and resources. GECI selection will also influence many other choices, for example, light source, filters, camera, image analysis algorithms, etc. A good rule of thumb is to characterize several GECIs in the context of a specific application, as they each have different strengths and weaknesses. Here, we discuss several practical criteria that users may wish to consider.

Two- versus one-FP GECIs

In general, FRET-based two-FP sensors have higher basal brightness (since they are typically based on intact FPs) and are less sensitive to motion artifacts or expression level differences (due to ratioing of the donor and acceptor fluorescence). In addition, FRET-based GECIs are better suited for applications in which precise quantification of $[Ca^{2+}]$ is desired. However, they require simultaneous measurement of fluorescence in two channels, slowing down measurement, compounding errors, and consuming large spectral bandwidth. Differential photobleaching, bleed-through between donor and acceptor FPs, and other artifacts may confound measurement and analysis (Piston and Kremers, 2007).

Intensity-based single-FP GECIs usually have superior dynamic range, kinetics, SNR, and photostability when compared to FRET-based GECIs (Table 1) (Mao et al., 2008; Tian et al., 2009). The smaller size of single-FP GECIs makes them more suitable for protein fusions to target specific cellular compartments and for viral payloading. They may also be used in conjunction with other labels or probes for multicolor imaging, since the spectral bandwidth is preserved. All FPs are intrinsically pH-sensitive, and this will affect both one- and two-FP GECIs. pH sensitivity may be exacerbated for one-FP GECIs based on circularly permuted FPs. A comparison between these two GECI classes is shown in Fig. 3.

Calcium affinity, kinetics, and dynamic range

The fluorescence response of a GECI to calcium transients is critically dependent on its affinity, kinetics, and dynamic range (Hires et al., 2008). To achieve optimal SNR, it is essential to match these parameters with the calcium dynamics of the system under study. In addition, a linear relationship between fluorescence response and target stimulus facilitates quantitative measurement of calcium transients (Sasaki et al., 2008). Cell-to-cell variability of GECI responses (e.g., see Fig. 7 in Borghuis et al., 2011; Fig. 6 in Tian et al., 2009) may complicate quantitation of responses across cell populations. For events inducing small calcium transients, such as sparse APs, a GECI with high affinity and fast on-kinetics is preferred (Hires et al., 2008). A slow decay rate (off-kinetics) may facilitate signaling integration, leading to better detection of lower or sparse APs (Wallace et al., 2008) but will confound interpretation of spike trains with high frequency. Indicator decay rates are not always the limiting factor for observed kinetics. For instance, dendritic GCaMP3 transients were roughly twice as fast as somatic responses in the same retinal ganglion cell (Borghuis et al., 2011). For events with large calcium transients, such as

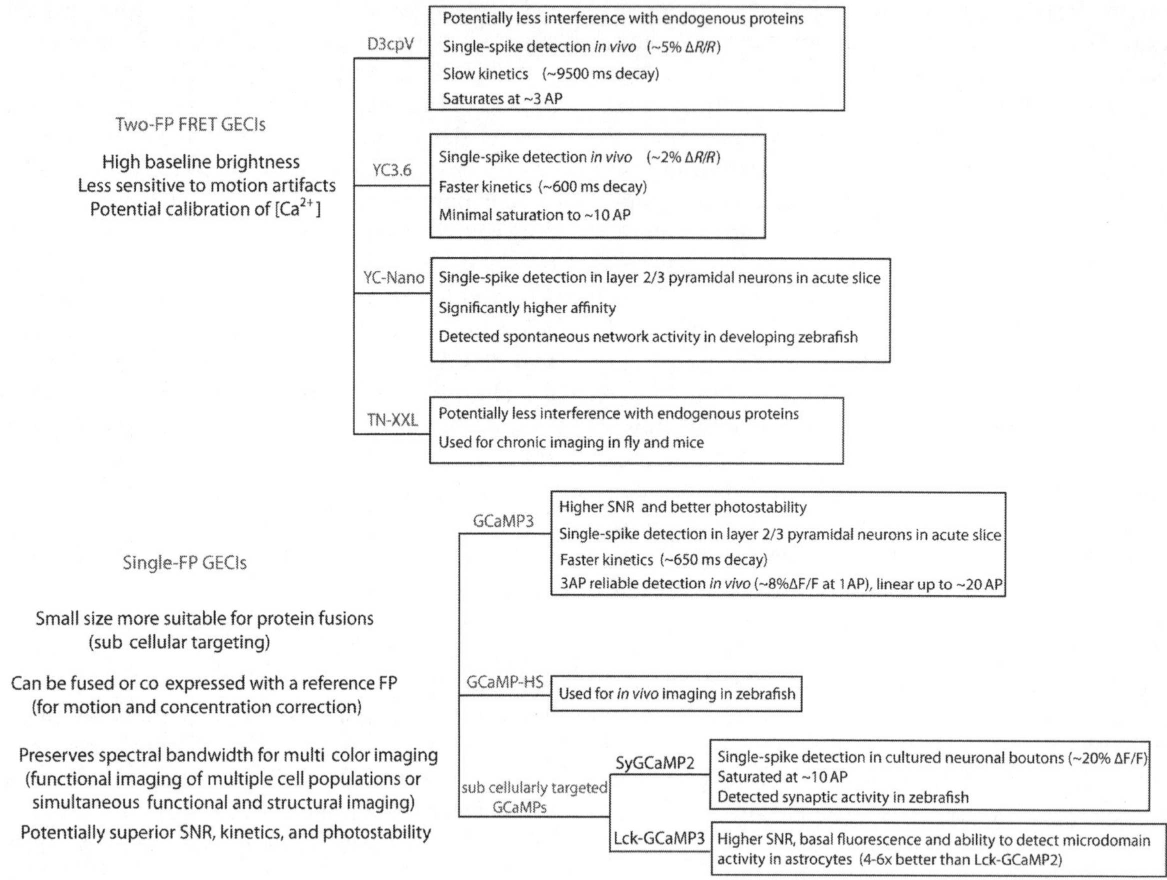

Fig. 3. A comparison between two-FP- or single-FP-based GECI classes. (For a color version of this figure, the reader is referred to the Web version of this chapter.)

in fast-spiking neurons, a low-affinity sensor will provide a greater effective dynamic range.

GECI expression level

The expression level of a GECI influences its performance in cells and organisms, in a complex manner. Very low GECI expression will preclude even sensor visualization. As [GECI] increases, the number of photons detected increases, but the amount of buffered Ca^{2+} will increase, which decreases $\Delta F/F$. These two opposing effects give rise to an "inverted U"-shaped plot for SNR as a function of [GECI], with a simulated optimum of $\sim$5–20μM for best SNR (Hires et al., 2008). In addition, higher [GECI] will tend to spread transients out (Helmchen et al., 1996), confounding kinetic analysis. On the other hand, it will facilitate imaging with lower laser power and/or shorter exposure times; this favorably impacts imaging speed and phototoxicity. In addition, GECI components (e.g., CaM, M13pep) may be sequestered by endogenous proteins, thus resulting in nonfunctional indicators that can increase background fluorescence (Hasan et al., 2004; Heim and Griesbeck, 2004; Tian et al., 2009). Such endogenous interference may also perturb cell-signaling pathways and lead to behavioral phenotypes as discussed above (Tian et al., 2009).

Balancing all these effects, we find that it is usually best to shoot for the lowest GECI expression level that allows imaging with tolerably low laser power. To accomplish this, users may wish to explore multiple promoters, regulatory sequences, and transduction methods (e.g., different viral serotypes, electroporation techniques, or transgenic lines). Following both *in utero* electroporation of a *CAG* promoter construct and AAV infection with a *synapsin-1* promoter, we found [GCaMP3] levels to be on the order of ~10 μM (Hires et al., 2008; Tian et al., 2009, 2011). This falls in the predicted range with high SNR with low cyto- and phototoxicity.

Subcellular targeting

One of the advantages of GECIs over small molecules is the ability to target them to specific subcellular locations. GECIs can fill some cellular compartments, such as dendrites and axons, without fusion to targeting peptides or proteins. For instance, expressing GCaMP3 in cortical layer 5 neurons (via stereotactic viral injection) allowed for apical dendritic imaging far away (~600–800 μm) from their somas of origin (Mittmann et al., 2011). Alternatively, GECIs can be targeted to specific subcellular compartments via protein fusions or signaling peptides. Different subcellular compartments experience unique Ca^{2+} fluxes; thus matching GECI properties to the calcium dynamics in the target environment is essential. For example, a low-affinity GECI may be less useful for probing small calcium transients in the cytosol, but highly effective at measuring large calcium transients in the endoplasmic reticulum and mitochondria (Miyawaki et al., 1997). Targeting a GECI to subcellular locations where calcium transients tend to be larger, such as synaptic terminals (Dreosti et al., 2009) or near the plasma membrane (Shigetomi et al., 2010a,b), can improve SNR without altering the sensor itself. Subcellular targeting may, however, negatively impact photostability of the sensor under laser-scanning illumination due to decreased protein mobility (L. Petreanu and K. Svoboda, personal communication). Fusion to essential proteins may also interfere with their evolved functions (T. Ryan, personal communication). Therefore, it is important to consider the impact of such manipulations on both the sensor and its environment before settling on a final targeting strategy.

Imaging neural activity with GECIs *in vivo*

Through iterative cycles of optimization, GECIs have been improved to the point that they are useful for *in vivo* neuronal imaging. State-of-the-art GECIs, such as D3cpV, YC3.60, YC-Nano, GCaMP-HS, and GCaMP3, have been widely used to report neural activity in living zebrafish (Muto et al., 2011), worms (Tian et al., 2009), flies (Cheng et al., 2010; Chiappe et al., 2010; Heim et al., 2007; Seelig et al., 2010; Tian et al., 2009), and mice (Dombeck et al., 2010; Lütcke et al., 2010; Mittmann et al., 2011; O'Connor et al., 2010; Tian et al., 2009; Wallace et al., 2008).

To make imaging data meaningful for quantitative neural activity measurements, calibration of GECI responses to ground truth measurements of spike trains in a representative set of neurons is necessary. In the mammalian brain, such calibration can be performed via cell-attached recording of spontaneous or sensory-evoked APs (Tian et al., 2009). Sensor response may vary from cell to cell, presumably due to variations in expression level, which can complicate interpretation. Such cell-to-cell variability is most apparent using *in utero* electroporation and to a lesser extent viral transduction. Transgenics, especially knock-ins, may alleviate this variability to some degree. Image processing is a necessary but tedious step, as signals must be assigned (segmented) to specific cells. This is inherently difficult and is further complicated by low probe SNR, dense labeling, motion artifacts, neuropil signal, probe photobleaching, tissue autofluorescence, hemodynamics, light scattering, artifacts from probe cytotoxicity, illumination phototoxicity, etc. A number of algorithmic improvements have been recently published (Greenberg and Kerr, 2009; Mukamel et al., 2009; Valmianski et al., 2010). Following segmentation, firing rate changes can be extracted from imaging traces via deconvolution (Yaksi and Friedrich, 2006), machine learning (Sasaki et al., 2008; D. Huber, D. Gutnisky, and K. Svoboda, personal communication), or Monte Carlo simulation (Vogelstein et al., 2009), among others.

The noise level *in vivo* greatly challenges the SNR of the sensor, which depends on the imaging configuration, surgical procedures (e.g., cranial window implantation or skull thinning (Yang et al., 2010), probe brightness, expression level, and imaging depth, among many other parameters. For reliably detecting sparse AP firing *in vivo*, especially single APs, further GECI engineering will be necessary. The kinetics of currently available GECIs is too slow to reconstruct high-frequency firing events. Potential cytotoxicity caused by long-term, high-level GECI expression can also make chronic *in vivo* imaging difficult. On the other hand, low-level GECI expression not only compromises SNR but also requires longer exposure times or higher excitation intensity, which can lead to sensor bleaching or tissue damage. To balance high SNR with minimal cytotoxicity, expression cassettes that are inducible and reversible or that hold [GECI] steady for long periods of time would facilitate signal calibration and further reduce toxicity concerns in chronic imaging. Finally, development of optics allowing faster, deeper, and higher-resolution imaging will greatly increase GECI utility. Recently, the applications of adaptive optics to two-photon microscopy to correct aberrations (Ji et al., 2010) and regenerative amplification to increase light penetration (Mittmann et al., 2011) have significantly increased the depth at which imaging may be reliably performed.

To target GECIs to genetically defined neuronal populations, a variety of genetic methods can be used. For example, transgenic lines expressing "drivers" (e.g., Cre recombinase or the Gal4 transcriptional activator) in specific cell types (with tissue-specific promoters) can be combined with "reporter" strains or viruses containing Cre- or Gal4/UAS-dependent GECIs (for review of such two-component systems, see Luo et al., 2008; Simpson, 2009). Large numbers of *Drosophila* Gal4 lines and Cre mice are currently available. Alternatively, a small but growing number of *cis*-regulatory elements (e.g., *synapsin-1*, *CaMKIIα*) can target transgenes delivered by viral infection or electroporation, even in non-genetic-model organisms. Through the combination of promoter selection and inherent viral serotype tropism, AAV vectors enable regional expression via local viral injection and have been broadly used to deliver sensors or other molecular tools to cell types of interest for mapping, monitoring, and manipulating neural circuitry (Betley and Sternson, 2011; Borghuis et al., 2011).

Expression of genetically encoded sensors may potentially perturb the physiology of the system studied, which is one of the major challenges for *in vivo* imaging, especially for applications where chronic imaging is required, for example, learning or disease progression. To facilitate signal calibration, protein sensors should be expressed at steady-state levels in individual neurons for long periods of time without altering physiology. At the same time, sensor performance should be constant to enable quantitative analysis. Optimizing sensors for *in vivo* application is difficult, although progress has been made (Tian et al., 2009). For each application, the timing and magnitude of sensor expression should be optimized to balance signal and cytotoxicity, by testing multiple promoters, regulatory sequences, and transduction methods (e.g., screening different viral serotypes, integration methods, copy number, etc.). Careful experiments are required to determine the effects of long-term GECI expression on both single-cell physiology and circuit function.

Outlook

The primary focus of current GECI engineering efforts is to further increase the SNR for robust detection of low firing rates (ideally single APs) in a variety of systems *in vivo*. Engineering high-affinity sensors for probing small stimuli while preserving fast kinetics is challenging (Mank et al., 2006). Further optimization of the GCaMP, Cameleon, and TnC sensor formats will undoubtedly improve signals in the short term. Alternatively, novel GECI scaffolds, with different (perhaps faster) calcium-binding proteins, may be employed to overcome this limitation.

Additionally, expanding the color spectrum of GECIs, particularly into the red and near-infrared, will open up many new applications in neuroscience, due to deeper tissue penetration, decreased autofluorescence, and less phototoxicity (Shcherbo et al., 2010). Also, many transgenic mouse lines already contain GFP (or equivalent), which precludes easy use of current GECIs for functional imaging. Red-shifted single-FP GECIs would not only facilitate deep imaging but also enable functional imaging from multiple cell populations and/or subcellular compartments, and make it easier to use with light-gated proteins such as channelrhodopsin-2 (Zhao et al., 2011). With the constant improvement of red and far-red FPs (Lin et al., 2009; Shcherbo et al., 2009, 2010; Shu et al., 2009), viable red-shifted GECIs should be possible.

The next generation of neural activity probes will be faster and more sensitive than the current one. It will also feature different colors of GECIs that can be used in concert. Mechanisms of GECI cytotoxicity must be elucidated; more "bio-orthogonal" sensor elements may help with this. Concomitant with this, improving indicator SNR will enable expression at lower levels, which reduces concerns. Other indicator improvements will include photoactivatable versions that can be turned on in selected populations and those that integrate [Ca^{2+}] over time. A defined toolkit of expression systems, such as transgenic mice and a library of viral serotype–promoter combinations, will further facilitate technology uptake. These advancements, together with improvement in the areas of lasers, cameras, imaging processing and deconvolution algorithms, and computation, will make long-term, high-resolution *in vivo* functional imaging in large, defined cell populations a reality.

Reagent availability

GCaMP3 plasmid DNA (*CMV* promoter for expression in mammalian cells) is available from Addgene (addgene.org): #22692. A membrane-fused version (Lck) is also available: #26974. Live AAV virus and pAAV constructs for both GCaMP3 and a Cre-dependent FLEX-GCaMP3 are available from the U Penn Vector Core, driven by a *synapsin-1* promoter (http://www.med.upenn.edu/gtp/vectorcore/Catalogue.shtml). *Drosophila* strains expressing GCaMP3 under *UAS* control are available from the Bloomington Stock Collection (#32116, #32234–32237; inserts are available on the X, 2L, and 3L chromosomes). A Cre-dependent GCaMP3 reporter mouse (*ROSA26* locus; *CAG* promoter) has been deposited at Jackson Labs (#014538). The Janelia Farm GECI Project continues to develop the GCaMP scaffold, as well as other GECIs. Information on the Project can be found at http://www.janelia.org/team-project/geci-project.

Note added in proof

Recently, blue- and red- shifted single-FP GECIs have been demonstrated (Zhao et al., 2011).

References

Akerboom, J., Rivera, J. D., Guilbe, M. M., Malave, E. C., Hernandez, H. H., Tian, L., et al. (2009). Crystal structures of the GCaMP calcium sensor reveal the mechanism of fluorescence signal change and aid rational design. *The Journal of Biological Chemistry*, *284*, 6455–6464.

Aubin-Horth, N., Landry, C. R., Letcher, B. H., & Hofmann, H. A. (2005). Alternative life histories shape brain gene expression profiles in males of the same population. *Proceedings of the Royal Society B*, *272*, 1655–1662.

Baird, G. S., Zacharias, D. A., & Tsien, R. Y. (1999). Circular permutation and receptor insertion within green fluorescent proteins. *Proceedings of the National Academy of Sciences of the United States of America*, *96*, 11241–11246.

Betley, J. N., & Sternson, S. M. (2011). Adeno-associated viral vectors for mapping, monitoring, and manipulating neural circuits. *Human Gene Therapy*, *22*, 669–677.

Borghuis, B. G., Tian, L., Xu, Y., Nikonov, S. S., Vardi, N., Zemelman, B. V., et al. (2011). Imaging light responses of targeted neuron populations in the rodent retina. *The Journal of Neuroscience*, *31*, 2855–2867.

Chalasani, S. H., Chronis, N., Tsunozaki, M., Gray, J. M., Ramot, D., Goodman, M. B., et al. (2007). Dissecting a circuit for olfactory behaviour in Caenorhabditis elegans. *Nature*, *450*, 63–70.

Cheng, L. E., Song, W., Looger, L. L., Jan, L. Y., & Jan, Y. N. (2010). The role of the TRP channel NompC in Drosophila larval and adult locomotion. *Neuron*, *67*, 373–380.

Chiappe, M. E., Seelig, J. D., Reiser, M. B., & Jayaraman, V. (2010). Walking modulates speed sensitivity in Drosophila motion vision. *Current Biology*, *20*, 1470–1475.

Del Bene, F., Wyart, C., Robles, E., Tran, A., Looger, L., Scott, E. K., et al. (2010). Filtering of visual information in the tectum by an identified neural circuit. *Science*, *330*, 669–673.

Diez-Garcia, J., Matsushita, S., Mutoh, H., Nakai, J., Ohkura, M., Yokoyama, J., et al. (2005). Activation of cerebellar parallel fibers monitored in transgenic mice expressing a fluorescent Ca2+ indicator protein. *The European Journal of Neuroscience*, *22*, 627–635.

Dombeck, D. A., Harvey, C. D., Tian, L., Looger, L. L., & Tank, D. W. (2010). Functional imaging of hippocampal place cells at cellular resolution during virtual navigation. *Nature Neuroscience*, *13*, 1433–1440.

Dreosti, E., Odermatt, B., Dorostkar, M. M., & Lagnado, L. (2009). A genetically encoded reporter of synaptic activity in vivo. *Nature Methods*, *6*, 883–889.

Fine, R., Lehman, W., Head, J., & Blitz, A. (1975). Troponin C in brain. *Nature*, *258*, 260–267.

Gourine, A. V., Kasymov, V., Marina, N., Tang, F., Figueiredo, M. F., Lane, S., et al. (2010). Astrocytes control breathing through pH-dependent release of ATP. *Science*, *329*, 571–575.

Greenberg, D. S., & Kerr, J. N. (2009). Automated correction of fast motion artifacts for two-photon imaging of awake animals. *Journal of Neuroscience Methods*, *176*, 1–15.

Hasan, M. T., Friedrich, R. W., Euler, T., Larkum, M. E., Giese, G., Both, M., et al. (2004). Functional fluorescent Ca2+ indicator proteins in transgenic mice under TET control. *PLoS Biology*, *2*, e163.

He, J., Ma, L., Kim, S., Nakai, J., & Yu, C. R. (2008). Encoding gender and individual information in the mouse vomeronasal organ. *Science*, *320*, 535–538.

Heim, N., Garaschuk, O., Friedrich, M. W., Mank, M., Milos, R. I., Kovalchuk, Y., et al. (2007). Improved calcium imaging in transgenic mice expressing a troponin C-based biosensor. *Nature Methods*, *4*, 127–129.

Heim, N., & Griesbeck, O. (2004). Genetically encoded indicators of cellular calcium dynamics based on troponin C and green fluorescent protein. *The Journal of Biological Chemistry*, *279*, 14280–14286.

Helmchen, F., Imoto, K., & Sakmann, B. (1996). Ca2+ buffering and action potential-evoked Ca2+ signaling in dendrites of pyramidal neurons. *Biophysical Journal*, *70*, 1069–1081.

Hendel, T., Mank, M., Schnell, B., Griesbeck, O., Borst, A., & Reiff, D. F. (2008). Fluorescence changes of genetic calcium indicators and OGB-1 correlated with neural activity and calcium in vivo and in vitro. *The Journal of Neuroscience*, *28*, 7399–7411.

Hires, S. A., Tian, L., & Looger, L. L. (2008). Reporting neural activity with genetically encoded calcium indicators. *Brain Cell Biology*, *36*, 69–86.

Horikawa, K., Yamada, Y., Matsuda, T., Kobayashi, K., Hashimoto, M., Matsu-ura, T., et al. (2010). Spontaneous network activity visualized by ultrasensitive Ca(2+) indicators, yellow Cameleon-Nano. *Nature Methods*, *7*, 729–732.

Ji, N., Milkie, D. E., & Betzig, E. (2010). Adaptive optics via pupil segmentation for high-resolution imaging in biological tissues. *Nature Methods*, *7*, 141–147.

Lin, M. Z., McKeown, M. R., Ng, H. L., Aguilera, T. A., Shaner, N. C., Campbell, R. E., et al. (2009). Autofluorescent proteins with excitation in the optical window for intravital imaging in mammals. *Chemistry & Biology*, *16*, 1169–1179.

Luo, L., Callaway, E. M., & Svoboda, K. (2008). Genetic dissection of neural circuits. *Neuron*, *57*, 634–660.

Lütcke, H., Murayama, M., Hahn, T., Margolis, D. J., Astori, S., zum Alten Borgloh, S. M., et al. (2010). Optical recording of neuronal activity with a genetically-encoded calcium indicator in anesthetized and freely moving mice. *Frontiers in Neural Circuits*, *4*, 9.

Mank, M., Reiff, D. F., Heim, N., Friedrich, M. W., Borst, A., & Griesbeck, O. (2006). A FRET-based calcium biosensor with fast signal kinetics and high fluorescence change. *Biophysical Journal*, *90*, 1790–1796.

Mao, T., O'Connor, D. H., Scheuss, V., Nakai, J., & Svoboda, K. (2008). Characterization and subcellular targeting of GCaMP-type genetically-encoded calcium indicators. *PLoS One*, *3*, e1796.

Mittmann, W., Wallace, D. J., Czubayko, U., Herb, J. T., Schaefer, A. T., Looger, L. L., et al. (2011). Two-photon calcium imaging of evoked activity from L5 somatosensory neurons in vivo. *Nature Neuroscience*, *14*, 1089–1093.

Miyawaki, A. (2011). Development of probes for cellular functions using fluorescent proteins and fluorescence resonance energy transfer. *Annual Review of Biochemistry*, *80*, 357–373.

Miyawaki, A., Llopis, J., Heim, R., McCaffery, J. M., Adams, J. A., Ikura, M., et al. (1997). Fluorescent indicators for Ca2+ based on green fluorescent proteins and calmodulin. *Nature*, *388*, 882–887.

Mukamel, E. A., Nimmerjahn, A., & Schnitzer, M. J. (2009). Automated analysis of cellular signals from large-scale calcium imaging data. *Neuron*, *63*, 747–760.

Muto, A., Ohkura, M., Kotani, T., Higashijima, S., Nakai, J., & Kawakami, K. (2011). Genetic visualization with an improved GCaMP calcium indicator reveals spatiotemporal activation of the spinal motor neurons in zebrafish. *Proceedings of the National Academy of Sciences of the United States of America*, *108*, 5425–5430.

Nagai, T., Sawano, A., Park, E. S., & Miyawaki, A. (2001). Circularly permuted green fluorescent proteins engineered to sense Ca2+. *Proceedings of the National Academy of Sciences of the United States of America*, *98*, 3197–3202.

Nagai, T., Yamada, S., Tominaga, T., Ichikawa, M., & Miyawaki, A. (2004). Expanded dynamic range of fluorescent indicators for Ca(2+) by circularly permuted yellow fluorescent proteins. *Proceedings of the National Academy of Sciences of the United States of America*, *101*, 10554–10559.

Nakai, J., Ohkura, M., & Imoto, K. (2001). A high signal-to-noise Ca(2+) probe composed of a single green fluorescent protein. *Nature Biotechnology*, *19*, 137–141.

O'Connor, D. H., Peron, S. P., Huber, D., & Svoboda, K. (2010). Neural activity in barrel cortex underlying vibrissa-based object localization in mice. *Neuron*, *67*, 1048–1061.

Ohkura, M., Matsuzaki, M., Kasai, H., Imoto, K., & Nakai, J. (2005). Genetically encoded bright Ca2+ probe applicable for dynamic Ca2+ imaging of dendritic spines. *Analytical Chemistry*, *77*, 5861–5869.

Palmer, A. E., Giacomello, M., Kortemme, T., Hires, S. A., Lev-Ram, V., Baker, D., et al. (2006). Ca2+ indicators based on computationally redesigned calmodulin-peptide pairs. *Chemical Biology*, *13*, 521–530.

Palmer, A. E., Jin, C., Reed, J. C., & Tsien, R. Y. (2004). Bcl-2-mediated alterations in endoplasmic reticulum Ca2+ analyzed with an improved genetically encoded fluorescent sensor. *Proceedings of the National Academy of Sciences of the United States of America*, *101*, 17404–17409.

Palmer, A. E., Qin, Y., Park, J. G., & McCombs, J. E. (2011). Design and application of genetically encoded biosensors. *Trends in Biotechnology*, *29*, 144–152.

Pedelacq, J. D., Cabantous, S., Tran, T., Terwilliger, T. C., & Waldo, G. S. (2006). Engineering and characterization of a superfolder green fluorescent protein. *Nature Biotechnology*, *24*, 79–88.

Petreanu, L., Huber, D., Sobczyk, A., & Svoboda, K. (2007). Channelrhodopsin-2-assisted circuit mapping of long-range callosal projections. *Nature Neuroscience*, *10*, 663–668.

Petreanu, L., Mao, T., Sternson, S. M., & Svoboda, K. (2009). The subcellular organization of neocortical excitatory connections. *Nature*, *457*, 1142–1145.

Piston, D. W., & Kremers, G. J. (2007). Fluorescent protein FRET: The good, the bad and the ugly. *Trends in Biochemical Sciences*, *32*, 407–414.

Pologruto, T. A., Yasuda, R., & Svoboda, K. (2004). Monitoring neural activity and [Ca2+] with genetically encoded Ca2+ indicators. *The Journal of Neuroscience*, *24*, 9572–9579.

Reiff, D. F., Plett, J., Mank, M., Griesbeck, O., & Borst, A. (2010). Visualizing retinotopic half-wave rectified input to the motion detection circuitry of Drosophila. *Nature Neuroscience*, *13*, 973–978.

Roisen, F. J., Wilson, F. J., Yorke, G., Inczedy-Marcsek, M., & Hirabayashi, T. (1983). Immunohistochemical localization of troponin-C in cultured neurons. *Journal of Muscle Research and Cell Motility*, *4*, 163–175.

Romoser, V. A., Hinkle, P. M., & Persechini, A. (1997). Detection in living cells of Ca2+–dependent changes in the fluorescence emission of an indicator composed of two green fluorescent protein variants linked by a calmodulin-binding sequence. A new class of fluorescent indicators. *Journal of Chemical Biology*, *272*, 13270–13274.

Ruta, V., Datta, S. R., Vasconcelos, M. L., Freeland, J., Looger, L. L., & Axel, R. (2010). A dimorphic pheromone circuit in Drosophila from sensory input to descending output. *Nature*, *468*, 686–690.

Sasaki, T., Takahashi, N., Matsuki, N., & Ikegaya, Y. (2008). Fast and accurate detection of action potentials from somatic calcium fluctuations. *Journal of Neurophysiology*, *100*, 1668–1676.

Scanziani, M., & Hausser, M. (2009). Electrophysiology in the age of light. *Nature*, *461*, 930–939.

Schevzov, G., Gunning, P., Jeffrey, P. L., Temm-Grove, C., Helfman, D. M., Lin, J. J., et al. (1997). Tropomyosin localization reveals distinct populations of microfilaments in neurites and growth cones. *Molecular and Cellular Neurosciences*, *8*, 439–454.

Seelig, J. D., Chiappe, M. E., Lott, G. K., Dutta, A., Osborne, J. E., Reiser, M. B., et al. (2010). Two-photon calcium imaging from head-fixed Drosophila during optomotor walking behavior. *Nature Methods*, *7*, 535–540.

Shcherbo, D., Murphy, C. S., Ermakova, G. V., Solovieva, E. A., Chepurnykh, T. V., Shcheglov, A. S., et al. (2009). Far-red fluorescent tags for protein imaging in living tissues. *The Biochemical Journal*, *418*, 567–574.

Shcherbo, D., Shemiakina, I. I., Ryabova, A. V., Luker, K. E., Schmidt, B. T., Souslova, E. A., et al. (2010). Near-infrared fluorescent proteins. *Nature Methods*, *7*, 827–829.

Shigetomi, E., Kracun, S., & Khakh, B. S. (2010). Monitoring astrocyte calcium microdomains with improved membrane targeted GCaMP reporters. *Neuron Glia Biology*, *16*, 1–9 [Epub ahead of print].

Shigetomi, E., Kracun, S., Sofroniew, M. V., & Khakh, B. S. (2010). A genetically targeted optical sensor to monitor calcium signals in astrocyte processes. *Nature Neuroscience*, *13*, 759–766.

Shindo, A., Hara, Y., Yamamoto, T. S., Ohkura, M., Nakai, J., & Ueno, N. (2010). Tissue-tissue interaction-triggered calcium elevation is required for cell polarization during Xenopus gastrulation. *PLoS One*, *5*, e8897.

Shu, X., Royant, A., Lin, M. Z., Aguilera, T. A., Lev-Ram, V., Steinbach, P. A., et al. (2009). Mammalian expression of infrared fluorescent proteins engineered from a bacterial phytochrome. *Science*, *324*, 804–807.

Simpson, J. H. (2009). Mapping and manipulating neural circuits in the fly brain. *Advances in Genetics*, *65*, 79–143.

Souslova, E. A., Belousov, V. V., Lock, J. G., Stromblad, S., Kasparov, S., Bolshakov, A. P., et al. (2007). Single fluorescent protein-based Ca2+ sensors with increased dynamic range. *BMC Biotechnology*, *7*, 37.

Svoboda, K., & Yasuda, R. (2006). Principles of two-photon excitation microscopy and its applications to neuroscience. *Neuron*, *50*, 823–839.

Tallini, Y. N., Ohkura, M., Choi, B. R., Ji, G., Imoto, K., Doran, R., et al. (2006). Imaging cellular signals in the heart in vivo: Cardiac expression of the high-signal Ca2+ indicator GCaMP2. *Proceedings of the National Academy of Sciences of the United States of America*, *103*, 4753–4758.

Tian, L., Hires, S. A., & Looger, L. L. (2011). Imaging neuronal activity with genetically encoded calcium indicators. In F. Helmchen & A. Konnerth (Eds.), *Imaging in neuroscience: A laboratory manual* (pp. 63–73). Cold Spring Harbor, NY: Cold Spring Harbor Laboratory Press.

Tian, L., Hires, S. A., Mao, T., Huber, D., Chiappe, M. E., Chalasani, S. H., et al. (2009). Imaging neural activity in worms, flies and mice with improved GCaMP calcium indicators. *Nature Methods*, *6*, 875–881.

Valmianski, I., Shih, A. Y., Driscoll, J. D., Matthews, D. W., Freund, Y., & Kleinfeld, D. (2010). Automatic identification of fluorescently labeled brain cells for rapid functional imaging. *Journal of Neurophysiology*, *104*, 1803–1811.

Vogelstein, J. T., Watson, B. O., Packer, A. M., Yuste, R., Jedynak, B., & Paninski, L. (2009). Spike inference from calcium imaging using sequential Monte Carlo methods. *Biophysical Journal*, *97*, 636–655.

Wallace, D. J., Zum Alten Borgloh, S. M., Astori, S., Yang, Y., Bausen, M., Kugler, S., et al. (2008). Single-spike detection in vitro and in vivo with a genetic Ca(2+) sensor. *Nature Methods*, *5*, 797–804.

Wang, Q., Shui, B., Kotlikoff, M. I., & Sondermann, H. (2008). Structural basis for calcium sensing by GCaMP2. *Structure*, *16*, 1817–1827.

Yaksi, E., & Friedrich, R. W. (2006). Reconstruction of firing rate changes across neuronal populations by temporally deconvolved Ca2+ imaging. *Nature Methods*, *3*, 377–383.

Yang, G., Pan, F., Parkhurst, C. N., Grutzendler, J., & Gan, W. B. (2010). Thinned-skull cranial window technique for long-term imaging of the cortex in live mice. *Nature Protocols*, *5*, 201–208.

Zhao, Y., Araki, S., Wu, J., Teramoto, T., Chang, Y-F., Nakano, M., et al. (2011). An expanded palette of genetically encoded Ca^{2+} indicators. *Science*, *333*, 1888–1891.

T. Knöpfel and E. Boyden (Eds.)
Progress in Brain Research, Vol. 196
ISSN: 0079-6123

CHAPTER 6

Manipulating cellular processes using optical control of protein–protein interactions

Chandra L. Tucker*

Department of Pharmacology, University of Colorado School of Medicine, Aurora, CO, USA

Abstract: Tools for optical control of proteins offer an unprecedented level of spatiotemporal control over biological processes, adding a new layer of experimental opportunity. While use of light-activated cation channels and anion pumps has already revolutionized neurobiology, an emerging class of more general optogenetic tools may have similar transformative effects. These tools consist of light-dependent protein interaction modules that allow control of target protein interactions and localization with light. Such tools are modular and can be applied to regulate a wide variety of biological activities. This chapter reviews the different properties of light-induced dimerization systems, based on plant phytochromes, cryptochromes, and light-oxygen-voltage domain proteins, exploring advantages and limitations of the different systems and practical considerations related to their use. Potential applications of these tools within the neurobiology field, including light control of various signaling pathways, neuronal activity, and DNA recombination and transcription, are discussed.

Keywords: optogenetics; protein–protein interactions; dimerizers; light control; photoreceptors; phytochrome; cryptochrome; LOV domain; inducible.

In recent years, coincident with the development of advanced imaging techniques and fluorescent protein tools that allow observation of cellular processes, there has been a growing interest in technologies allowing artificial control of biological pathways. Exogenous factors such as chemicals that permit acute inhibition or activation of pathways can allow the study of effects of protein or pathway activation and inhibition in the absence of genetic compensation that can occur when pathways are perturbed on a longer timescale. To this end, novel chemical–genetic strategies have been developed that allow activation or inhibition of a wide range of cellular functions within seconds or minutes by addition of membrane-permeable small molecules. Such chemical tools have been immensely valuable, but chemicals cannot provide spatial resolution, are not rapidly reversible, and

*Corresponding author.
Tel.: 303-724-6337; Fax: 303-724-3663
E-mail: chandra.tucker@ucdenver.edu

DOI: 10.1016/B978-0-444-59426-6.00006-9

can be difficult to deliver *in vivo*. Another exogenous factor that can be conditionally applied to cells is light, which allows precise spatial resolution and fast temporal resolution. Systems for modulating biological activity with light using photocaged chemical compounds have been in use for decades (Adams and Tsien, 1993; Callaway and Katz, 1993; Ellis-Davies, 2007; Pettit et al., 1997; Wang and Augustine, 1995; Wieboldt et al., 1994), where they provide precise spatial control. However, chemicals can be difficult to deliver into cells and tissues and do not always have specific sites for attachment of photolabile moieties. Other approaches such as CALI (chromophore-assisted light inactivation) (Jay, 1988) have also been used that can allow knockdown of protein levels in specific locations illuminated with light, but these systems only allow inactivation of activity. The above systems are also typically not photoreversible, though there are some exceptions, notably covalently bound photoswitchable affinity labels that contain azobenzene groups (Banghart et al., 2004; Fortin et al., 2008; Janovjak et al., 2010; Volgraf et al., 2006).

In the past several years, a number of genetically encoded systems that allow modulation of protein activity using light have been developed. The first generation of these tools to be generally adopted have been cation channels or anion pumps such as channelrhodopsin-2 and halorhodopsin, which are normally found, respectively, in the green algae *Chlamydomonas reinhardtii* and the halobacteria *Natronomonas pharaonis* (Boyden et al., 2005; Han and Boyden, 2007; Li et al., 2005; Nagel et al., 2003; Zhang et al., 2007). In a short time, these tools have revolutionized neurobiology, giving researchers the ability to precisely stimulate or inhibit neural activity in a local, spatially defined manner (Deisseroth, 2011; Gradinaru et al., 2010; Peron and Svoboda, 2011). More recently, heterologous proteins from light-sensing organisms have given rise to an emerging class of broad-use tools that may be used for optical control of a variety of enzymes and intracellular signaling pathways. These tools consist of modular light-responsive photoreceptor domains that are engineered to allosterically control protein activity or mediate protein–protein interactions with light. This review will focus on the latter of these emerging technologies, light-stimulated protein–protein interactions (e.g., protein dimerization).

The concept of controlling biological activity via protein–protein interactions is well documented in nature. Systems allowing artificial control of protein interactions using small molecules (chemical inducers of dimerization) have been in use for almost two decades, where they have been used to inducibly control diverse biological processes (Belshaw et al., 1996; Farrar et al., 1996; Inoue et al., 2005; Spencer et al., 1993). With chemical dimerizers, two proteins are brought together with addition of a bridging chemical ligand (Fig. 1a). Attachment of target proteins to the dimerizing modules allows conditional interaction of target domains of interest upon addition of a small molecule. With light, the same general strategy is used, but the dimerizing molecules are a photoreceptor protein and a second protein that interacts with the photoreceptor protein in a light-dependent manner (Fig. 1a). Similar to chemical dimerizer systems, the light-interacting modules can be attached to target proteins or domains, allowing conditional interaction of target proteins of interest with light (Fig. 1b), but with spatial control and greatly enhanced temporal control. Many general mechanisms for regulation of protein activity have been used with chemical dimerizers (Fegan et al., 2010), and these would be expected to also be effective with light-induced dimerizers. These approaches include inducible recruitment of a protein to a specific subcellular location (resulting in protein activation or inactivation), reconstitution of two halves of a split protein or enzyme, and activation of downstream signaling pathways initiated by homo- or heterodimerization (Fig. 1b).

The use of light offers many advantages over chemical regulation. Unlike chemicals, which require some time to be taken up into cells and are not easily removed, application of light is nearly instantaneous. Light-dependent conformational

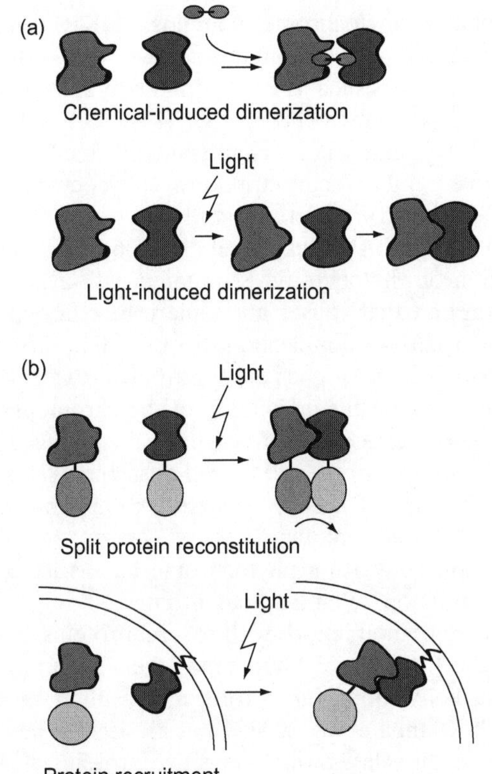

Fig. 1. Strategies for induction of protein dimerization. (a) Small-molecule or light-induced dimerizers conditionally interact with light. At top, two noninteracting proteins are forced together by a tethered compound that bridges the proteins. At bottom, a photoreceptor protein and a second protein are unable to interact in the dark, but light alters the conformation of the photoreceptor and enables interaction. (b) Schemes for regulation of target proteins with light. Target proteins or domains of interest (circles) can be fused to the dimerizer modules, allowing inducible dimerization of target proteins. In the top figure, the target proteins are inactive fragments of a split protein, which when dimerized result in reconstitution of functional protein. Alternatively, inducible dimerization can be used to force proteins to particular subcellular locations, where they may be active or sequestered (bottom figure). (For color version of this figure, the reader is referred to the Web version of this chapter.)

changes in most photoreceptor proteins occur with subsecond kinetics, and delivery is not limited by the cell membrane or affected by differences in factors affecting small molecule delivery to tissues. Indeed, several light-activated dimerizers show response rates more or less equivalent to the rates of protein diffusion in whole cells (Kennedy et al., 2010; Levskaya et al., 2009).

Light application is also quickly reversible, as light can be immediately extinguished. Light-stimulated photoreceptor proteins may remain in a stimulated state for time frames that range from seconds to hours, but eventually spontaneously revert back to their dark states through the process of dark reversion. Some photoreceptor protein interactions such as those with phytochromes and PIF family proteins can be conformationally switched on or off with different wavelengths of light (Ni et al., 1999; Shimizu-Sato et al., 2002), and dimerizers based on these allow fine optical control of both protein association and protein dissociation (Leung et al., 2008; Levskaya et al., 2009; Shimizu-Sato et al., 2002; Tyszkiewicz and Muir, 2008). Dimerization systems that naturally revert on the time frame of minutes or that are inducibly reversible can also be regulated with light pulses or light intensity, thus allowing dose-dependent input to a particular pathway (Kennedy et al., 2010; Levskaya et al., 2009). Thus, one major advantage of light control is that it can be used to activate or inhibit a biological process on a timecourse and magnitude defined by the user.

Perhaps the biggest advantage of using light to control biological activity is the ability for precise spatial regulation. While there are a number of genetic approaches allowing spatial regulation of activity in specific cells using cell-targeted promoters, light offers the ability for spatial regulation in single cells or user-defined cell clusters. In addition, the use of light offers the prospect for local activation of pathways or proteins in specific subcellular regions. This concept has already been exploited in neurobiology with caged chemicals, which can be locally activated at dendrites (Bagal et al., 2005; Matsuzaki et al., 2001, 2004; Smith et al., 2003). While, as with caged compounds, light scattering can limit ability for local activation, two-photon activation has been demonstrated with at least one light-activated dimerizer technology (Kennedy et al., 2010), and this technology will facilitate local activation as well as allow improved penetration of light into deep tissues.

Phytochromes, cryptochromes, and LOV domains—Modular photosensory domains

Optogenetic tools are based on existing modular photosensory domains derived from plants, algae, and other light-sensing organisms. Thus, a review of these technologies requires an introduction to light-sensing protein domains and mechanisms of light activation. There are six main classes of photosensory proteins: phytochromes, cryptochromes, light-oxygen-voltage (LOV) domain proteins, BLUF domain proteins, xanthopsins, and rhodopsins (Moglich et al., 2010)—this review will focus on the first three. Shown in Fig. 2a are representative phytochrome, cryptochrome, and LOV domain photoreceptor proteins. All photoreceptors bind one or more chromophores, small molecule cofactors that are involved in photon absorption. Light absorption causes a physical change in the chromophore that is coupled to a structural change in the photoreceptor protein. However, the specific events and mechanisms involved in photoexcitation differ among the photoreceptors and vary from creation of a covalent bond with LOV domain proteins to isomerization around a carbon–carbon double bond with phytochromes (for an in-depth review of photoreceptor photochemistry, see Moglich et al., 2010; Zoltowski and Gardner, 2011).

Phytochromes mediate responses to red (660 nm) and far-red (730nm) light in plants and are also found in bacteria and cyanobacteria. The bacterial and cyanobacterial phytochromes maintain high sequence homology with the plant proteins, sharing a common domain structure of an amino-terminal photosensory domain with three conserved structural domains, a PAS domain, a GAF domain, and a PHY domain (Rockwell et al., 2006; Fig. 2a). The C-terminal domain of phytochromes contains a HKRD (histidine kinase-related domain) but can also contain other domains such as PAS domains as shown with *Arabidopsis* phytochrome B (Fig. 2a). Phytochromes contain a covalently linked linear tetrapyrrole (bilin) chromophore, phycocyanobilin (PCB) in cyanobacteria and phytochromobilin (PΦB) in plants (Rockwell et al., 2006). Light stimulation of phytochrome is initiated with a photon absorption that causes an isomerization around a carbon–carbon double bond in the bilin chromophore (Fig. 2b; Moglich et al., 2010; Rockwell et al., 2006). This results in rotation of the chromophore and a corresponding structural change in phytochrome from a "Pr" red-absorbing state to a "Pfr" far-red absorbing state. Absorption of a second photon of far-red light reverts the bilin cofactor, converting phytochrome back to the Pr conformation. Plant phytochromes show light-dependent interactions with several proteins, notably the Pfr form of PhyB binds the PIF family of basic helix–loop–helix proteins (Castillon et al., 2007; Khanna et al., 2004; Ni et al., 1999).

A second class of light-sensitive proteins, which also play a major role in plant photoresponses, are cryptochromes. Cryptochromes are blue-light-responsive photoreceptors and are structurally similar to DNA photolyases, which repair breaks in DNA caused by UV damage. Cryptochrome family members contain an N-terminal photolyase homology (PHR) domain of approximately 500 amino acids that mediates light responses and divergent C-terminal domains (Fig. 2a; Lin and Shalitin, 2003). Cryptochromes bind a flavin chromophore (FAD) (Fig. 2c) and are thought to also bind folate as a second light-harvesting chromophore. Folate binds DNA photolyase (Sancar and Sancar, 1988) and, though it was not observed in the crystal structure of the photosensory domain of *Arabidopsis* Cry1 (Brautigam et al., 2004), is speculated to be present in plant cryptochrome based on biochemical studies (Hoang et al., 2008; Malhotra et al., 1995). As with phytochromes, several light-dependent interactions of plant and *Drosophila*

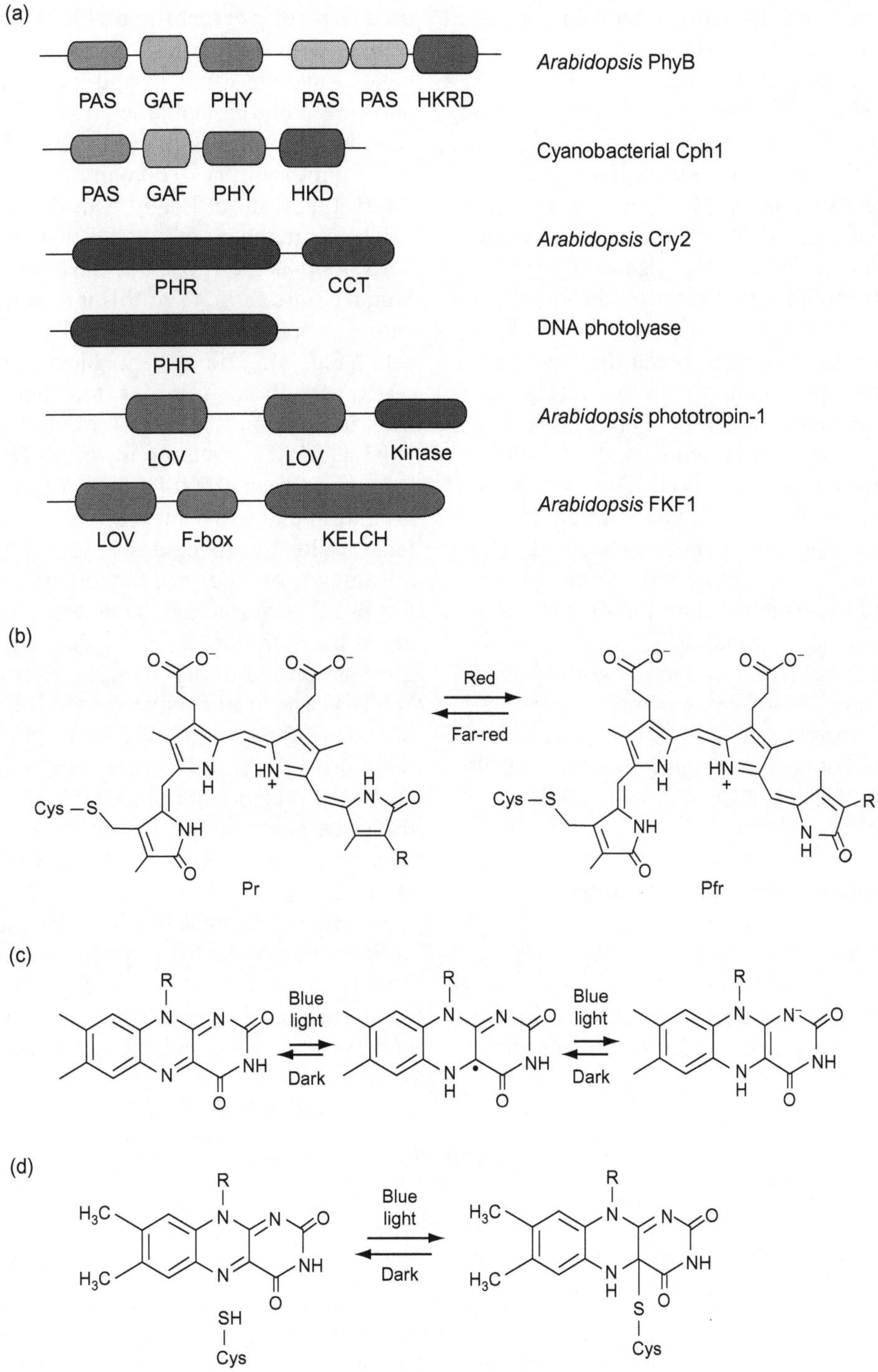

(a)

PAS GAF PHY PAS PAS HKRD — *Arabidopsis* PhyB

PAS GAF PHY HKD — Cyanobacterial Cph1

PHR CCT — *Arabidopsis* Cry2

PHR — DNA photolyase

LOV LOV Kinase — *Arabidopsis* phototropin-1

LOV F-box KELCH — *Arabidopsis* FKF1

(b)

Red / Far-red

Cys — S, Pr

Cys — S, Pfr

(c)

Blue light / Dark

Blue light / Dark

(d)

Blue light / Dark

SH — Cys

S — Cys

cryptochromes have been identified (Liu et al., 2008, 2011; Peschel et al., 2009).

A third light-responsive domain is the LOV domain, which is a member of the conserved Per-ARNT-Sim (PAS) domain family (Moglich et al., 2010). LOV domains are smaller (~110 amino acids) photosensory units than phytochromes or cryptochromes and are often found as modular elements of larger multidomain signaling proteins. LOV domain proteins bind a flavin chromophore (FMN or FAD), which is stimulated by blue light to form a covalent linkage between the flavin ring and a nearby cysteine in the protein (Fig. 2d; Zoltowski and Gardner, 2011). This causes in a change in chromophore structure and a resultant change in protein confirmation. The conformational changes involved in photoactivation of several LOV domains have been well studied. The LOV2 domain of *Avina sativa* phototropin 1 binds a C-terminal Jα helix in the dark but dissociates in light, resulting in unwinding of the Jα helix (Harper et al., 2003). These large, light-triggered structural changes have been exploited for allosteric control of protein activity by strategic insertion of an LOV domain (Moglich et al., 2009; Strickland et al., 2008, 2010; Wu et al., 2009).

Phytochrome-based chemical dimerizers

The PhyB–PIF3 system

The first engineered system that allowed inducible control of a protein interaction with light used a plant phytochrome (PhyB) and its interacting partner to bring together a transcription factor binding domain and activation domain, allowing light-inducible control of DNA transcription (Shimizu-Sato et al., 2002). The N-terminal photosensory domain of *Arabidopsis* PhyB binds in red light (in the Pfr state) to PIF3, a member of the PIF family of basic helix–loop–helix proteins (Ni et al., 1999). Shimizu-Sato et al. used this interaction to reconstitute a split Gal4 transcription factor in yeast cells (Fig. 3). The system allowed light control of transcription, wherein red light stimulated the interaction between PhyB–Gal4BD and PIF3–Gal4AD, resulting in robust transcriptional induction (over 1000-fold induction over background in one experiment). Remarkably, this system could be toggled on and off by taking advantage of the red/far-red reversal of the PhyB–PIF3 interaction. Thus while red light triggered transcription, far-red light triggered protein dissociation, resulting in transcriptional shutoff. While yeast do not make the PΦB chromophore necessary for photostimulation of PhyB, this chromophore (or typically an analogous bilin, PCB, that is extracted from Spirulina) can be added to the yeast medium, where it is taken up into cells and assembles to form holo-PhyB (Shimizu-Sato et al., 2002).

Several additional studies demonstrated the utility of the PhyB–PIF3 system to control protein activity. Tyszkiewicz and Muir, also working in yeast, generated a split intein, a protein that undergoes a self-splicing reaction, that was

Fig. 2. Light-responsive protein domains and cofactors. (a) Domain structure of representative phytochrome, cryptochrome, and LOV domain containing proteins from plants and light-sensing organisms. Domain abbreviations are as follows: PAS, Per-ARNT-Sim domain; GAF, GAF domain; PHY, phytochrome domain; HKRD, histidine kinase-related domain; PHR, DNA photolyase homology-related domain; CCT, cryptochrome C-terminal domain; LOV, light-oxygen-voltage domain; KELCH, KELCH motif. (b–d) Chromophores and light-induced photochemical changes for phytochrome bilin (b), LOV domain flavin (c), or cryptochrome flavin (d). (b) The bilin chromophore that binds phytochrome undergoes a *Z*/*E* isomerization around a C—C double bond upon stimulation with red light, resulting in a change from the Pr to Pfr form. (c) In *Arabidopsis* cryptochromes 1 and 2, light stimulation of FAD results in a semiquinone intermediate, followed by further reduction to FADH. (d) Light absorption by an LOV domain-binding flavin stimulates formation of a covalent bond between the flavin ring and a nearby conserved cysteine in the protein. (For color version of this figure, the reader is referred to the Web version of this chapter.)

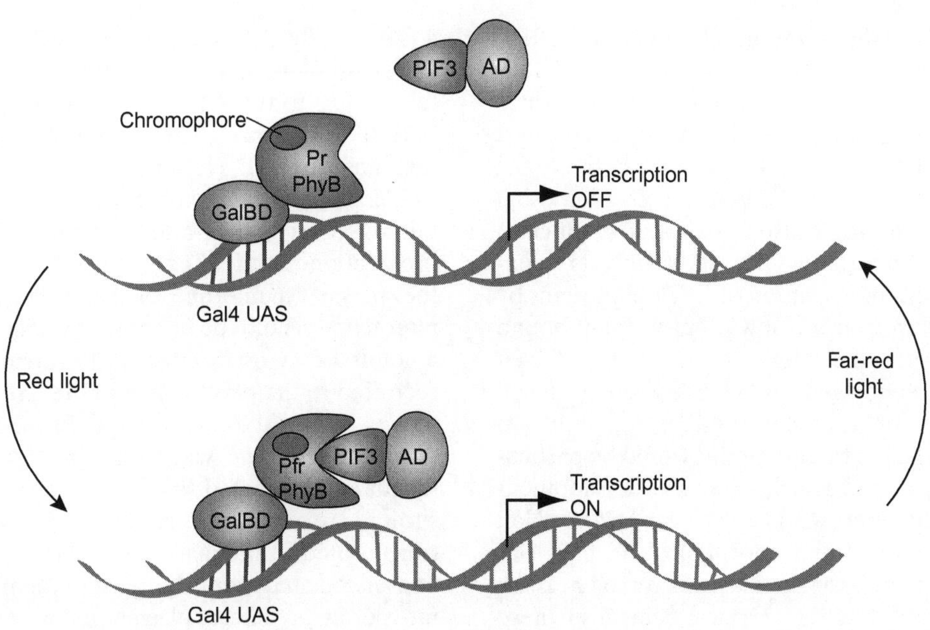

Fig. 3. Light-regulated induction of transcription. In the transcriptional activation system described by Shimizu-Sato et al., interaction between PhyB and PIF3 was used to reconstitute a split Gal4 transcription factor. In this case, PhyB was fused to a Gal4-binding domain (GalBD) and PIF3 fused to a Gal4 activation domain (Gal4AD). In dark or far-red light, PhyB does not interact with PIF3 and transcription remains off. With red light, PhyB interacts with PIF3, reconstituting a functional Gal4 transcription factor and allowing transcriptional induction. (For color version of this figure, the reader is referred to the Web version of this chapter.)

brought together by PhyB and PIF3 to allow light-dependent protein splicing (Tyszkiewicz and Muir, 2008). Their initial experiments used full-length PIF3, which not only triggered light-dependent protein splicing but also suffered from high background levels of splicing in the dark. Use of a truncated N-terminal domain of PIF3 (APB domain, aa 1–100) that had been shown to be sufficient for binding to PhyB (Khanna et al., 2004) resulted in reduced background levels of splicing in the dark, while retaining the same level of splicing in the light. In addition to reduced levels of background activation, the PIF3APB domain is much smaller than full-length PIF3, simplifying its use as a modular domain in engineering strategies. Importantly, this work also demonstrated reversibility. Not only could the PhyB–PIF3APB interaction be switched off with far-red light, but also the reconstituted split intein, once brought together by the dimerizers, could reversibly dissociate (Tyszkiewicz and Muir, 2008).

A third study used the PhyB–PIF3APB interaction to recruit a GDP-bound form of the Rho family GTPase Cdc42 to its effector, the Wiskott–Aldrich syndrome protein (WASP), allowing activation of WASP (Leung et al., 2008). While GDP-bound Cdc42 has poor affinity for WASP, prior studies had shown that sufficiently high (saturating) concentrations of Cdc42 (GDP) can activate WASP (Leung and Rosen, 2005). Leung et al. found that tethering PhyB and PIF3APB to GDP-bound Cdc42 and WASP could increase their interaction upon light-activated dimerization of PhyB and PIF3, allowing light-dependent activation of WASP. While these studies were carried out *in vitro*, since the bilin chromophore can be taken up by yeast and mammalian cells (Levskaya et al., 2009; Shimizu-Sato et al., 2002), WASP activation using this mechanism would also be expected to function in live cells. Importantly, this study suggests a general approach for activation of GTPase proteins by forcing interaction of low-affinity GDP-bound versions with their effector proteins. Given the general importance of GTPase proteins in cell morphology, polarity, and motility, the ability to precisely regulate these proteins would be particularly useful in areas such as neural development (e.g., neurite outgrowth) (Govek et al., 2005; Hall and Lalli, 2010) and morphological plasticity (Harvey et al., 2008; Hayashi et al., 2004; Murakoshi et al., 2011). Optical control of these pathways in neuronal cells is discussed in more detail in a subsequent section.

The PhyB–PIF6 system

One of the biggest advances in using PhyB for light-triggered dimerization did not involve use of the PIF3 or PIF3APB interaction, but a different PIF family protein, PIF6 (Levskaya et al., 2009). Using an interaction between PhyB and PIF6, Levskaya et al. elegantly demonstrated that this system could be used in mammalian cells to recruit proteins to specific subcellular locations and to conditionally regulate protein activity. Importantly, this work was the first demonstration of the fast temporal control, precise subcellular spatial control, and reversibility afforded by light-activated dimerizers. As mammalian cells do not synthesize the bilin chromophore required for photostimulation of PhyB, these studies showed that, as in yeast (Shimizu-Sato et al., 2002), exogenous chromophore (PCB) applied to the medium could be taken up into mammalian cells, where it assembles with apo-PhyB to form holo-PhyB.

To initially characterize the PhyB–PIF6 system, Levskaya et al. attached a yellow fluorescent protein (YFP) to the APB domain of PIF6 (residues 1–100) (Fig. 4a). They then fused different domains of PhyB (residues 1–450, 1–650, 1–908, full length) to the monomeric red fluorescent protein mCherry, which also contained a C-terminal prenylation motif (CaaX box) for targeting to the plasma membrane. Thus, the PHYB–PIF6 interaction could be assayed by visualizing the amount of cytosolic PIF–YFP versus PIF–YFP recruited to the plasma membrane. All PhyB constructs were able to recruit PIF–YFP to the plasma membrane with red light, but only full-length and truncated (aa 1–908) versions of PhyB showed reversible recruitment, in which PIF6 translocated to the plasma membrane in red light and dissociated from the plasma membrane with far-red light. These plasma membrane recruitment studies allowed visualization of the fast binding and dissociation kinetics of the interaction. As shown in Fig. 4b, recruitment to the plasma membrane occurs within seconds, with time constants (τ=1.3±0.1 s for recruitment and τ=4±1 s for membrane release) that are calculated to be near the limits of diffusion. Due to such fast kinetics, the group was able to demonstrate precise subcellular spatial control of membrane recruitment: by globally illuminating the entire cell with infrared (>750 nm) light and focusing a red light at a specific points in the cell, PIF could be rapidly recruited to user-specified regions of the cell. Importantly, any of the PhyB–PIF6 complex that diffused out of the defined red light spot would be quickly reverted by the infrared mask.

To demonstrate the biological utility of their system, Levskaya et al. used the PhyB–PIF6 system to recruit the catalytic domains of guanine nucleotide exchange factors (GEFs) that activate Rho family small GTPases (Fig. 4c and d). Previous experiments with chemical dimerizers had shown that inducible activation of Rho GTPases can be accomplished by recruitment of the catalytic (DH–PH) domain of Rho GEFs to the

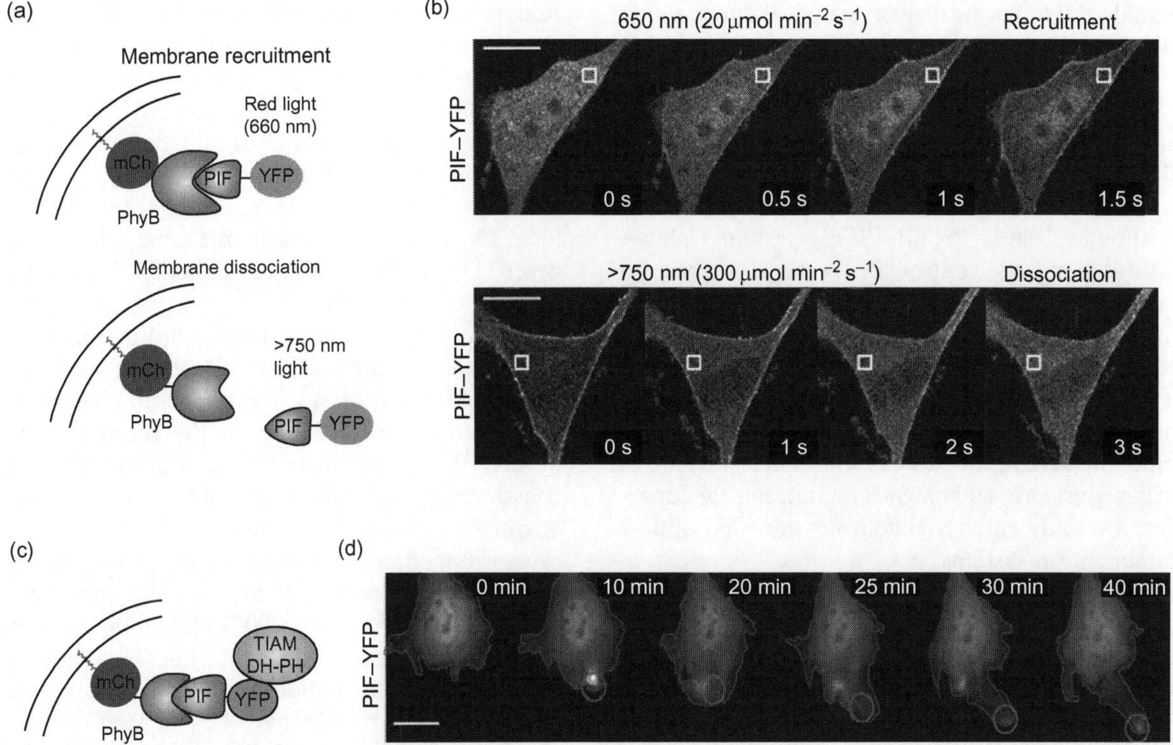

Fig. 4. Membrane recruitment using PhyB–PIF6 dimerizers. (a) Schematic showing constructs used for recruitment of PIF6–YFP to the membrane. (b) Fluorescence images reveal rapid translocation of PIF–YFP between membrane and cytosol with red and infrared light. (c) Schematic showing constructs used for membrane recruitment of the Rho GEF Tiam DH–PH domain. (d) Localized membrane recruitment results in formation of lamellipodia extrusions in NIH3T3 cells. Samples were globally irradiated with a 750nm light source, while a 650nm laser was focused on the region of the cell indicated with a circle. The cell body is outlined in color. Figures adapted with permission from Macmillan Publishers Ltd., *Nature* (Levskaya et al., 2009), copyright 2009. (For color version of this figure, the reader is referred to the Web version of this chapter.)

plasma membrane (Inoue et al., 2005). Levskaya et al. used a similar strategy, fusing the DH–PH domains of several Rho family GEF proteins (the Rac GEF Tiam, the Cdc42 GEF intersectin, and the Rho GEF Tim) to YFP–PIF3 (Fig. 4c). These constructs were then coexpressed with the membrane-localized PhyB–mCherry–CaaX construct, such that light application would recruit the GEFs to the plasma membrane where they exert their effects. As shown in Fig. 4d with the Rac GEF Tiam, application of red light to a specific subcellular region resulted in localized Rac activation, causing protrusion of the plasma membrane at the illuminated site on a timescale of tens of minutes. Again, this work provided a critical demonstration of the ability of light-activated dimerizers to regulate protein activity in a precise spatially restricted manner. In addition, these studies showed that, as with chemical dimerizers, proteins can be efficiently activated by light dimerizers via membrane recruitment. As a wide variety of proteins of importance to neurobiologists and cell biologists are regulated by membrane binding, this should provide a general strategy for light regulation of protein activity by removing endogenous membrane targeting signals and replacing them with optical dimerization domains for conditional membrane recruitment.

Light-activated dimerizers derived from flavin-binding proteins

Although the PHYB–PIF3 system allows light-dependent control of protein interaction and dissociation with red/far-red light, one major limitation for use in tissues or whole organisms is the requirement for addition of a bilin ligand. Two different light-activated dimerizer systems have been generated using photoreceptors that bind flavin chromophores, which are present in organisms across all kingdoms, and thus far represent the only entirely genetically encoded optical dimerization systems.

LOV domain-based dimerizers: FKF–GIGANTIA

The first blue-light-controlled dimerization system developed was based on an interaction between an LOV domain containing protein from *Arabidopsis*, FKF1 and a second protein, GIGANTIA (GI) (Yazawa et al., 2009). As demonstration of membrane recruitment, Yazawa et al. attached GI to the plasma membrane (via a membrane-targeted mCherry protein) and fused FKF to YFP. Wild-type FKF1 showed light-dependent interaction with GI but retained significant affinity for GI even in the dark. They then identified a G128D mutant of FKF1 that reduced the dark interaction but maintained the same degree of interaction in the light as the wild-type protein. The timecourse of FKF–GI interaction was much slower than PhyB–PIF3, requiring up to 30min for maximal membrane translocation. Also, unlike the PhyB–PIF3 interaction which could be reversed by far-red light, the FKF–GI interaction cannot be inducibly dissociated with light and possesses very slow spontaneous reversal kinetics: FKF and GI remained associated for >1.5h. While the slow interaction and reversal timecourse will preclude use of this system for many studies that require fast temporal control, it is possible that the poor activation kinetics may be improved by optimizing expression levels and other parameters. The slow reversal, in turn, may be ideal for some applications, including a light-triggered DNA recombinase, discussed in a later section.

As a biological demonstration of the FKF1–GI dimerization system, Yazawa et al. used the G128D–FKF–GI interaction to recruit a constitutively active form of Rac to the plasma membrane. Previous experiments showed that a constitutively active, GTP-bound form of Rac missing its prenylation motif is inactive in the cytosol but, if it is brought to the plasma membrane, can be functional and activate downstream effectors (Inoue et al., 2005). Illumination with blue light for 5min resulted in an 87% increase in lamellipodia formation after 30min, indicating activation of Rac pathways. As a second demonstration of light regulation, they used the dimerizers to reconstitute a split Gal4–VP16 bipartite transcription factor in mammalian cells, similar to the experiments that were done by Shimizu-Sato et al. using PhyB–PIF3 in yeast. However, activation was much less robust than that seen in yeast: using G128D–FKF–GI, they observed a 4.5-fold increase in transcriptional activation with light (Yazawa et al., 2009), as compared with a 1000-fold increase with PhyB–PIF3 in yeast (Shimizu-Sato et al., 2002). Further study of this system may allow a better understanding of ways to engineer this system for improved transcriptional activation.

Cryptochrome-based dimerizers: CRY2–CIB

A second blue-light-controlled dimerization system was recently developed that is based on an interaction between the *Arabidopsis* photoreceptor cryptochrome 2 (CRY2) and a putative transcription factor, CIB1 (Kennedy et al., 2010). CRY2 and CIB1 interact in plants in blue light but not dark, and prior studies had shown that an N-terminal truncation of CIB1 is sufficient for binding to CRY2 (Liu et al., 2008). Using a similar membrane recruitment approach as used with PhyB–PIF and FKF–GI, Kennedy et al. tethered the truncated domain of CIB1 (CIBN) to the plasma membrane and expressed a CRY2–mCherry fusion in the cytosol (Fig. 5a). Illumination of cells with a single (100ms) pulse of blue-light results in rapid translocation of CRY2–mCherry to the plasma membrane (τ=0.7s) (Fig. 5b and c). All photoreceptors have natural dark reversion rates, where after photostimulation they revert in darkness to an unstimulated state with varying kinetics. The CRY2–CIBN interaction completely (>90%) dissociates over a timecourse of approximately 10–12min (τ=5.5min) (Fig. 5d), consistent with the known reversion rate for a closely related cryptochrome, *Arabidopsis* CRY1 (Bouly et al., 2007).

To demonstrate functionality, Kennedy et al. used the CRY2–CIBN dimerizers for two biological applications, generating systems for light-inducible control of protein transcription in yeast cells and Cre-mediated DNA recombination in mammalian cells (Fig. 6). The reversibility observed in the translocation experiments allowed tight dose-dependent control, and they demonstrated use of light pulses to induce transcription with light (Fig. 6a and b). They also used the dimerizers to activate a split Cre recombinase enzyme (Fig. 6c and d) that was nonfunctional in the dark but could be reconstituted with blue light to allow recombination at loxP sites. While essentially no light-independent recombinase activity was seen, they were able to obtain dose-dependent light activation of Cre, with ~3% of cells showing recombination after delivery of three pulses of blue light over 15min, and ~20% of wild-type activity after delivery of light pulses for 24h (Fig. 6d). While the reconstituted Cre enzyme activity was not robust, its activity was tightly light controlled and similar to that seen in a previously reported split Cre reconstituted with a chemical dimerizer (Jullien et al., 2003), indicating that light-dimerization systems can function at similar levels as chemical dimerization systems. In addition, this was the first split enzyme (other than a transcription factor, which has very modular domains) that could be functionally reconstituted with light. Thus, this work provides a proof of principle for regulation of other enzymes by light using similar strategies. In the case of split Cre, as reversibility is not required, a system with a longer-lived interaction (i.e., slower reversal kinetics) may improve performance, allowing more robust levels of Cre recombinase activity to be achieved with just a single pulse of light.

As the kinetics of the CRY–CIB system is quite fast and no exogenous ligand is required, this dimerization system may be well suited for use in model organisms and *in vivo* studies. As blue light poorly penetrates tissues, Kennedy et al. also showed that the CRY–CIB system can be activated by two-photon excitation in the range of 820–980nm, with optimal stimulation at 860 nm. The ability to activate this, and possibly other systems, by two-photon microscopy enables their use in tissue slices and *in vivo*.

Allosteric control of protein function using light

Engineering allosteric control is also a useful means of controlling protein function, and although this is not the main focus of this review, there have been a number of groundbreaking studies exploring placement of light-responsive domains into proteins for allosteric control of protein function with light (see Moglich and Moffat, 2010; Rana and Dolmetsch, 2010 for reviews). For example, by replacing a heme-binding PAS domain in the *Bradyrhizobium japonicum* histidine kinase protein FixL with an LOV domain from *Bacillus subtilis* YtvA, a light-activated histidine kinase was generated that could be used to control gene expression in bacteria (Moglich

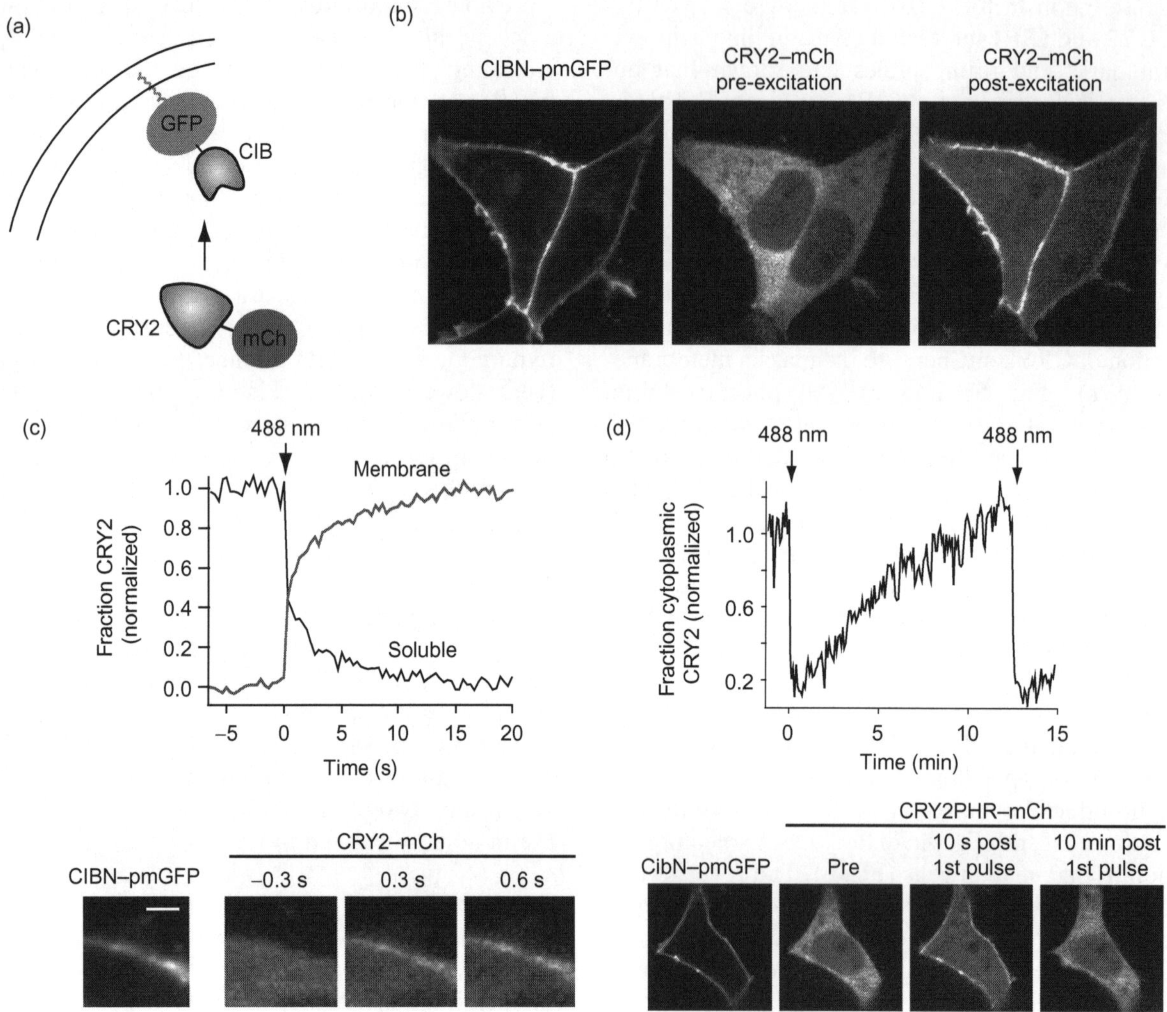

Fig. 5. Blue light-triggered translocation of CRY2–mCherry in mammalian cells. (a) Schematic showing fusion proteins used. (b) Fluorescence images of CIBN–pmEGFP and CRY2–mCh coexpressed in HEK293T cells. CRY2–mCh was imaged before light excitation and 20s after a 100-ms pulse of blue light (488nm). (c, d) The graph above and pictures below show the timecourse of CRY2–mCh recruitment to (c) or dissociation from (d) the plasma membrane after a blue-light pulse (488nm). The graphs show quantification of CRY2–mCh or CRY2PHR–mCh in the cytoplasm and/or at the plasma membrane, as labeled. Each fraction was normalized between 0 and 1. The photos show localization of mCh in the cell at indicated times before and after pulsing. Figure adapted from one originally published in Macmillan Publishers Ltd., *Nature Methods*, Kennedy et al. (2010), copyright 2010. (For color version of this figure, the reader is referred to the Web version of this chapter.)

et al., 2009). The *A. sativa* phototropin 1 LOV2 domain, which undergoes unwinding of its Jα helix when photostimulated (Harper et al., 2003), has been used in numerous engineering projects. In one study, it was used *in vitro* to confer light dependence to binding of an *Escherichia coli* Trp repressor to DNA (Strickland et al., 2008). While the engineered protein showed only

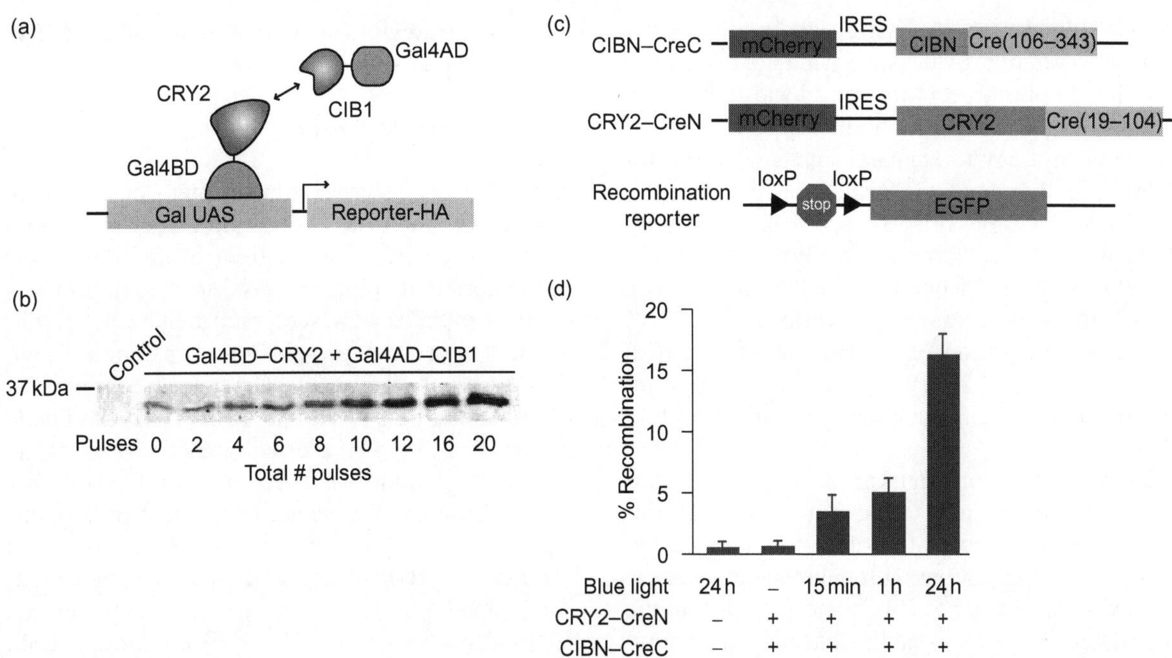

Fig. 6. Light-induced activation of transcription and DNA recombination. (a) Schematic of split Gal4 modules expressed in yeast cells containing a gene encoding a hemagglutinin (HA)-tagged reporter protein under control of a galactose-inducible promoter. UAS, upstream activating sequence. (b) Immunoblot analysis of the HA-tagged reporter in response to blue-light pulses (10 s pulses, 1.7 mW, 8 min apart). Control, lysates from cells expressing only the reporter. (c) Schematic showing the two split Cre recombinase constructs (CIBN–CreC and CRY2–CreN) and the reporter construct. IRES, internal ribosome entry site. (d) Cre reporter recombination measured 48 h after transfection of HEK293T cells with the Cre reporter and indicated constructs. Cells were exposed to blue-light pulses (450 nm, 4.5 mW) for the indicated durations or kept in the dark (–). Error bars, SD (n=3) from three independent experiments. Figure adapted from one originally published in Macmillan Publishers Ltd., *Nature Methods*, Kennedy et al. (2010), copyright 2010. (For color version of this figure, the reader is referred to the Web version of this chapter.)

approximately fivefold higher affinity for DNA in light in initial studies (Strickland et al., 2008), a follow-up study greatly improved the dynamic range of the light response, resulting in a protein showing 64-fold higher affinity for DNA in light (Strickland et al., 2010).

One of the most promising studies of light-regulated allosteric control of protein function was the development of a light-activated Rac GTPase (PA-Rac1) that showed robust function in mammalian cells (Wu et al., 2009). Wu et al. fused the LOV2–Jα helix to a constitutively active Rac protein (containing additional mutations to eliminate binding to GTPase activating proteins), such that in the dark Rac was prevented from interacting with effector proteins by LOV–Jα, but the light-triggered conformational change of LOV–Jα allowed interaction with downstream effectors. PA-Rac1 activation was reversible, with a half-life for spontaneous dark reversion of $\sim$ 45 s. Thus, prolonged activation requires near-constant light input (or very closely spaced pulses). Because of its rapid reversibility, this system can be used for very local activation with subcellular precision, since activated protein diffusing away from the site of illumination is quickly inactivated by spontaneous reversion to the dark state. Thus, while not as spatially precise as the

PhyB–PIF6 system, the fast decay kinetics of the LOV–Jα system provides an advantage over more long-lived photoreceptors for local activation studies. In further experiments, Wu et al. used the same strategy to engineer light regulation for Cdc42, another small GTPase, indicating that this approach may be broadly applicable to other proteins in the GTPase family (Wu et al., 2009). PA-Rac1 was subsequently used to induce cell polarization in *Drosophila* (Wang et al., 2010) and direct migration of neutrophils in zebrafish (Yoo et al., 2010), providing important demonstrations of functionality of this tool *in vivo*.

In addition to engineered proteins, light-sensing organisms have a number of natural light-responsive signaling proteins that could be used in optogenetic applications with minimal engineering. For example, the genes *PACα* and *PACβ* from the flagellate *Euglena gracilis* encode a photo-activatable adenylyl cyclase controlled by BLUF domains (Iseki et al., 2002). These proteins have been promising for use as optogenetic tools (Nagahama et al., 2007; Schroder-Lang et al., 2007) but also have had problems when expressed in heterologous systems due to their large size (~100kDa), poor expression, and significant light-independent adenylyl cyclase activity. Recently, two different groups identified a smaller BLUF domain regulated adenylyl cyclase from a different organism (*blaC*/*bPAC* gene from *Beggiatoa*) and demonstrated its utility to control cAMP levels in heterologous systems with light (Ryu et al., 2010; Stierl et al., 2010). In addition to its smaller size, this cyclase has minimal activity in the dark and is strongly activated with light. In an additional engineering twist, as adenylyl cyclases can be converted to guanylyl cyclases and vice versa by mutagenesis (Sunahara et al., 1998; Tucker et al., 1998), one of the groups engineered the adenylyl cyclase to generate a light-responsive guanylyl cyclase (Ryu et al., 2010). Given the importance of cAMP and cGMP as second messengers in signaling pathways, these tools could be useful for acutely perturbing these pathways in heterologous systems.

General strategies for regulating protein activity with dimerizers

Subcellular recruitment

One strategy for regulating protein activity with dimerizers that has been successful with both chemical dimerizers and light-activated dimerizers is membrane recruitment. As discussed in previous sections of this review, successful membrane recruitment of fluorescent proteins has been demonstrated in three of the light dimerizer technologies: PhyB–PIF6, FKF1–GI, and CRY2–CIB (Kennedy et al., 2010; Levskaya et al., 2009; Yazawa et al., 2009). While fluorescent proteins do not have elaborate structural requirements to maintain activity, this approach has also been successfully demonstrated with the PhyB–PIF6 and FKF1–GI systems to activate small GTPase pathways (Levskaya et al., 2009; Yazawa et al., 2009). Both groups used approaches that were previously successful with chemical dimerizers, in which they substituted the light-triggered protein–protein interaction modules for the chemically stimulated modules, a result that suggests that target proteins that have been successfully regulated with chemical dimerizers will work with light dimerizers without extensive engineering. While only GTPase activation has been demonstrated using recruitment strategies with light-induced dimerizers, this approach should be generalizable with a variety of membrane-acting proteins, with nuclear proteins, and with other proteins that function only at specific subcellular locations.

Membrane recruitment strategies offer particular promise for local perturbation of proteins in specific subcellular regions, such as individual dendritic spines. In such applications, photo-stimulated proteins that require constant light input may be more useful than long-lived systems, as protein recruited to the plasma membrane will quickly spread by lateral diffusion to other parts of the cell. For example, PA-Rac1, with a dark reversion rate of ~45s, can be used for local activation with constant light input, as protein that diffuses from the activation region rapidly inactivates (Wu et al., 2009). Photoreversal techniques such as those employed with the PhyB–PIF6 system using full-field illumination with far-red light (to dissociate the interaction) combined with focal illumination with red light (to promote interaction only at a specific region of a cell) can be used for even better resolution of the spatial pattern of membrane recruitment.

Assembly of split proteins

The use of protein interactions to assemble split protein fragments is another general strategy that can be used to confer light regulation to proteins of interest. In this approach, a protein is split into two inactive fragments that do not associate on their own but assemble with the help of dimerization domains to form an active protein. The list of proteins that can be conditionally controlled in this manner is growing and includes Gal4 transcriptional activator (Fields and Song, 1989), ubiquitin (Johnsson and Varshavsky, 1994), dihydrofolate reductase (Pelletier et al., 1998), β-lactamase (Galarneau et al., 2002; Wehrman et al., 2002), GFP (Wilson et al., 2004), luciferase (Remy and Michnick, 2006), TEV protease (Wehr et al., 2006), cytosine deaminase (Ear and Michnick, 2009), and Cre recombinase (Jullien et al., 2003). One main constraint has been that identification of optimal protein split sites often requires considerable screening. In addition, the lengths of linkers separating the proteins of interest from the dimerizers are also important, and thus screening different length linkers is often necessary to optimize split protein activity. Depending on the application, dimerizers with longer or shorter reversion rates may be desirable for split protein reconstitution. For example, a single light pulse will trigger prolonged activity of the cryptochrome, FKF1 or PhyB systems which may be advantageous for experiments where one would need to elicit large effects with minimal light stimulation (e.g., transcription or DNA recombination).

Applications of light-responsive proteins in neurobiology

The ability to activate proteins in spatially restricted patterns and on a rapid, user-defined timescale will allow an unprecedented level of control of biological processes. When combined with advanced optical reporters, these approaches will allow the ability to evaluate the consequences of perturbation of specific pathways in select cells at defined points in time. While such tools will be generally useful for scientists in all areas of cell and developmental biology, such systems will be particularly relevant to neurobiologists, as they offer the ability to dissect effects of local perturbation of cell function, as well as the ability to elicit effects on the second and millisecond timescales that are relevant to many processes in the nervous system. In addition, as advanced optical methods and optogenetic tools such as channelrhodopsin are already in use by many neurobiologists, additional hardware for optical control of protein activity will be easily assimilated, particularly for flavin-based photoreceptor systems whose optical spectra overlap considerably with that of channelrhodopsin. Conversely, with additional light sources, systems such as PhyB that absorb in the red and far-red spectrum regions could be used in conjunction with channelrhodopsin.

Control of neuronal morphology with light

The complex morphological architecture of neurons is a large determinant of neuronal function and connectivity. In all cell types, the actin cytoskeleton and its regulation by small GTPases control cell shape, size, and ability to migrate. A number of approaches for light activation of small GTPases have been demonstrated, using either dimerization technologies or allosteric control. In neurons, small GTPases play critical roles in establishment of cell polarity, neurite outgrowth, and activity-triggered morphological changes that occur in dendritic spines. Rho family GTPases regulate neuronal cell morphology including collapse of growth cones and neurite contraction for Rho, and neurite outgrowth for Rac and Cdc42 (Govek et al., 2005; Hall and Lalli, 2010; Tada and Sheng, 2006). Small GTPases have also been shown to be involved in dendrite morphological changes (Luo et al., 1996; Nakayama et al., 2000; Tada and Sheng, 2006) and synaptic plasticity (Harvey et al., 2008; Hayashi et al., 2004; Murakoshi et al., 2011). As small GTPases act centrally in cells to regulate a variety of critical processes, use of light-activated GTPases should allow interrogation of the roles of various GTPases at specific times during development, in specific cell clusters, or even within subcellular domains such as individual dendritic spines. While strategies for membrane recruitment have already been demonstrated with GTPases, activation of many other proteins should be possible using similar recruitment strategies. For example, several groups have used chemical dimerizers to recruit $PI(4,5)P_2$ 5-phosphatase to the cell membrane, where it acts to deplete PIP_2, allowing the effect of acute PIP_2 depletion to be examined independently of other effects (Heo et al., 2006; Suh et al., 2006; Varnai et al., 2006). In one such study, depletion of $PI(4,5)P_2$ while simultaneously recording membrane currents was found to rapidly regulate KCNQ channel activity (Suh et al., 2006). Use of light-activated dimerizers in similar experiments would allow study of the effects of phospholipid composition at specific times in specific subcellular domains of neurons.

Genetic manipulation with light

A second class of proteins that should be useful in neuronal studies, in particular developmental studies and cell circuit tracing, are light-activated DNA recombinases. A first-generation light-activated Cre recombinase showed strong light dependence and no background but the overall levels of activation were modest (Kennedy et al., 2010). Further refinement of this system, perhaps by using longer-lived photorecepters and random mutagenesis and screening, will be needed. A robust light-activated Cre recombinase would be useful for a number of studies in the brain, in particular for regulating genes (permanently turning on or off activity) involved in neuronal development and plasticity, and for tracing cellular lineages in user-specified cells or cell clusters. As an example, the genome of individual neurons participating in a defined circuit (visualized using calcium or voltage dyes) could be optically manipulated to test the effect of any gene on circuit dynamics and ultimately behavior in a completely noninvasive experiment. In another scenario, using light to turn on a gene for horseradish peroxidase or another reporter could be used to permanently mark cells *in vivo* "on the fly" during circuit visualization for further *post hoc* ultrastructural analysis by electron microscopy (Bock et al., 2011; Briggman et al., 2011). A light-inducible Cre recombinase also has advantages over current inducible systems such as the tamoxifen-induced CreERt system (Metzger and Feil, 1999), which has no spatial regulation and requires addition of tamoxifen, which causes secondary biological effects in many tissues potentially complicating experimental interpretation (see Kellen, 1996).

Regulation of protein transcription by light has been demonstrated with dimerizer technologies (Kennedy et al., 2010; Shimizu-Sato et al., 2002; Yazawa et al., 2009), and these approaches will be useful to neurobiologists. To this end, it will be important to develop more robust systems for light-induced transcriptional regulation in mammalian cells, as the only system demonstrated in mammalian cells (Yazawa et al., 2009) provides only low levels of transcriptional activation. Although regulation of proteins at the level of transcription is slow, and thus does not take advantage of the fast temporal resolution offered by light, such approaches would be suitable for studies of neuronal development, neurodegeneration, and general cell biology that involve longer time frames. One advantage of light regulation of transcription is the ability to toggle transcription on and off on a user-defined time frame. Thus, not only can protein expression be induced at a particular time, but it can also be shut off. Such regulation may be useful for the study of factors that are expressed at specific timepoints during development.

Light-controlled signaling

Light-activated dimerizers will be useful for activation of a number of signaling proteins that are naturally activated by heterodimerization or homodimerization. For example, caspase cascades, which have recently been shown to play a role in synaptic depression (Li et al., 2010), are regulated by dimerization and have been successfully activated with chemical dimerizers (Fan et al., 1999; MacCorkle et al., 1998). Receptor protein tyrosine kinases such as trkA are regulated by receptor dimerization and have been functionally activated by chemical dimerizers, allowing activation of downstream pathways independent of NGF (Alfa et al., 2009). As neurotrophin receptors such as trkA control neurite outgrowth, use of such a strategy with light dimerizers could allow light-triggered axonal growth in neurons. In the future, therapeutic uses for light dimerizers can even be envisioned, such as light guidance of axonal growth (using fiber optic delivery) for directing neuronal process migration following CNS trauma.

Functional circuit studies

While there are a number of robust optogenetic tools for neuronal silencing such as halorhodopsin (Han and Boyden, 2007; Zhang et al., 2007), these tools require direct light application for the duration of neuronal inactivation. Light-induced dimerizers could provide a complementary tool set for long-term (hours to days) neuronal silencing following a single pulse of light. One approach to this would be to use light-induced dimerizers that interfere with exocytosis of neurotransmitter vesicles. This has already been accomplished using chemical dimerizers, which were used to perturb SNARE proteins involved in synaptic vesicle exocytosis, resulting in cessation of presynaptic neurotransmitter release (Karpova et al., 2005). Optical control of neurotransmitter release would have an advantage over chemical methods in that light could be delivered to specific boutons, allowing silencing of a subset of user-defined presynaptic terminals from an individual neuron. Alternatively, a global and longer-lived approach for silencing neurons could be taken such as light-triggered transcription of botulinum or tetanus toxins, endoproteinases that target the SNARE machinery required for neurotransmitter release. In this case, the soma of individual cells could be illuminated to induce expression of toxin molecules that would presumably block neurotransmitter release from every axon terminal. This approach has been demonstrated in neurons by placing tetanus toxin expression under tetracycline control to demonstrate inducible and reversible silencing of hippocampal CA3 neurons (Nakashiba et al., 2008). Longer-term activation or silencing will be particularly useful for studies of animal behavior, as the light source would not need to continuously target a given subset of neurons. In these cases, a neuron or cluster of neurons could be stimulated during surgery and then the animal monitored without the light source over the ensuing hours or days for behavioral changes.

Limitations and practical considerations

Light-activated dimerizers have tremendous potential, and their general adoption will require systems that are robust, stable, show little or no light-independent activity, and have a wide dynamic range. The systems that are described here are already useful for a number of applications; however, additional engineering will improve the function of these domains considerably.

For heterologous use, photosensory domains present a number of engineering challenges. The photosensory domains of phytochromes and cryptochromes are quite large: the N-terminus of PhyB used for the PhyB–PIF6 studies is 908 amino acids; the PHR domain of CRY2 used for the CRY–CIB interaction is 498 amino acids. The large size of these constructs will make packaging into many types of viral vectors challenging. In addition, phytochrome and cryptochrome domains have strict conformational requirements. For example, phytochrome B does not tolerate N-terminal fusions (Shimizu-Sato et al., 2002) and CRY2 also shows some fusion site preference in some applications (Kennedy et al., 2010). LOV domain protein are much smaller and may work better as fusion proteins, though generating a system that is faster and more reversible than the current FKF1–GI dimerizers will be important.

Many photosensory proteins, including phytochromes and cryptochromes, themselves homodimerize (Rockwell et al., 2006; Rosenfeldt et al., 2008; Sang et al., 2005), which may limit the use of these systems in regulating homodimeric interactions. For example, chemical dimerizers have been used to dimerize the intracellular domain of trkA, resulting in neurite outgrowth (Alfa et al., 2009); however, this experiment would not be possible with a photoreceptor module that constitutively homodimerizes as trkA would be activated independently of light. The dimerization region of phytochrome appears to be in the C-terminus, so this may not be a problem for C-terminally truncated versions of PhyB, which retain light-triggered Pif6 binding (e.g., PhyB 1–651 or 1–908). Some LOV domains appear to themselves dimerize, but others do not (see Zoltowski and Gardner, 2011 for a review). Monomeric forms of fluorescent reporter proteins such as RFP and bacteriophytochrome, which normally form multimers, have been successfully engineered (Campbell et al., 2002; Shu et al., 2009), and similar approaches can be used to engineer monomeric forms of cryptochromes and phytochromes, if needed.

Another practical concern when using light-regulated proteins is the overlap between wavelengths used for activation/control of biological processes and those used for observation. While the spectrum of fluorescent reporter proteins is rapidly expanding (Tsien, 2010), many fluorescent proteins have overlapping excitation spectra with phytochromes, cryptochromes, and LOV domain proteins. For example, imaging GFP at 488nm also activates PA-Rac1 (Yi Wu, personal communication) and the CRY2–CIB and FKF1–GI dimerization system (Kennedy et al., 2010). Thus, GFP cannot be used for observation of perturbations caused by light, as imaging GFP itself activates these systems. Similarly, the excitation spectra of certain genetically encoded reporters such as FRET reporters or GCaMP Ca^{2+} sensors overlaps with the flavin-binding photoreceptor-based systems, although there is also the possibility of identifying spectrally shifted variants or using nonstandard activation wavelengths.

Other concerns with working with photoactive proteins are background light (i.e., the need to keep samples in the dark during preparation) and light delivery. Although different photoactivation systems show different light sensitivities, exposure of samples to background light is a concern. This can be minimized by using nonactivating wavelengths of filtered light to illuminate a room during sample preparation. With the explosion of interest in channelrhodopsin and related optogenetic tools, there has been a parallel explosion in development of advanced systems and approaches for light delivery that allow penetration into deep parts of the brain (Adamantidis et al., 2007; Aravanis et al., 2007; Flusberg et al., 2008; Gradinaru et al., 2007). Such approaches will likely be easily adaptable for precise delivery of light to stimulate protein–protein interactions.

Summary

The concept of using light to control biological processes, as opposed to use in visualizing biological processes, is rapidly expanding, and technologies such as light-controlled ion channels and dimerizers are likely the start of a vast wave of new genetically encoded systems for light regulation. Allowing unprecedented temporal, spatial, and dosage control, such tools are expected to transform many fields including neurobiology, allowing more precise, nuanced studies of cellular events. One of the greatest challenges at present is to improve the performance of existing light-induced dimerizers. As with channelrhodopsins and halorhodopsins, improvements to the kinetics, spectral sensitivity, expression levels, and trafficking can be made to enhance these systems or alter their functionality for specific applications. Another challenge will be to develop general approaches to allow regulation of new target proteins of interest without extensive engineering. While light-stimulated dimerizers are quite modular, applications with specific target proteins still require significant engineering for optimal performance.

Each of the present set of light-inducible dimerization systems has distinct biological properties, with different optimal activation wavelengths, dark reversion rates, and activation kinetics. Other known or newly discovered light-interacting protein pairs could also eventually be integrated into dimerization systems, imparting improved or altered properties desirable for specific applications. Ideally, these different optogenetic tools will complement each other, allowing sophisticated manipulation of complex cellular circuits. By combining different optogenetic actuator technologies with visualization tools such as fluorescent protein sensors, complex cellular behaviors can be induced, precisely controlled, and visualized, allowing formerly intractable experiments to be conducted to address new questions.

Acknowledgments

I would like to thank Matthew Kennedy and reviewers Andreas Möglich and Takanari Inoue for providing critical and insightful comments on the chapter.

References

Adamantidis, A. R., Zhang, F., Aravanis, A. M., Deisseroth, K., & de Lecea, L. (2007). Neural substrates of awakening probed with optogenetic control of hypocretin neurons. *Nature*, *450*, 420–424.

Adams, S. R., & Tsien, R. Y. (1993). Controlling cell chemistry with caged compounds. *Annual Review of Physiology*, *55*, 755–784.

Alfa, R. W., Tuszynski, M. H., & Blesch, A. (2009). A novel inducible tyrosine kinase receptor to regulate signal transduction and neurite outgrowth. *Journal of Neuroscience Research*, *87*, 2624–2631.

Aravanis, A. M., Wang, L. P., Zhang, F., Meltzer, L. A., Mogri, M. Z., Schneider, M. B., et al. (2007). An optical neural interface: In vivo control of rodent motor cortex with integrated fiberoptic and optogenetic technology. *Journal of Neural Engineering*, *4*, S143–S156.

Bagal, A. A., Kao, J. P., Tang, C. M., & Thompson, S. M. (2005). Long-term potentiation of exogenous glutamate responses at single dendritic spines. *Proceedings of the National Academy of Sciences of the United States of America*, *102*, 14434–14439.

Banghart, M., Borges, K., Isacoff, E., Trauner, D., & Kramer, R. H. (2004). Light-activated ion channels for remote control of neuronal firing. *Nature Neuroscience*, *7*, 1381–1386.

Belshaw, P. J., Ho, S. N., Crabtree, G. R., & Schreiber, S. L. (1996). Controlling protein association and subcellular localization with a synthetic ligand that induces heterodimerization of proteins. *Proceedings of the National Academy of Sciences of the United States of America*, *93*, 4604–4607.

Bock, D. D., Lee, W. C., Kerlin, A. M., Andermann, M. L., Hood, G., Wetzel, A. W., et al. (2011). Network anatomy and in vivo physiology of visual cortical neurons. *Nature*, *471*, 177–182.

Bouly, J. P., Schleicher, E., Dionisio-Sese, M., Vandenbussche, F., Van Der Straeten, D., Bakrim, N., et al. (2007). Cryptochrome blue light photoreceptors are activated through interconversion of flavin redox states. *The Journal of Biological Chemistry*, *282*, 9383–9391.

Boyden, E. S., Zhang, F., Bamberg, E., Nagel, G., & Deisseroth, K. (2005). Millisecond-timescale, genetically targeted optical control of neural activity. *Nature Neuroscience*, *8*, 1263–1268.

Brautigam, C. A., Smith, B. S., Ma, Z., Palnitkar, M., Tomchick, D. R., Machius, M., et al. (2004). Structure of the photolyase-like domain of cryptochrome 1 from Arabidopsis thaliana. *Proceedings of the National Academy of Sciences of the United States of America*, *101*, 12142–12147.

Briggman, K. L., Helmstaedter, M., & Denk, W. (2011). Wiring specificity in the direction-selectivity circuit of the retina. *Nature*, *471*, 183–188.

Callaway, E. M., & Katz, L. C. (1993). Photostimulation using caged glutamate reveals functional circuitry in living brain slices. *Proceedings of the National Academy of Sciences of the United States of America*, *90*, 7661–7665.

Campbell, R. E., Tour, O., Palmer, A. E., Steinbach, P. A., Baird, G. S., Zacharias, D. A., et al. (2002). A monomeric red fluorescent protein. *Proceedings of the National Academy of Sciences of the United States of America*, *99*, 7877–7882.

Castillon, A., Shen, H., & Huq, E. (2007). Phytochrome interacting factors: Central players in phytochrome-mediated light signaling networks. *Trends in Plant Science*, *12*, 514–521.

Deisseroth, K. (2011). Optogenetics. *Nature Methods*, *8*, 26–29.

Ear, P. H., & Michnick, S. W. (2009). A general life-death selection strategy for dissecting protein functions. *Nature Methods*, *6*, 813–816.

Ellis-Davies, G. C. (2007). Caged compounds: Photorelease technology for control of cellular chemistry and physiology. *Nature Methods*, *4*, 619–628.

Fan, L., Freeman, K. W., Khan, T., Pham, E., & Spencer, D. M. (1999). Improved artificial death switches based on caspases and FADD. *Human Gene Therapy*, *10*, 2273–2285.

Farrar, M. A., Alberol-Ila, J., & Perlmutter, R. M. (1996). Activation of the Raf-1 kinase cascade by coumermycin-induced dimerization. *Nature*, *383*, 178–181.

Fegan, A., White, B., Carlson, J. C., & Wagner, C. R. (2010). Chemically controlled protein assembly: Techniques and applications. *Chemical Reviews*, *110*, 3315–3336.

Fields, S., & Song, O. (1989). A novel genetic system to detect protein-protein interactions. *Nature*, *340*, 245–246.

Flusberg, B. A., Nimmerjahn, A., Cocker, E. D., Mukamel, E. A., Barretto, R. P., Ko, T. H., et al. (2008). High-speed, miniaturized fluorescence microscopy in freely moving mice. *Nature Methods*, *5*, 935–938.

Fortin, D. L., Banghart, M. R., Dunn, T. W., Borges, K., Wagenaar, D. A., Gaudry, Q., et al. (2008). Photochemical control of endogenous ion channels and cellular excitability. *Nature Methods*, *5*, 331–338.

Galarneau, A., Primeau, M., Trudeau, L. E., & Michnick, S. W. (2002). Beta-lactamase protein fragment complementation assays as in vivo and in vitro sensors of protein–protein interactions. *Nature Biotechnology*, *20*, 619–622.

Govek, E. E., Newey, S. E., & Van Aelst, L. (2005). The role of the Rho GTPases in neuronal development. *Genes & Development*, *19*, 1–49.

Gradinaru, V., Thompson, K. R., Zhang, F., Mogri, M., Kay, K., Schneider, M. B., et al. (2007). Targeting and readout strategies for fast optical neural control in vitro and in vivo. *The Journal of Neuroscience*, *27*, 14231–14238.

Gradinaru, V., Zhang, F., Ramakrishnan, C., Mattis, J., Prakash, R., Diester, I., et al. (2010). Molecular and cellular approaches for diversifying and extending optogenetics. *Cell*, *141*, 154–165.

Hall, A., & Lalli, G. (2010). Rho and Ras GTPases in axon growth, guidance, and branching. *Cold Spring Harbor Perspectives in Biology*, *2*, a001818.

Han, X., & Boyden, E. S. (2007). Multiple-color optical activation, silencing, and desynchronization of neural activity, with single-spike temporal resolution. *PLoS One*, *2*, e299.

Harper, S. M., Neil, L. C., & Gardner, K. H. (2003). Structural basis of a phototropin light switch. *Science*, *301*, 1541–1544.

Harvey, C. D., Yasuda, R., Zhong, H., & Svoboda, K. (2008). The spread of Ras activity triggered by activation of a single dendritic spine. *Science*, *321*, 136–140.

Hayashi, M. L., Choi, S. Y., Rao, B. S., Jung, H. Y., Lee, H. K., Zhang, D., et al. (2004). Altered cortical synaptic morphology and impaired memory consolidation in forebrain-specific dominant-negative PAK transgenic mice. *Neuron*, *42*, 773–787.

Heo, W. D., Inoue, T., Park, W. S., Kim, M. L., Park, B. O., Wandless, T. J., et al. (2006). PI(3,4,5)P3 and PI(4,5)P2 lipids target proteins with polybasic clusters to the plasma membrane. *Science*, *314*, 1458–1461.

Hoang, N., Bouly, J. P., & Ahmad, M. (2008). Evidence of a light-sensing role for folate in Arabidopsis cryptochrome blue-light receptors. *Molecular Plant*, *1*, 68–74.

Inoue, T., Heo, W. D., Grimley, J. S., Wandless, T. J., & Meyer, T. (2005). An inducible translocation strategy to rapidly activate and inhibit small GTPase signaling pathways. *Nature Methods*, *2*, 415–418.

Iseki, M., Matsunaga, S., Murakami, A., Ohno, K., Shiga, K., Yoshida, K., et al. (2002). A blue-light-activated adenylyl cyclase mediates photoavoidance in Euglena gracilis. *Nature*, *415*, 1047–1051.

Janovjak, H., Szobota, S., Wyart, C., Trauner, D., & Isacoff, E. Y. (2010). A light-gated, potassium-selective glutamate receptor for the optical inhibition of neuronal firing. *Nature Neuroscience*, *13*, 1027–1032.

Jay, D. G. (1988). Selective destruction of protein function by chromophore-assisted laser inactivation. *Proceedings of the National Academy of Sciences of the United States of America*, *85*, 5454–5458.

Johnsson, N., & Varshavsky, A. (1994). Split ubiquitin as a sensor of protein interactions in vivo. *Proceedings of the National Academy of Sciences of the United States of America*, *91*, 10340–10344.

Jullien, N., Sampieri, F., Enjalbert, A., & Herman, J. P. (2003). Regulation of Cre recombinase by ligand-induced complementation of inactive fragments. *Nucleic Acids Research*, *31*, e131.

Karpova, A. Y., Tervo, D. G., Gray, N. W., & Svoboda, K. (2005). Rapid and reversible chemical inactivation of synaptic transmission in genetically targeted neurons. *Neuron*, *48*, 727–735.

Kellen, J. A. (1996). *Tamoxifen, beyond the antiestrogen.* Boston: Birkhauser.

Kennedy, M. J., Hughes, R. M., Peteya, L. A., Schwartz, J. W., Ehlers, M. D., & Tucker, C. L. (2010). Rapid blue-light-mediated induction of protein interactions in living cells. *Nature Methods, 7*, 973–975.

Khanna, R., Huq, E., Kikis, E. A., Al-Sady, B., Lanzatella, C., & Quail, P. H. (2004). A novel molecular recognition motif necessary for targeting photoactivated phytochrome signaling to specific basic helix-loop-helix transcription factors. *The Plant Cell, 16*, 3033–3044.

Leung, D. W., Otomo, C., Chory, J., & Rosen, M. K. (2008). Genetically encoded photoswitching of actin assembly through the Cdc42-WASP-Arp2/3 complex pathway. *Proceedings of the National Academy of Sciences of the United States of America, 105*, 12797–12802.

Leung, D. W., & Rosen, M. K. (2005). The nucleotide switch in Cdc42 modulates coupling between the GTPase-binding and allosteric equilibria of Wiskott-Aldrich syndrome protein. *Proceedings of the National Academy of Sciences of the United States of America, 102*, 5685–5690.

Levskaya, A., Weiner, O. D., Lim, W. A., & Voigt, C. A. (2009). Spatiotemporal control of cell signalling using a light-switchable protein interaction. *Nature, 461*, 997–1001.

Li, X., Gutierrez, D. V., Hanson, M. G., Han, J., Mark, M. D., Chiel, H., et al. (2005). Fast noninvasive activation and inhibition of neural and network activity by vertebrate rhodopsin and green algae channelrhodopsin. *Proceedings of the National Academy of Sciences of the United States of America, 102*, 17816–17821.

Li, Z., Jo, J., Jia, J. M., Lo, S. C., Whitcomb, D. J., Jiao, S., et al. (2010). Caspase-3 activation via mitochondria is required for long-term depression and AMPA receptor internalization. *Cell, 141*, 859–871.

Lin, C., & Shalitin, D. (2003). Cryptochrome structure and signal transduction. *Annual Review of Plant Biology, 54*, 469–496.

Liu, H., Yu, X., Li, K., Klejnot, J., Yang, H., Lisiero, D., et al. (2008). Photoexcited CRY2 interacts with CIB1 to regulate transcription and floral initiation in Arabidopsis. *Science, 322*, 1535–1539.

Liu, B., Zuo, Z., Liu, H., Liu, X., & Lin, C. (2011). Arabidopsis cryptochrome 1 interacts with SPA1 to suppress COP1 activity in response to blue light. *Genes & Development, 25*, 1029–1034.

Luo, L., Hensch, T. K., Ackerman, L., Barbel, S., Jan, L. Y., & Jan, Y. N. (1996). Differential effects of the Rac GTPase on Purkinje cell axons and dendritic trunks and spines. *Nature, 379*, 837–840.

MacCorkle, R. A., Freeman, K. W., & Spencer, D. M. (1998). Synthetic activation of caspases: Artificial death switches. *Proceedings of the National Academy of Sciences of the United States of America, 95*, 3655–3660.

Malhotra, K., Kim, S. T., Batschauer, A., Dawut, L., & Sancar, A. (1995). Putative blue-light photoreceptors from Arabidopsis thaliana and Sinapis alba with a high degree of sequence homology to DNA photolyase contain the two photolyase cofactors but lack DNA repair activity. *Biochemistry, 34*, 6892–6899.

Matsuzaki, M., Ellis-Davies, G. C., Nemoto, T., Miyashita, Y., Iino, M., & Kasai, H. (2001). Dendritic spine geometry is critical for AMPA receptor expression in hippocampal CA1 pyramidal neurons. *Nature Neuroscience, 4*, 1086–1092.

Matsuzaki, M., Honkura, N., Ellis-Davies, G. C., & Kasai, H. (2004). Structural basis of long-term potentiation in single dendritic spines. *Nature, 429*, 761–766.

Metzger, D., & Feil, R. (1999). Engineering the mouse genome by site-specific recombination. *Current Opinion in Biotechnology, 10*, 470–476.

Moglich, A., Ayers, R. A., & Moffat, K. (2009). Design and signaling mechanism of light-regulated histidine kinases. *Journal of Molecular Biology, 385*, 1433–1444.

Moglich, A., & Moffat, K. (2010). Engineered photoreceptors as novel optogenetic tools. *Photochemical and Photobiological Sciences, 9*, 1286–1300.

Moglich, A., Yang, X., Ayers, R. A., & Moffat, K. (2010). Structure and function of plant photoreceptors. *Annual Review of Plant Biology, 61*, 21–47.

Murakoshi, H., Wang, H., & Yasuda, R. (2011). Local, persistent activation of Rho GTPases during plasticity of single dendritic spines. *Nature, 472*, 100–104.

Nagahama, T., Suzuki, T., Yoshikawa, S., & Iseki, M. (2007). Functional transplant of photoactivated adenylyl cyclase (PAC) into Aplysia sensory neurons. *Neuroscience Research, 59*, 81–88.

Nagel, G., Szellas, T., Huhn, W., Kateriya, S., Adeishvili, N., Berthold, P., et al. (2003). Channelrhodopsin-2, a directly light-gated cation-selective membrane channel. *Proceedings of the National Academy of Sciences of the United States of America, 100*, 13940–13945.

Nakashiba, T., Young, J. Z., McHugh, T. J., Buhl, D. L., & Tonegawa, S. (2008). Transgenic inhibition of synaptic transmission reveals role of CA3 output in hippocampal learning. *Science, 319*, 1260–1264.

Nakayama, A. Y., Harms, M. B., & Luo, L. (2000). Small GTPases Rac and Rho in the maintenance of dendritic spines and branches in hippocampal pyramidal neurons. *The Journal of Neuroscience, 20*, 5329–5338.

Ni, M., Tepperman, J. M., & Quail, P. H. (1999). Binding of phytochrome B to its nuclear signalling partner PIF3 is reversibly induced by light. *Nature, 400*, 781–784.

Pelletier, J. N., Campbell-Valois, F. X., & Michnick, S. W. (1998). Oligomerization domain-directed reassembly of active dihydrofolate reductase from rationally designed fragments. *Proceedings of the National Academy of Sciences of the United States of America, 95*, 12141–12146.

Peron, S., & Svoboda, K. (2011). From cudgel to scalpel: Toward precise neural control with optogenetics. *Nature Methods*, *8*, 30–34.

Peschel, N., Chen, K. F., Szabo, G., & Stanewsky, R. (2009). Light-dependent interactions between the Drosophila circadian clock factors cryptochrome, jetlag, and timeless. *Current Biology*, *19*, 241–247.

Pettit, D. L., Wang, S. S., Gee, K. R., & Augustine, G. J. (1997). Chemical two-photon uncaging: A novel approach to mapping glutamate receptors. *Neuron*, *19*, 465–471.

Rana, A., & Dolmetsch, R. E. (2010). Using light to control signaling cascades in live neurons. *Current Opinion in Neurobiology*, *20*, 617–622.

Remy, I., & Michnick, S. W. (2006). A highly sensitive protein-protein interaction assay based on *Gaussia luciferase*. *Nature Methods*, *3*, 977–979.

Rockwell, N. C., Su, Y. S., & Lagarias, J. C. (2006). Phytochrome structure and signaling mechanisms. *Annual Review of Plant Biology*, *57*, 837–858.

Rosenfeldt, G., Viana, R. M., Mootz, H. D., von Arnim, A. G., & Batschauer, A. (2008). Chemically induced and light-independent cryptochrome photoreceptor activation. *Molecular Plant*, *1*, 4–14.

Ryu, M. H., Moskvin, O. V., Siltberg-Liberles, J., & Gomelsky, M. (2010). Natural and engineered photoactivated nucleotidyl cyclases for optogenetic applications. *The Journal of Biological Chemistry*, *285*, 41501–41508.

Sancar, A., & Sancar, G. B. (1988). DNA repair enzymes. *Annual Review of Biochemistry*, *57*, 29–67.

Sang, Y., Li, Q. H., Rubio, V., Zhang, Y. C., Mao, J., Deng, X. W., et al. (2005). N-terminal domain-mediated homodimerization is required for photoreceptor activity of Arabidopsis CRYPTOCHROME 1. *The Plant Cell*, *17*, 1569–1584.

Schroder-Lang, S., Schwarzel, M., Seifert, R., Strunker, T., Kateriya, S., Looser, J., et al. (2007). Fast manipulation of cellular cAMP level by light in vivo. *Nature Methods*, *4*, 39–42.

Shimizu-Sato, S., Huq, E., Tepperman, J. M., & Quail, P. H. (2002). A light-switchable gene promoter system. *Nature Biotechnology*, *20*, 1041–1044.

Shu, X., Royant, A., Lin, M. Z., Aguilera, T. A., Lev-Ram, V., Steinbach, P. A., et al. (2009). Mammalian expression of infrared fluorescent proteins engineered from a bacterial phytochrome. *Science*, *324*, 804–807.

Smith, M. A., Ellis-Davies, G. C., & Magee, J. C. (2003). Mechanism of the distance-dependent scaling of Schaffer collateral synapses in rat CA1 pyramidal neurons. *The Journal of Physiology*, *548*, 245–258.

Spencer, D. M., Wandless, T. J., Schreiber, S. L., & Crabtree, G. R. (1993). Controlling signal transduction with synthetic ligands. *Science*, *262*, 1019–1024.

Stierl, M., Stumpf, P., Udwari, D., Gueta, R., Hagedorn, R., Losi, A., et al. (2010). Light modulation of cellular cAMP by a small bacterial photoactivated adenylyl cyclase, bPAC, of the soil bacterium Beggiatoa. *The Journal of Biological Chemistry*, *286*, 1181–1188.

Strickland, D., Moffat, K., & Sosnick, T. R. (2008). Light-activated DNA binding in a designed allosteric protein. *Proceedings of the National Academy of Sciences of the United States of America*, *105*, 10709–10714.

Strickland, D., Yao, X., Gawlak, G., Rosen, M. K., Gardner, K. H., & Sosnick, T. R. (2010). Rationally improving LOV domain-based photoswitches. *Nature Methods*, *7*, 623–626.

Suh, B. C., Inoue, T., Meyer, T., & Hille, B. (2006). Rapid chemically induced changes of PtdIns(4,5)P2 gate KCNQ ion channels. *Science*, *314*, 1454–1457.

Sunahara, R. K., Beuve, A., Tesmer, J. J., Sprang, S. R., Garbers, D. L., & Gilman, A. G. (1998). Exchange of substrate and inhibitor specificities between adenylyl and guanylyl cyclases. *The Journal of Biological Chemistry*, *273*, 16332–16338.

Tada, T., & Sheng, M. (2006). Molecular mechanisms of dendritic spine morphogenesis. *Current Opinion in Neurobiology*, *16*, 95–101.

Tsien, R. Y. (2010). Nobel lecture: Constructing and exploiting the fluorescent protein paintbox. *Integrative Biology: Quantitative Biosciences from Nano to Macro*, *2*, 77–93.

Tucker, C. L., Hurley, J. H., Miller, T. R., & Hurley, J. B. (1998). Two amino acid substitutions convert a guanylyl cyclase, RetGC-1, into an adenylyl cyclase. *Proceedings of the National Academy of Sciences of the United States of America*, *95*, 5993–5997.

Tyszkiewicz, A. B., & Muir, T. W. (2008). Activation of protein splicing with light in yeast. *Nature Methods*, *5*, 303–305.

Varnai, P., Thyagarajan, B., Rohacs, T., & Balla, T. (2006). Rapidly inducible changes in phosphatidylinositol 4,5-bisphosphate levels influence multiple regulatory functions of the lipid in intact living cells. *The Journal of Cell Biology*, *175*, 377–382.

Volgraf, M., Gorostiza, P., Numano, R., Kramer, R. H., Isacoff, E. Y., & Trauner, D. (2006). Allosteric control of an ionotropic glutamate receptor with an optical switch. *Nature Chemical Biology*, *2*, 47–52.

Wang, S. S., & Augustine, G. J. (1995). Confocal imaging and local photolysis of caged compounds: Dual probes of synaptic function. *Neuron*, *15*, 755–760.

Wang, X., He, L., Wu, Y. I., Hahn, K. M., & Montell, D. J. (2010). Light-mediated activation reveals a key role for Rac in collective guidance of cell movement in vivo. *Nature Cell Biology*, *12*, 591–597.

Wehr, M. C., Laage, R., Bolz, U., Fischer, T. M., Grunewald, S., Scheek, S., et al. (2006). Monitoring regulated protein-protein interactions using split TEV. *Nature Methods*, *3*, 985–993.

Wehrman, T., Kleaveland, B., Her, J. H., Balint, R. F., & Blau, H. M. (2002). Protein-protein interactions monitored in mammalian cells via complementation of beta-lactamase enzyme fragments. *Proceedings of the National Academy of Sciences of the United States of America, 99*, 3469–3474.

Wieboldt, R., Gee, K. R., Niu, L., Ramesh, D., Carpenter, B. K., & Hess, G. P. (1994). Photolabile precursors of glutamate: Synthesis, photochemical properties, and activation of glutamate receptors on a microsecond time scale. *Proceedings of the National Academy of Sciences of the United States of America, 91*, 8752–8756.

Wilson, C. G., Magliery, T. J., & Regan, L. (2004). Detecting protein-protein interactions with GFP-fragment reassembly. *Nature Methods, 1*, 255–262.

Wu, Y. I., Frey, D., Lungu, O. I., Jaehrig, A., Schlichting, I., Kuhlman, B., et al. (2009). A genetically encoded photoactivatable Rac controls the motility of living cells. *Nature, 461*, 104–108.

Yazawa, M., Sadaghiani, A. M., Hsueh, B., & Dolmetsch, R. E. (2009). Induction of protein-protein interactions in live cells using light. *Nature Biotechnology, 27*, 941–945.

Yoo, S. K., Deng, Q., Cavnar, P. J., Wu, Y. I., Hahn, K. M., & Huttenlocher, A. (2010). Differential regulation of protrusion and polarity by PI3K during neutrophil motility in live zebrafish. *Developmental Cell, 18*, 226–236.

Zhang, F., Wang, L. P., Brauner, M., Liewald, J. F., Kay, K., Watzke, N., et al. (2007). Multimodal fast optical interrogation of neural circuitry. *Nature, 446*, 633–639.

Zoltowski, B. D., & Gardner, K. H. (2011). Tripping the light fantastic: Blue-light photoreceptors as examples of environmentally modulated protein-protein interactions. *Biochemistry, 50*, 4–16.

T. Knöpfel and E. Boyden (Eds.)
Progress in Brain Research, Vol. 196
ISSN: 0079-6123

CHAPTER 7

Two-photon optogenetics

Dan Oron[†], Eirini Papagiakoumou[‡], F. Anselmi[‡] and Valentina Emiliani[‡,*]

[†] *Department of Physics of Complex Systems, Weizmann Institute of Science, Rehovot, Israel*
[‡] *Neurophysiology and New Microscopies Laboratory, Wavefront Engineering Microscopy Group, CNRS UMR 8154, INSERM U603, Paris Descartes University, Paris Cedex, France*

Abstract: The use of optogenetics, the technology that combines genetic and optical methods to monitor and control the activity of specific cell populations, is now widely adopted in neuroscience. The development of optogenetic tools, such as natural photosensitive ion channels and pumps or calcium- and voltage-sensitive proteins, has been growing tremendously during the past 10 years, thanks to the improvement of their performances in terms of facilitating light stimulation. To this aim, efficient illumination methods are also needed. The most common way to photostimulate optogenetic tools has been, so far, widefield illumination with visible light. However, the necessity of addressing the complexity of brain architecture has recently imposed switching to the use of two-photon excitation, which provides a better spatial specificity and deeper penetration in scattering tissue. Two-photon excitation is still challenging, due to intrinsic characteristics of optogenetic tools (e.g., the low conductivity of light-sensitive channels), and efficient illumination methods are therefore essential for advancing in this domain. Here, we present a review on the existing two-photon optical approaches for photoactivation of optogenetic tools, and future perspectives for the widespread implementation of these techniques.

Keywords: optogenetics; photoactivation; light patterning; digital holography; temporal focusing; generalized phase contrast.

Introduction

To achieve an efficient control of brain activity with light, precise spatiotemporal stimulation of neuronal structures is a fundamental requirement. In this respect, optogenetic tools (e.g., light-gated ion channels and pumps for photoactivation, or calcium- and voltage-sensitive fluorescent proteins, C/VSFPs, for functional imaging) have an important advantage over synthetic optical reporters because their expression can be restricted to a specific neuronal population by genetic targeting (Kramer et al., 2009). This has allowed loosening some constraints on illumination specificity so that

*Corresponding author.
Tel.: +33-142-864253; Fax: +33-142-864255
E-mail: valentina.emiliani@parisdescartes.fr

DOI: 10.1016/B978-0-444-59426-6.00007-0

many neurobiological studies have been performed with one-photon (1P) widefield excitation, which has the advantage of technical simplicity. In the case of photoactivation, the low-power density necessary for 1P stimulation of optogenetic ion channels and pumps (e.g., $1\,\text{mW}\,\text{mm}^{-2}$ for generating action potentials with the light-gated ion channel channelrhodopsin-2—ChR2; Aravanis et al., 2007) has permitted to stimulate deep into scattering brain tissues with widefield visible light, both *in vitro* (Arenkiel et al., 2007; Petreanu et al., 2009; Wang et al., 2007) and *in vivo* (Cardin et al., 2009; Gradinaru et al., 2009; Witten et al., 2010). As for functional fluorescence imaging, widefield illumination, together with the use of rapid CCD cameras, allows fast (>kHz) optical recordings and the possibility of averaging signals spatially over large surfaces (e.g., a cell soma or a portion of dendrite), thus increasing the number of collected photons. This is particularly important in the case of voltage imaging (Peterka et al., 2011), where dynamics are often in the order of the millisecond, and the collected fluorescent signal can be small (relative fluorescent increase ~1% per 100mV for VSFPs in brain slices; Akemann et al., 2010; Mutoh et al., 2011). Indeed, optogenetic voltage imaging in neuronal preparations has been mostly performed with 1P widefield excitation, allowing to record neuronal activity both *in vitro* and *in vivo* (Akemann et al., 2010; Chanda et al., 2005).

Despite the impact of these results, widefield illumination for optogenetics presents some important limitations. When used for photoactivation, it does not allow stimulating a subpopulation of genetically identical neurons (a configuration which could be useful, e.g., to mimic sparse physiological activity patterns) or targeting subcellular compartments, such as dendritic and axonal branches, or dendritic spines. Similarly, in widefield imaging, the signal from a region of interest is contaminated by aspecific fluorescence coming from out-of-focus planes and blurred by scattering, making it sometimes difficult to isolate contributions from a single-cell body or dendrite, and decreasing the signal-to-noise ratio (SNR).

To address these issues, spatial shaping of the illumination light (light patterning) is required. For 1P photoactivation, laser scanning approaches have been used to map functional neuronal connectivity, by successively stimulating different ChR2-expressing targets in brain slices (Petreanu et al., 2007, 2009; Wang et al., 2007). Alternatively, extended illumination patterns have been generated by using micro-Light Emitting Diode (LED) arrays as light sources or by inserting Digital Micromirror Devices (DMDs) in the widefield excitation path (intensity modulation). In these cases the patterns are imaged onto the sample plane by a telescope (generally formed by the microscope objective and tube lens): illumination profiles are generated by creating "dark regions" on the excitation field, either by deviating part of the excitation light outside the optical path (DMDs) or by turning off the corresponding emitters (micro-LED arrays). Intensity modulation techniques have been used both *in vitro* and *in vivo*, for example, to photoactivate ChR2-expressing retinal ganglion cells (Farah et al., 2007; Grossman et al., 2010), to study the central pattern generator for locomotion (Wyart et al., 2009), and to probe odor coding mechanisms in the olfactory bulb (Dhawale et al., 2010).

Still, these approaches remain limited by intrinsic drawbacks of 1P illumination. In particular, the precision of excitation is hindered by the lack of axial resolution, which is often incompatible with single-cell selectivity or with stimulating thin neuronal processes. Moreover, visible light is highly scattered by living tissues, thus limiting the penetration depth and, in the case of imaging, decreasing the SNR.

Two-photon (2P) optogenetics has the potentiality to solve these problems, thanks to the inherent optical sectioning capability of 2P excitation (2PE) and the increased robustness to scattering of the longer wavelength, used in that case.

Two-photon optogenetics: Illumination methods

A common challenge for 2P optogenetics is the need to enlarge the excitation surface. For photoactivation, this problem was first highlighted by Rickgauer and Tank, in the case of ChR2 (Rickgauer and Tank, 2009). Indeed, ChR2 has low single channel conductance (~80fS; Feldbauer et al., 2009). As a consequence, the depolarization induced by photostimulating ChR2 channels comprised into the volume of a standard 2P diffraction-limited spot is not big enough to exceed the threshold for action potential generation in neurons. Moreover, the relatively high 2P absorption cross section of ChR2 (~260Goeppert–Mayer units at 920nm; Rickgauer and Tank, 2009) and the long lifetime of the excited states (~10ms) cause fast saturation of the channels, preventing the increment of neuronal depolarization by simply raising the excitation power. Similarly, for 2P voltage imaging, the membrane surface, and so the number of reporter molecules, which can be excited by a 2P spot is small and, consequently, the SNR is low (*shot-noise* $\propto$ square root of the number of collected fluorescent photons).

To improve the efficiency of 2P stimulation, the solution is then to extend the excitation area. This might be achieved with different approaches that can be divided in two main groups: scanning and parallel excitation techniques. In the first case, the excitation area is increased by rapidly steering the laser beam through different positions on the target structure, while, in the second case, larger excitation spots are generated by modulating the phase of the illumination laser beam.

Scanning approaches

The first demonstration of 2P activation of ChR2 was performed by Tank's group (Rickgauer and Tank, 2009). Two configurations to increase the excitation area were compared. In the first one, the laser spot diameter was enlarged by underfilling the back aperture of the microscope objective. This was effective in increasing the size of ChR2-evoked currents but caused a degradation of axial resolution, since it reduced the effective numerical aperture (NA) of the microscope objective. In the second case, the 2PE spot was scanned across different positions on the target cell. In this configuration, ChR2-evoked currents from regions stimulated sequentially would sum up, contributing to the final response. The scan time, T_s, was limited by the decay time constant, τ_d, of ChR2 currents ($I^*(T_s)/I^*_{max}=(1-e^{-n})/n$, with $n=T_s/\tau_d$, where I^* is the evoked photocurrent and I^*_{max} is the maximal available current through ChR2 channels in the conductive state (*) (Rickgauer and Tank, 2009). A combination of the two approaches led to the first successful attempt to stimulate action potentials by 2P activation of cultured hippocampal neurons transfected with ChR2 (Fig. 1). The microscope objective was slightly underfilled to give an effective NA of 0.2–0.5 (corresponding to an axial profile covering, approximately, the thickness of the cell). Then, the excitation spot was scanned along spiral trajectories on the neuronal soma: an action potential could be generated with a scan time of ~30ms. This is still limited in terms of temporal resolution: in particular, the technique is not suitable for "simultaneous" (within a few milliseconds) stimulation of multiple cells, as it is necessary to mimic fast spontaneous neuronal circuit activity. Nevertheless, these results opened the way to 2P optogenetics.

To date, no data are available, to our knowledge, on 2P imaging with VSFPs, but some studies have been performed with synthetic voltage-sensitive dyes (VSDs). In these works, as in the case of ChR2, rapid laser scanning was used to extend the area of excited membrane. For example, 2P laser scanning over a short line (three points ~0.5µm, 100µs per point) was used to perform voltage imaging on small (µm) axonal terminals in a neurohypophysis preparation *in vitro* (Fisher et al., 2008): action potentials and action potential trains were visualized in single trials, with a relative fluorescence increase up to 10% (di-3-ANEPPDHQ dye, 10–20mW at the sample).

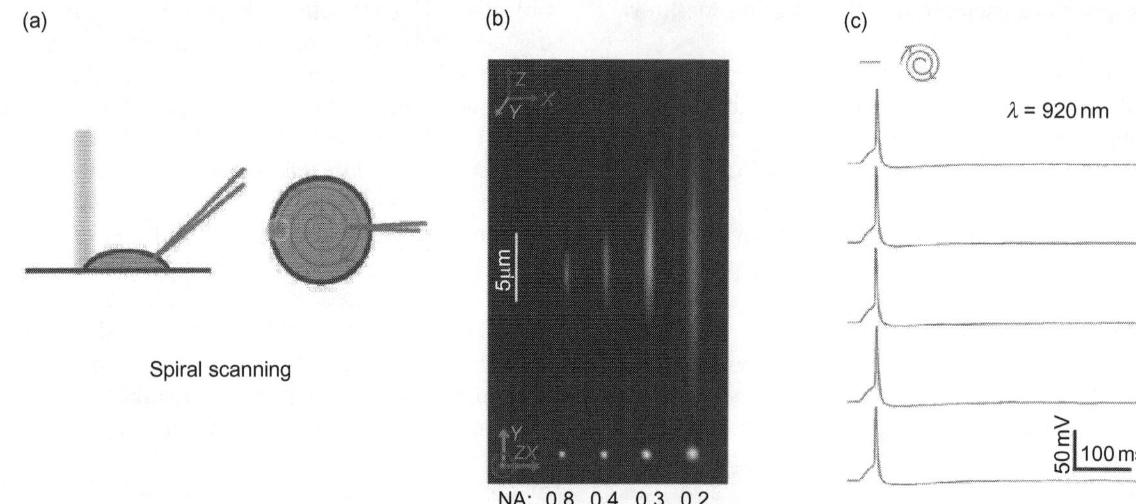

Fig. 1. First demonstration of action potential generation by 2P activation of ChR2 in neurons. (a) Schematic depiction of geometry of whole-cell scanning 2P stimulation stimulation, shown in side and top view. (b) Measured 2P intensity profiles of the point spread function (PSF) at several effective NA values, along the optical axis z (top), and in the lateral (x–y) plane (bottom). (c) Current–clamp recordings of membrane voltage changes, stimulated by using 32-ms spiral scans with a 2PE spot with PSF corresponding to the case of NA=0.3. Overline indicates stimulation time. Adapted from Rickgauer and Tank (2009). (For color version of this figure, the reader is referred to the Web version of this chapter.)

2P voltage imaging (ANNIE-6 dye) was also used with living mice, to monitor global changes of membrane potential in the upper layers of the somatosensory cortex, following whisker deflection, or during the transition from the anesthetized to the awake state (Kuhn et al., 2008): the laser spot was scanned over a line crossing an entire cortical column (~200 μm, 0.8 ms); the signal was averaged spatially and temporally (over 400 trials, relative fluorescence increase ≤1%). These preliminary results seem to indicate that 2P voltage imaging on enlarged areas is a promising research line to be pursued and, possibly, extended to the use of VSFPs.

Parallel approaches

Lateral shaping: Digital holography and generalized phase contrast method

For 2P parallel light patterning, methods based on intensity modulation of the illumination light are far too inefficient, so the only option is to use phase modulation. This can be achieved either by the use of static beam multiplexing (Nikolenko et al., 2007) or of reconfigurable liquid crystal devices. In the latter, two approaches have been proposed: one is based on the principle of digital holography (DH; Curtis et al., 2002) and the second on the generalized phase contrast (GPC) method (Glückstad and Mogensen, 2001).

One-photon and two-photon DH Originally proposed for generating multiple optical tweezers (Curtis et al., 2002; Reicherter et al., 1999), the experimental scheme for DH consists in computing, with an iterative algorithm, the phase pattern at the rear aperture of the objective that permits to reproduce the desired target intensity in the objective focal plane. The calculated phase hologram is addressed to a liquid crystal spatial light modulator (LC-SLM) that is designed to impose the phase modulation onto the input beam's wavefront. After propagation through the objective, the beam is focused onto an intensity pattern, reproducing the desired template (Fig. 2a).

For multiple-spot generation, a simple algorithm called "gratings and lenses" (Leach et al., 2006; Liesener et al., 2000) can be used. This consists in imposing a grating effect to control the lateral position of each spot and a lens effect to control the axial position (Fig. 3). This algorithm is of easy implementation but does not permit to control the relative intensity of the spots at the focal plane (Leach et al., 2006; Liesener et al., 2000) and significant intensity inhomogeneity can be observed for the generation of a high number of spots.

To improve multiple-spot light distribution or to generate arbitrary two-dimensional (2D) intensity distributions, the holographic phase at the SLM is normally calculated with another algorithm called the Gerchberg and Saxton algorithm (Gerchberg and Saxton, 1972; Fig. 2b), or with an improved version of it known as Gerchberg–Saxton weighted (GSW) algorithm, which is particularly suited for optimizing uniformity in multiple diffraction-limited spots (Di Leonardo et al., 2007).

With these approaches, generation of multiple photoactivation spots in 2D (Lutz et al., 2008; Nikolenko et al., 2008, 2010) or three-dimensional (3D) patterns (Anselmi et al., 2011; Daria et al., 2009; Yang et al., 2011) has been demonstrated. It has also been shown that, by designing the target intensity profile on the base of a fluorescence picture, light excitation can be shaped to perfectly match a specific subcellular process (Dal Maschio et al., 2010; Lutz et al., 2008), a cell soma (Zahid et al., 2010) or a defined group of cells (Dal Maschio et al., 2010; Zahid et al., 2010; Fig. 4). Interestingly, for 2D-shaped patterns, the spatial phase distribution of a holographic wavefront permits also improved axial resolution (Fig. 5).

An important parameter to be taken into account when designing a holographic system is the maximum lateral size of the excitation area. According to previous papers (e.g., Golan et al., 2009) we can adopt for that the term field of view (FOV), normally used for imaging systems. In DH, the FOV is mainly limited by the SLM pixel size, d_{SLM}, which sets the maximum achievable deflection angle and gives rise to a position (x,y) dependent diffraction efficiency, $\delta(x,y)$, defined as the ratio between the intensity, I_{spot}, redirected into the desired target spot(s) and the light intensity, I_{tot}, incident on the LC-SLM, that is, $\delta(x,y)=I_{spot}(x,y)/I_{tot}$ (we neglect for simplicity the losses due to the optical elements of the light path). Inside the FOV, $\delta(x,y)$ decreases proportionally to $\delta(x,y)_{DH}=(\sin X/X)^2(\sin Y/Y)^2$ (Golan et al., 2009; Yang et al., 2011), where $X=((\pi f_2 2d_{SLM}/\lambda f_1 f_{obj})x)$, $Y=((\pi f_2 2d_{SLM}/\lambda f_1 f_{obj})y)$, and reaches a zero value at $x_{max}=y_{max}=((\lambda f_1/2d_{SLM})(f_{obj}/f_2))$, where λ is the illumination wavelength, f_1/f_2 is the telescope magnification (typically $\sim$1.5), f_{obj} is the objective focal length, and $1/2d_{SLM}$ is the maximum SLM spatial frequency (typically $\sim$25 lp/mm). In 2PE, one has to take into account the quadratic dependence of the signal from the excitation density so that the generated signal decreases to zero proportionally to $(\sin X/X)^4(\sin Y/Y)^4$; for example, for a diffraction efficiency $\geq$50% ($1/2d_{SLM}\sim$10lp/mm for 2PE), λ=900nm, and an objective of 60×, 40×, or 20×, the FOV_{2PE} ($FOV=2x_{max}\times 2y_{max}$) in 2PE is $\sim 80\times 80$, 120×120, or $240\times 240\,\mu m^2$, respectively.

To date, 1P and 2P holographic photostimulation have been applied for glutamate uncaging and imaging; however, we anticipate that the extension to optogenetics will be straightforward. By combining 1P or 2P holographic photoactivation with electrophysiological recordings or calcium imaging, it has been possible to show that patterned excitation allows for a precise control of glutamate release in single (Anselmi et al., 2011; Lutz et al., 2008) and multiple (Yang et al., 2011) subcellular processes or in single and multiple neurons (Dal Maschio et al., 2010; Lutz et al., 2008; Nikolenko et al., 2008; Zahid et al., 2010; Fig. 6).

Alternatively, holographic light patterning was used to illuminate specific regions of interest (ROIs) within the FOV, while performing conventional galvo-steered uncaging of MNI-glutamate

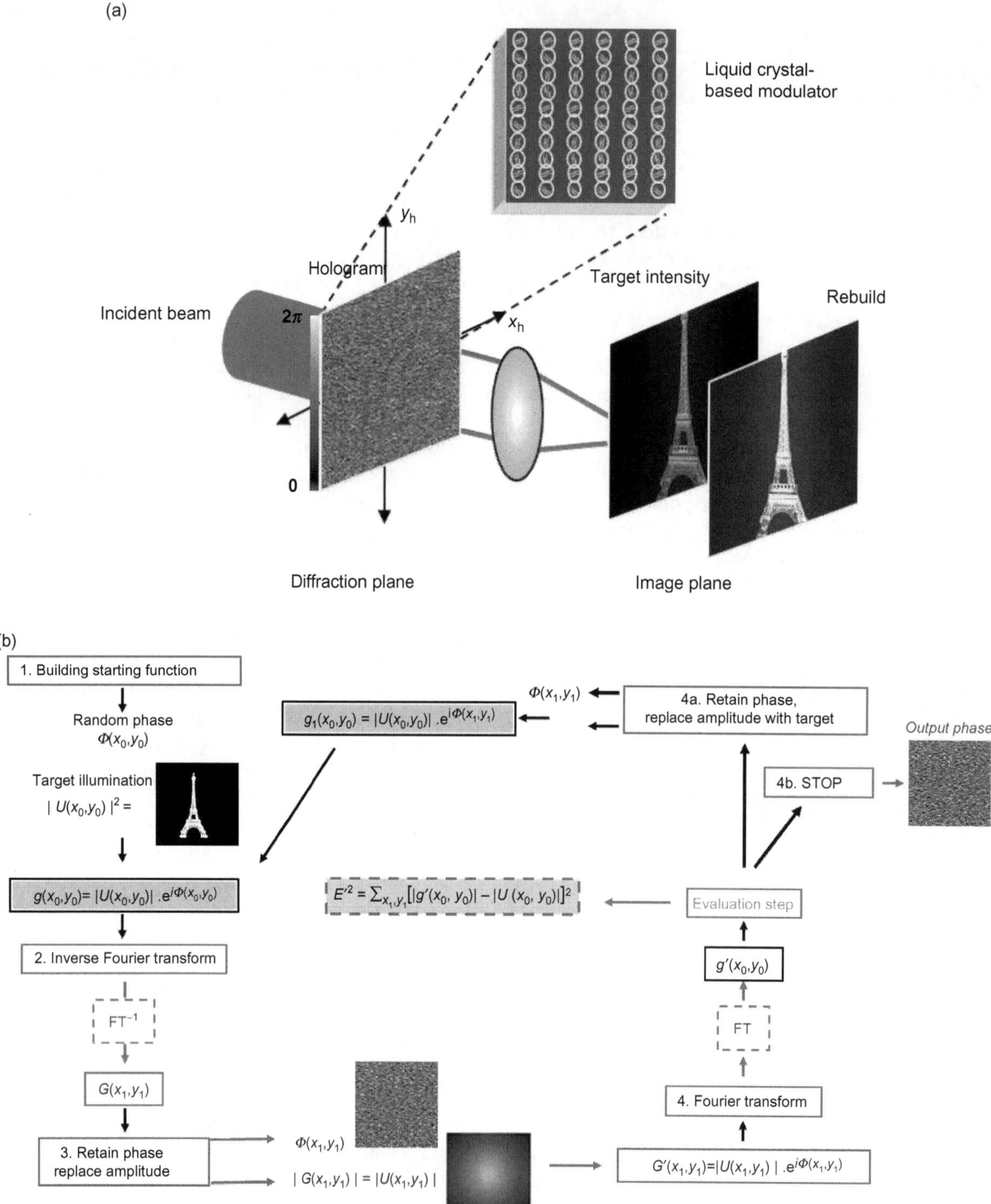

Fig. 2. Digital holography. (a) Diagram of the experimental scheme for DH. (b) Block diagram of the iterative Fourier transform algorithm (Gerchberg and Saxton algorithm) for calculating the phase-hologram addressed to the SLM. (For color version of this figure, the reader is referred to the Web version of this chapter.)

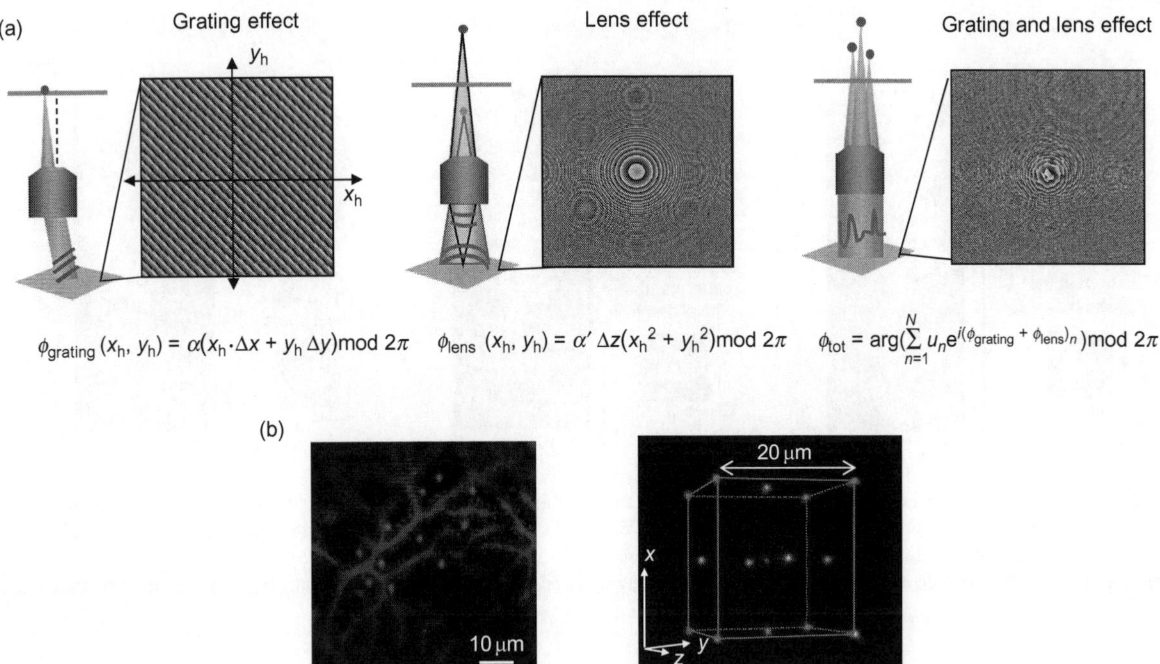

Fig. 3. Multispot generation. (a) Schematic diagram and phase hologram for displacing the spots laterally (left panel), axially (middle panel), and both laterally and axially (right panel) by imposing a grating effect, lens effect and combined grating, and lens effect, respectively, to the phase hologram. At the bottom of each panel, a description of the phase function is given for each case, in relation to the spatial coordinates x_h, y_h at the back focal plane of the objective lens. α and α' are constants, u_n is the amplitude of the nth spot, and mod 2π stands for 2π modulation. (b) Holographic diffraction-limited spots placed in two- (left panel) and three-dimensions (right panel) by using a grating and lenses and a GSW algorithm, respectively. Spots are visualized by exciting fluorescence on a thin fluorescent film by 2PE (left) and 1PE (right) (adapted from Yang et al., 2011). On the left panel, the fluorescence of the spots was superimposed to the fluorescence image of a Purkinje cell. (For color version of this figure, the reader is referred to the Web version of this chapter.)

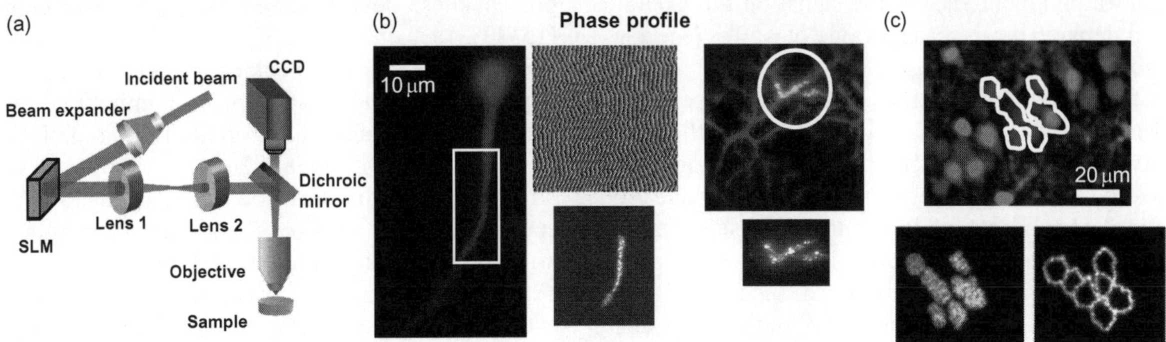

Fig. 4. Holographic excitation. (a) Layout of the experimental setup for DH. By designing the target intensity profile on the base of the fluorescence picture, light excitation can be shaped to perfectly match (b) a specific subcellular process (adapted from Lutz et al., 2008; Papagiakoumou et al., 2008) and (c) a cell soma, a defined group of cells or the extracellular space (adapted from Zahid et al., 2010). (For color version of this figure, the reader is referred to the Web version of this chapter.)

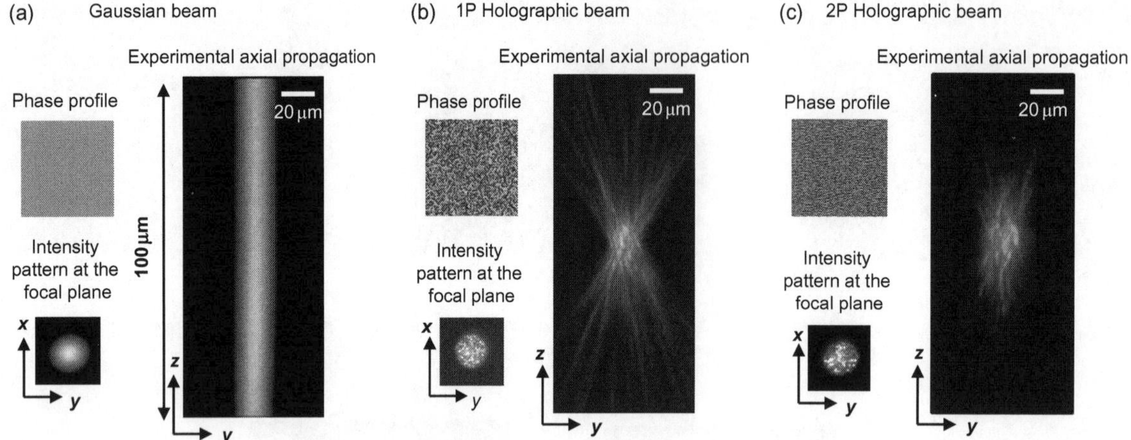

Fig. 5. Axial resolution of holographic spots. Axial propagations measured in a range of 100 μm around the focal plane of the objective for (a) a 20-μm-full width at half maximum (FWHM) Gaussian beam, (b) a 20-μm holographic spot in 1PE (axial resolution of 33μm at FWHM), and (c) a 20-μm holographic spot in 2PE (axial resolution of 25μm at FWHM). In each case, the corresponding phase profile at the back aperture of the objective and the *x–y* intensity profile at the sample plane are shown.

(Dal Maschio et al., 2010; Fig. 6). This configuration permits to integrate the calcium response on large ROIs and thus to significantly increase the SNR.

One limitation of DH is the inhomogeneous light distribution within the generated light patterns. This can be of the order of 15–20% for 1P excitation (1PE), while for 2PE, due to the quadratic dependence of the signal on the excitation density, it can reach a value of ~50% (see *x-y* light distribution in Figs. 3 and 4).

These fluctuations, also called intensity speckles, rise principally from the approximation in the iterative algorithm (since the problem of obtaining a desired intensity pattern in the Fourier plane by using only phase modulation does not usually have a solution, the output intensity is optimized by leaving a random phase distribution at the focal plane; Fig. 2b) and from the cross talk of adjacent pixels of the LC-SLM. Solutions based on using a rotating diffuser (Papagiakoumou et al., 2008) or phase mask shift-averaging (Golan and Shoham, 2009) permit averaging over the speckles and lead to a smoother spatial distribution. However, in the case of a diffuser, this causes a significant loss of light and a deterioration of the axial resolution. The shift-averaging methods require projecting a series of holograms and are thus limited by the refresh rate of the LC-SLM (60–200 Hz); this can be reduced by using binary ferroelectric LCs (which can get up to the kHz range). Unfortunately, at present these devices have a very poor efficiency (15–20%).

The presence of speckles is not critical in applications using holographic illumination for recording the integrated signal from a defined ROI (Dal Maschio et al., 2010) or for uncaging on large areas (the diffusion of the uncaged molecule quickly smooths out the speckle distribution after the end of the stimulation). Nevertheless, for applications requiring a precise control of light excitation on fine subcellular processes or a precise quantification of the excitation, one should consider using alternative schemes for light patterning, such as the GPC method described in the following section.

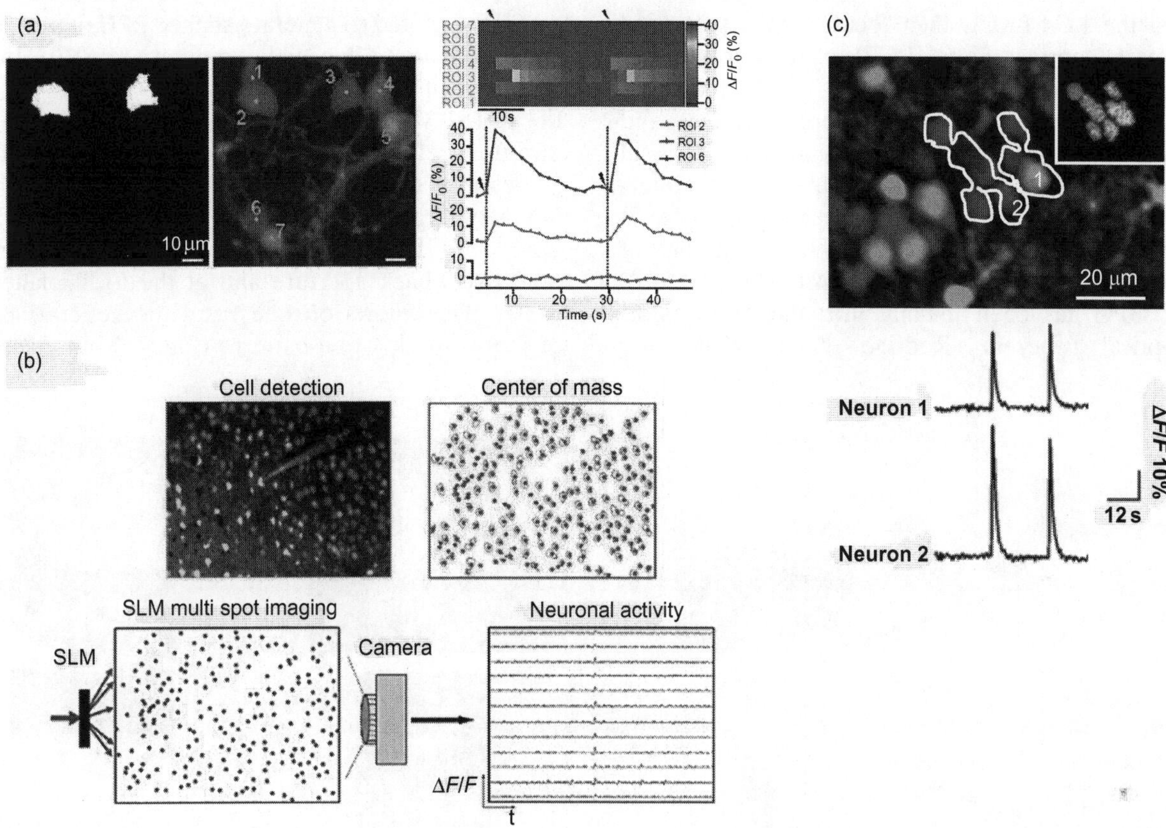

Fig. 6. Patterned excitation for glutamate uncaging. (a) 2P uncaging of MNI-glutamate performed simultaneously on cerebellar granule cells in culture (targeted cells are marked by red lines) with the image mask shown on the left, generated to shape the laser wavefront of the uncaging beam. Imaging is performed, simultaneously to uncaging, with conventional scanning microscopy at 0.54Hz. The time course of $\Delta F/F_0$ values of Fluo-4 fluorescence in the seven ROIs denoted in the fluorescence image with green, before and after the application of the holographic uncaging stimulus, are shown on the right. The arrows indicate the time of delivery of the photolysis stimulus (0.26mW/μm^2) (adapted from Dal Maschio et al., 2010). (b) SLM multispot imaging of Ca^{2+} transients induced by 2P uncaging of glutamate to activate cortical neurons in brain slices. After detection of the ROIs (red contours) on the fluorescence image of labeled neurons (obtained with conventional raster scanning), pixel centers of mass of each ROI are calculated, and their coordinates are used as a command image uploaded to the SLM software. The SLM illuminates all or a subset of the ROIs, and a widefield detector, with spatial resolution suitable to resolve individually illuminated objects, is used to record neuronal activity from all illuminated neurons simultaneously (adapted from Nikolenko et al., 2008). (c) Variations of intracellular Ca^{2+} concentration of two different target CA1 hippocampal neurons (1 and 2 on the fluorescence image) during 1P uncaging of MNI-glutamate with the pattern configuration of the inset (adapted from Zahid et al., 2010). (For interpretation of the references to color in this figure legend, the reader is referred to the Web version of this chapter.)

Principle of the GPC method The GPC method is based on an extension of the Zernike phase-contrast method (Zernike, 1955) into the domain of full range [0, 2π] phase modulation (Glückstad, 1996; Rhodes, 2009). Briefly, a desired target intensity map is converted into a binary [0, π] phase map that is used to modify the input beam wavefront via a LC-SLM. The beam modulated by the LC-SLM is then focused on a patterned phase contrast filter (PCF) plate that imposes appropriate phase retardation between the on-axis focused component (reference wave) and the higher-order Fourier components (signal wave, focused around the central spot). The interference between these two beams generates, at the focal plane of a second lens (the output plane), the original target intensity (Fig. 7a, top).

GPC has been initially introduced to realize optical tweezers (Rodrigo et al., 2004) and recently adapted to generate shaped 2PE patterns (Fig. 7a, bottom; Papagiakoumou et al., 2010). In the latter, it has been shown that, by addressing the LC-SLM with binary phase map conversion of a user-defined ROI in a fluorescence image, it is possible to precisely reproduce the original intensity target (Fig. 7b). Contrary to the case of DH, in GPC the optical wavefront at the objective's back aperture and at the focal plane is flat. This allows on one hand the generation of sharp, speckle-free patterns (Fig. 7b) but gives

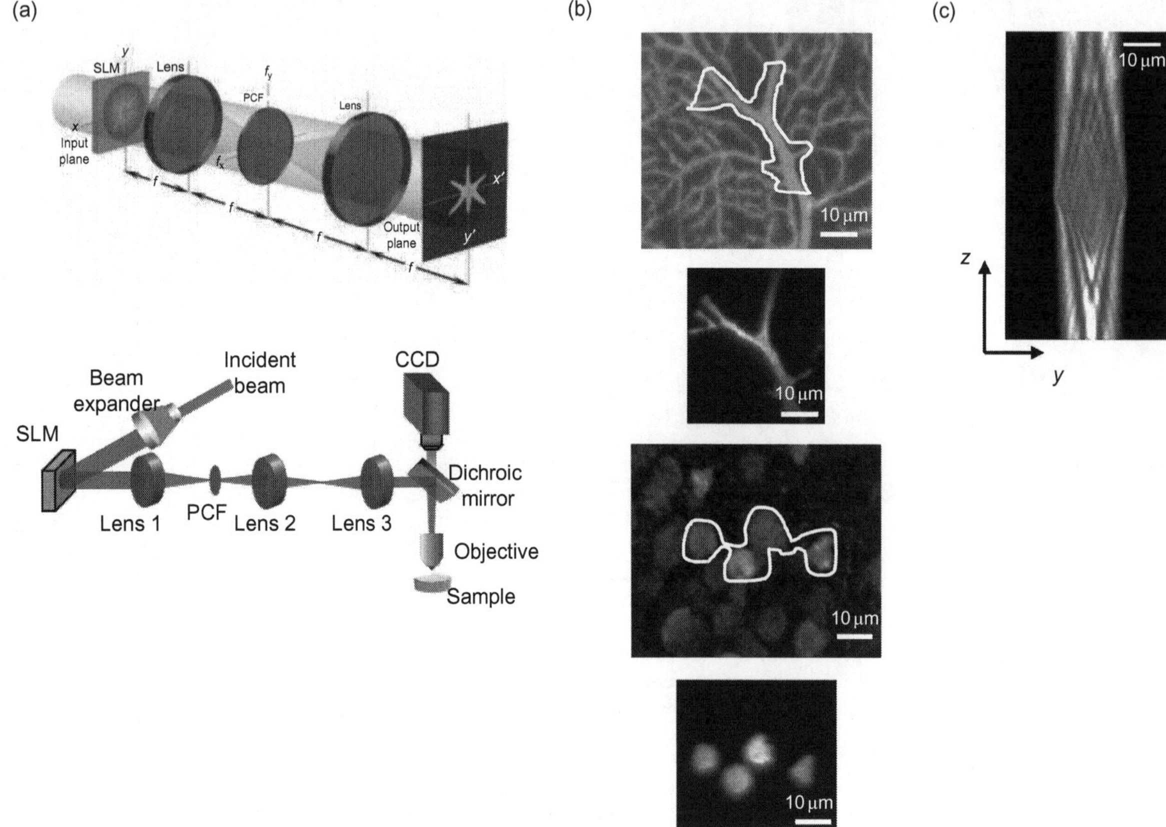

Fig. 7. The generalized phase contrast method. (a) Optical setup for the GPC (adapted from Palima and Glückstad, 2008) and layout of the corresponding experimental setup. (b) Shaped laser patterns generated with the GPC method (2PE) designed on the basis of a confocal fluorescence image of a Purkinje cell (up) and a widefield fluorescence image of CA1 hippocampal neurons loaded with oregon green bapta (down), in selected ROIs (white line) (adapted from Papagiakoumou et al., 2010). (c) *y–z* section of the measured axial propagation of a 20-μm-diameter spot, generated with GPC in 2PE. Note the diffraction pattern rising from the interference of the signal wave with the reference wave component. (For interpretation of the references to color in this figure legend, the reader is referred to the Web version of this chapter.)

rise, on the other hand, to the loss of optical confinement (the axial propagation of a GPC beam being very similar to the one of low-NA Gaussian beams; Fig. 7c).

In GPC, the FOV is given by the circular illuminated area at the SLM (of radius R_c) multiplied by the magnification of the GPC interferogram and the relay telescope $FOV_{GPC}=\pi(R_c(f_2/f_1)(f_{obj}/f))^2$, where f_1 and f_2 are the focal lengths of the corresponding lenses in the GPC setup (Fig. 7a, bottom) and f and f_{obj} are the focal lengths of the telescope forming the final intensity pattern at the objective's focal plane (Lens 3 and objective). Although, in this case, it is more appropriate to talk about a phase interferometric contrasting, we will keep for δ the denotation "diffraction efficiency" for similitude to the case of DH.

Two parameters determine the value of δ within the FOV. The first one is the ratio, $S=A_{spot}/FOV$, of the excitation pattern area, A_{spot}, to the FOV: it can be demonstrated that, for $S=0.25$ (i.e., $A_{spot}(max)=FOV/4$), δ reaches its theoretical maximum value of 100% (Palima and Glückstad, 2008). However, for smaller excitation areas, it decreases proportionally to the ratio $A_{spot}/A_{spot}(max)$.

Second, the spatial dependence of δ within the FOV is given by the central filtering size, η, defined as the ratio between the diameter of the PCF, R_1, to the main lobe of the Airy profile of the reference wave focused at the PCF plane, R_2, that is, the Fourier transform of the input circular aperture of radius R_c (Glückstad and Mogensen, 2001). This parameter determines the strength and the wavefront curvature of the reference wave at the GPC output aperture and therefore, implicitly, the diffraction efficiency δ in the FOV (Glückstad and Mogensen, 2001). In general, a value of η between 0.5 and 0.6 represents a good compromise between wavefront curvature and strength of the synthetic reference wave, that is, between FOV and intensity contrast. We can include the dependence of the diffraction efficiency on the two parameters, η and S, by defining the total diffraction efficiency for binary input phase GPC: $\delta_{GPC}=I_{spot}/I_{tot}=\delta_\eta 4(A_{spot}/FOV)$ for $A_{spot} \leq \frac{FOV}{4}$, where r is the radial coordinate at the excitation plane and ΔR is the excitation field radius.

Axial shaping: Temporal focusing

As is evident from Figs. 5 and 7c, the rules of diffraction practically determine a relation between the lateral excitation shape and its axial extent. For flat wavefronts (Figs. 4a and 6c), the axial extent scales quadratically with the lateral size (Goodman, 2005). Holographic excitation (Fig. 5b), which utilizes the full angular acceptance (NA) of the excitation lens, enables to achieve a linear scaling between the two, but at the price of adding significant speckle to the lateral pattern (Fig. 5b and c). These restrictions seem to be an insurmountable barrier for generating large area axially confined excitation patterns. For 2PE, however, various alternatives for illuminating large areas, while maintaining axial confinement, are available, taking advantage of temporal multiplexing (i.e., exciting different points within the FOV at different times). This can be achieved by fast scanning of a single excitation point, but the finite residence and traveling time required limit the number of positions that can be considered "simultaneously" activated on a biological timescale.

Provided that the excitation source is powerful enough, axially resolved spatial multiplexing can also be achieved by splitting the beam into several beamlets, time delayed (10ps) relative to one another and directed at different spatial locations (Fricke and Nielsen, 2005). Going to a large number of beamlets, though, their separate control becomes cumbersome, making this approach difficult to scale beyond several tens of beamlets.

Several years ago, temporal focusing (TF) has been suggested as a continuous, easily controlled, alternative to the use of a discrete number of beamlets for multipoint excitation. TF is a technique based on imaging the required excitation pattern onto the sample. The standard TF setup is depicted in Fig. 8 and consists of a grating which is imaged onto the sample via a $2f$–$2f$ telescope consisting of an achromatic lens and the microscope objective. Since this is a perfect imaging system, the pattern illuminating the grating is simply demagnified and relayed onto the objective focal plane. To understand why axial resolution is maintained in this configuration, it is instructive to consider the time-domain evolution of a short pulse impinging upon the grating. For ultrafast excitation, the incoming pulse of duration, τ, is pancake shaped, covering a relatively large transverse area, but having a thickness of only $c\tau$ (where c is the speed of light), corresponding to about 30μm for 100fs pulses, typically used for multiphoton excitation. Since light scattered from the grating is diffracted onto the first order, the illuminating pulse has to impinge upon the grating at an angle. Thus, different regions of the grating are *not illuminated simultaneously*. Transiently, only the intersection

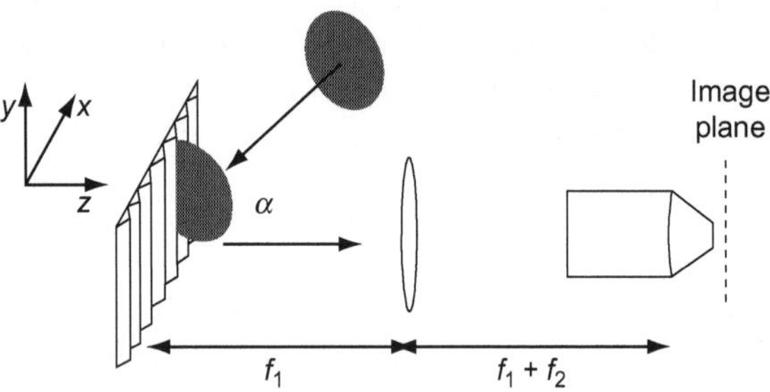

Fig. 8. Temporal focusing. The standard setup for temporal focusing consists of a grating, placed at one end of a $2f$–$2f$ telescope comprising an achromatic lens of focal length f_1 and an objective lens of effective focal length f_2 placed at a distance f_1+f_2 from one another. The ultrashort pulse impinges on the grating at an angle α such that the center wavelength is diffracted perpendicular to the grating, toward the optical axis of the lens system. As can be seen, only a thin line is instantaneously illuminated on the grating as the pancake-shaped ultrashort pulse crosses it. (For color version of this figure, the reader is referred to the Web version of this chapter.)

between the pancake-shaped pulse and the grating surface, which is an extended line along the grating grooves, is illuminated. Upon proper choice of the grating and telescope parameters, this line is imaged onto a diffraction-limited line in the objective focal plane. Considering all this, TF can be viewed as temporally multiplexed line-scanning multiphoton excitation. Indeed, for widefield temporally focused excitation, the axial resolution, $2\Delta z_0$, defined as the full width at half maximum (FWHM) of the axial intensity distribution, I_{2P}, has been shown to be equivalent to that of multiphoton line-scanning microscopy both for multiphoton absorption (Oron et al., 2005) and for coherent nonlinear processes such as multiharmonic generation (Oron and Silberberg, 2005): $I_{2P}\approx[1+(\Delta z/z_R)^2]^{-0.5}$; $2\Delta z_0=2\sqrt{3}z_R$, where z_R is the Rayleigh range.

Two clear consequences follow from the above description. The first is that as long as the pulse front is not distorted, TF affords to maintain the axial resolution regardless of the particular shape of the excitation pattern and that axial resolution should only slowly deteriorate for weakly distorted pulses. The second is that axial resolution is improved if the excitation pattern on the grating is shaped as a thin line perpendicular to the grating grooves (Tal et al., 2005; Zhu et al., 2005). In this case, the instantaneous intersection of the pulse with the grating is a small area, which is imaged onto a diffraction-limited spot in the objective focal plane, and the axial resolution would be equivalent to that of a conventional 2P microscope: $I_{2P}\approx[1+(\Delta z/z_R)^2]^{-1}$; $2\Delta z_0=2z_R$.

It should be noted here that a complementary description of TF in the frequency domain exists. It relies on the fact that the various colors comprising the ultrashort excitation pulse are diffracted by the grating toward different directions and thus propagate toward the objective focal plane at different angles. This leads to axially dependent spectral dispersion so that the pulse is temporally stretched outside the focal plane. As a result, multiphoton excitation is axially confined. While less intuitive than the time-domain description, the frequency domain one is sometimes useful to understand the physics behind the evolution of temporally focused pulses, for example, in the cases of remote focusing or of transmission through scattering media, as described below.

The frequency domain is also useful to provide an alternative description why TF enables to decouple the lateral and axial resolution. For a conventional Gaussian illumination, the Rayleigh range, z_R, depends quadratically on the excitation spot size, w_0, or on the inverse of the square of the beam NA:

$$z_R = \frac{\pi w_0^2}{2\lambda} \propto \lambda \text{NA}^2, \quad (1)$$

where $\text{NA}=s/2f$, f is the objective focal length, and s is the size of the illuminating beam at the objective back aperture (equal in the x and y directions). For a diffraction-limited spot, s is larger than the objective back aperture, D, (overfilling) and NA in Eq. (1) coincides to the objective NA, $\text{NA}_{\text{obj}}=D/2f$. Increasing the size w_0 of the spot at the focal plane requires decreasing s and consequently NA so that NA becomes less than NA_{obj} and the axial resolution decreases, as well. For a temporally focused beam $z_R=(2f^2/k\ (s^2+\alpha^2\Omega^2))$ (Durst et al., 2006), where k is the mean magnitude of the excitation wave vector and s and $\alpha\Omega$ are the spot sizes at the back aperture of the objective in the direction orthogonal and parallel to the grating linear dispersion, respectively. As for the case of conventional Gaussian beams, the generation of large excitation spots at the focal plane requires reducing s, but this affects the back aperture illumination only in one direction, the other being always equal to $\alpha\Omega$; for $\alpha\Omega>D$ and a large excitation area, $s\ll\alpha\Omega$ and $z_R\approx(2f^2/k(\alpha^2\Omega^2))=(\lambda/\text{NA}_{\text{obj}}{}^2)=\text{const}$. In this regime, the axial resolution depends only on the NA_{obj} independently of the excitation spot size, w_0. In other words, w_0 can be increased without sacrifying the NA of the system and therefore without losses in axial resolution. This leads to an effective decoupling of the axial and lateral beam parameters.

3D sculpting: Lateral and axial shaping

The fact that TF is practically an imaging technique greatly facilitates its use for generating complex spatial excitation patterns. As long as it does not lead to significant distortion of the pulse front, any phase modulation technique can be used to generate the required magnified image on the grating. This image will then be relayed onto the sample resulting in an axially resolved excitation pattern. In particular, both wavefront shaping methods described above, DH and GPC, can be easily integrated with TF.

TF was first combined with wavefront shaping by Papagiakoumou et al. (2008, 2009), who coupled it with a standard DH setup. A phase-only LC-SLM was placed at the focal plane of a lens positioned one focal length away from the grating. Thus, the image on the grating was the Fourier transform of the phase pattern applied to the SLM. To obtain the desired image, a Gerchberg–Saxton iterative optimization algorithm was used.

Two years later, TF was combined with GPC; in this case, the grating was placed at the output plane of the GPC interferometer (Fig. 9a; Papagiakoumou et al., 2010).

In both cases, an axial resolution close to the theoretical limit for TF has been demonstrated (Fig. 9b). It should be noted that due to the inherent distortion of the pulse front in the case of DH, there is a slight deterioration of the axial resolution as compared with GPC, which is well accounted for by theoretical calculations (Papagiakoumou et al., 2008, 2009).

The combination of GPC and TF has been used for efficient in-depth 2P activation of ChR2 in cultured neurons and brain slices (Papagiakoumou et al., 2010), enabling action potential generation with 2-ms temporal resolution (Fig. 10a) and ~5-μm axial resolution (Fig. 10b) and, for the first time, reliable generation of action potentials with the simultaneous excitation of multiple neurons (Fig. 10a) or multiple neuronal compartments (Papagiakoumou et al., 2010).

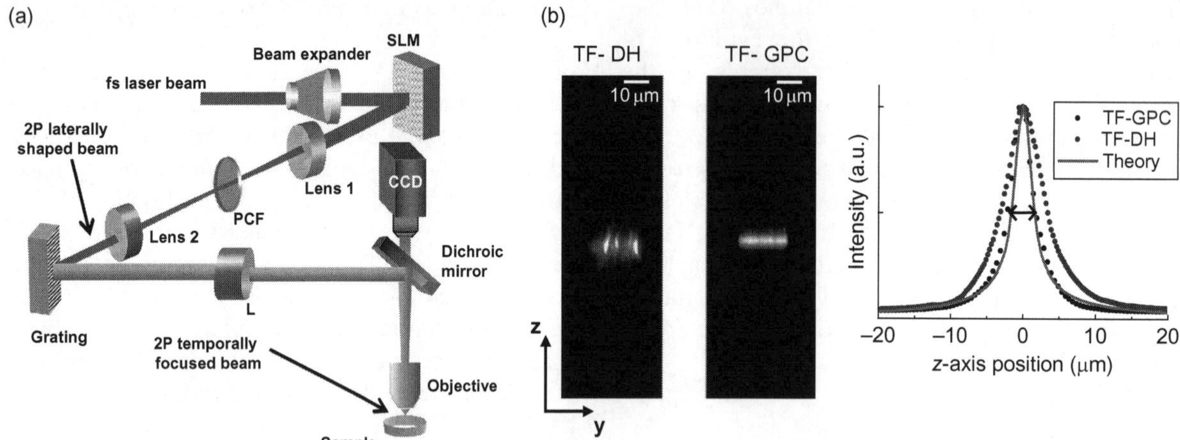

Fig. 9. Temporal focusing combined with spatial light patterning. (a) Layout of the experimental implementation of the GPC technique in combination with TF. (b) *y–z* section of the measured axial propagation of a 20-μm temporally focused holographic spot (left panel) and a 20-μm-GPC-generated temporally focused spot. The axial distribution of the integrated fluorescence intensity on the area of the spots is plotted on the diagram shown (adapted from Papagiakoumou et al., 2010). (For color version of this figure, the reader is referred to the Web version of this chapter.)

Alternatively to these approaches, enlarged and optical confined excitation area for 2P activation of ChR2 can be achieved by using a low-NA Gaussian beam in combination with TF, as demonstrated by Andrasfalvy et al. (2010). In this case, excitation patterns that match the shape of neuronal processes (Andrasfalvy et al., 2010) or of a targeted neuronal population (Losonczy et al., 2010) were achieved by integrating the technique into a laser scanning head. Action potentials from a single dendrite in acute brain slices could be triggered at ~150 μm depth by placing multiple spots along the dendrite (Andrasfalvy et al., 2010) with ~6 ms temporal resolution. The same approach has allowed to study the underlying mechanisms of theta phase precession by inducing spatiotemporal patterns that mimic the perisomatic inhibition on hippocampal pyramidal cells (Losonczy et al., 2010).

Comparison between the different approaches

Overall, each of the above approaches has certain advantages and limits and, as a result, each of them might prove to be best suited in a given experimental setting (see also Vaziri and Emiliani, 2011). These are briefly described in this paragraph and summarized in Table 1.

The spiral scanning approach proposed by Rickgauer and Tank (2009) can be easily implemented, since it only requires the introduction, in a conventional microscope, of an iris at the objective back aperture or of a variable beam expander in the excitation optical path, to reduce the effective NA of the objective. All the available power can be concentrated in a single spot scanned through the sample. The FOV is that of a conventional scanning microscope, that is, FOV$=(2f_{\text{scan}}\tan(\vartheta)M)^2$, where f_{scan} is the focal length of the scanning lens; ϑ, the galvo-scanning angle (typically $\pm 11°$); and M, the objective magnification; this leads, for example, to a FOV of $\sim 600\times 600\,\mu\text{m}^2$ for a 40× objective.

The main limitation of this approach is that a compromise between axial and temporal, T_s, resolution has to be found. The latter is given by $T_s=N(t_{\text{dwell}}+S_t)$, where N is the number of visited positions during the scan, which is given by the ratio between the total excited area of the

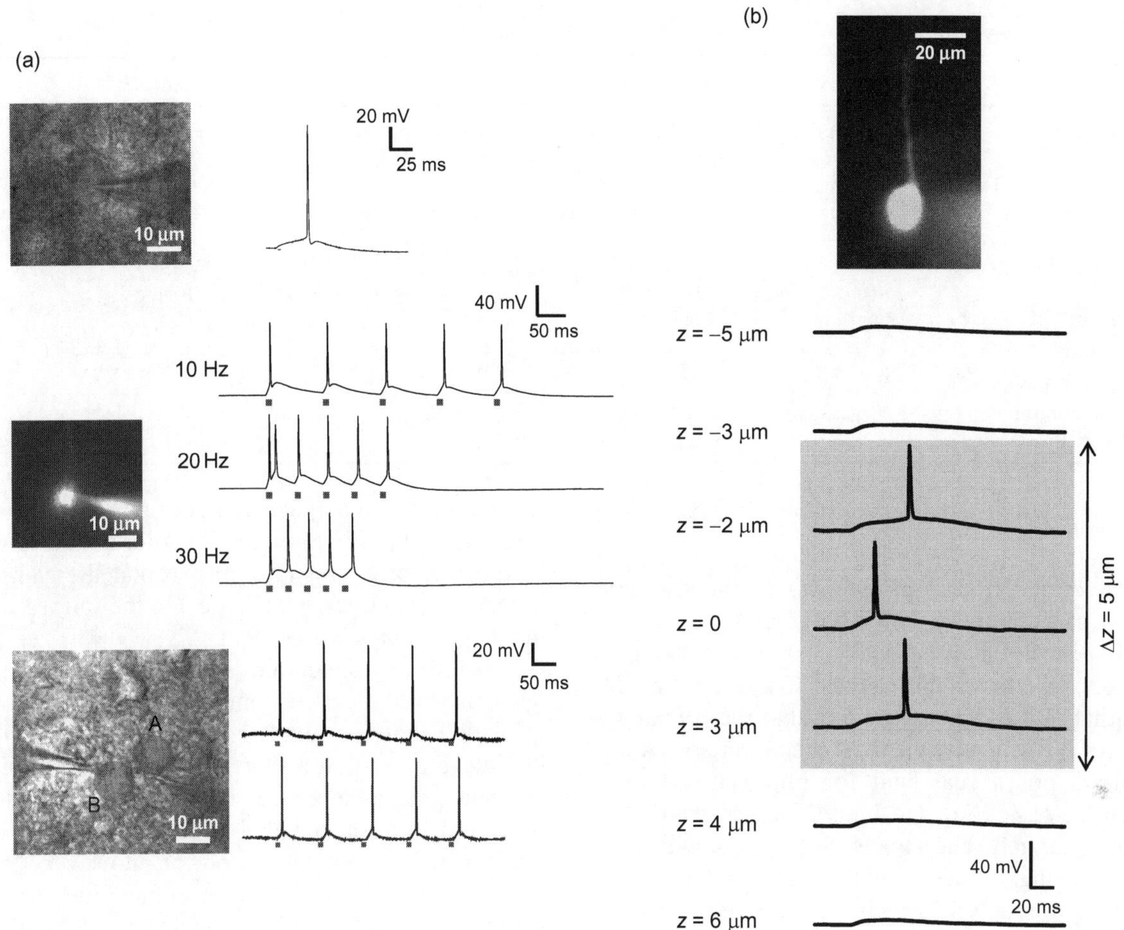

Fig. 10. TF-GPC enables efficient 2P excitation of ChR2 in brain slices. (a) Excitation (0.3–0.5 mW/µm^2) of layer V pyramidal neurons, ChR2-YFP positive, in cortical slices with temporally focused GPC spots of 15 µm with a 2-ms light stimulation (top), with 10-ms light stimulation in different firing frequencies (middle), and by simultaneous photoactivation (10 ms pulses) of two neurons (bottom). In each case, the neurons responded reliably to the stimulation by action potentials. (b) Fluorescence image of a neuron filled with Alexa Fluor 594 with superimposed shaped excitation profile covering the apical dendrite (red, top). Photo-depolarizations evoked by the excitation shape (0.3 mW/µm^2) with a 10-ms light stimulation at different z-axis positions (bottom). Adapted from Papagiakoumou et al. (2010). (For interpretation of the references to color in this figure legend, the reader is referred to the Web version of this chapter.)

pattern, A_{spot}, and the excitation spot size, w_0; t_{dwell} is the residence time at a given position (also known as dwell time); and S_t is the point to point scanning time. Increasing the temporal resolution requires increasing w_0, with consequent deterioration of the axial resolution: $R_z \propto w_0^2$. This approach is therefore particularly suitable for single-cell stimulation, but generation of high-frequency action potential trains or millisecond excitation of multiple cells is not possible.

Scanning of a temporally focused low-NA beam, as proposed by Andrasfalvy et al. (2010), allows for increasing w_0 without deteriorating the axial resolution, which remains equal to

Table 1. Summary of the illumination techniques for 2P photoactivation

	Spiral scanning	Scanning and TF	DH and TF	GPC and TF
Implementation	Conventional 2P microscope	2P microscope+ grating for TF	2P microscope+setup for DH and grating for TF	2P microscope+setup for GPC and grating for TF
Max excitation density	Laser power/w_0	Laser power/A_{spot}	Laser power/(FOV/4)	Laser power/(FOV/4)
Power losses		15–20% at grating	15–20% at grating 30% at SLM	15–20% at grating
FOV (for 40× objective)	$(2f_{scan}\tan(\vartheta)M)^2$ $\sim 600\times 600\,\mu m^2$	$(2f_{scan}\tan(\vartheta)M)^2$ $\sim 600\times 600\,\mu m^2$	$[2\times((\lambda f_{obj}f_1)/(2d_{SLM}f_2))]^2$ $\sim 120\times 120\,\mu m^2$ (at 900 nm)	$\pi\left(R_c\frac{f_2}{f_1}\frac{f_{obj}}{f}\right)^2$ $R_c \sim 100\,\mu m^2$
Diffraction efficiency	Constant	Constant	$\propto\left(\frac{\sin X}{X}\right)^4$	$\delta_\eta 4\left(\frac{A_{spot}}{FOV}\right)$
Axial resolution	$\propto \frac{\pi w_0^2}{2\lambda} \cong \frac{\lambda}{NA^2}$	$\propto \frac{2f^2}{k\left(s^2+\alpha^2\Omega^2\right)} \cong \frac{\lambda}{NA^2_{obj}}$	$\propto \frac{2f^2}{k\left(s^2+\alpha^2\Omega^2\right)} \cong \frac{\lambda}{NA^2_{obj}}$	$\propto \frac{2f^2}{k\left(s^2+\alpha^2\Omega^2\right)} \cong \frac{\lambda}{NA^2_{obj}}$
Temporal resolution	$N(t_{dwell}+S_t)$	$N(t_{dwell}+S_t)$	t_{dwell}	t_{dwell}
Temporal resolution or sequential patterning	$N(t_{dwell}+S_t)$	$N(t_{dwell}+S_t)$	$t_{dwell}+1/R_e$	$t_{dwell}+1/R_e$

$\frac{\lambda}{NA^2_{obj}}$ = const. The implementation of this approach is relatively easy (requiring adding a TF grating in the external optical path or inside the scanning head of a commercial 2P microscope).

In terms of light efficiency, one has to take into account the losses due to the diffraction at the TF grating (15–20%) and the fact that the remaining power is divided by the total excitation area A_{spot}. This is not a real limit for photoactivation of ChR2, considering the high 2P cross section of these channels. The situation is less clear with voltage imaging. Indeed, even if the 2P cross section of some synthetic VSDs has been measured (Fisher et al., 2005), these compounds have never been tested, to our knowledge, with extended excitation patterns. Besides, VSFPs have not yet been characterized with 2PE. The FOV in that case is the same as for the spiral scanning approach.

For single-spot excitation, this approach is limited to the shape of circular symmetry and needs readjusting the spot size with a variable telescope for experiments requiring different excitation spots. For multiple-cell excitation or excitation of arbitrary patterns, it requires scanning the excitation beam. In this case, the expression for the temporal resolution is the same as for the spiral scanning approach: $T_s=N(t_{dwell}+S_t)$. Nevertheless, the use of a temporally focused beam allows for larger values of w_0 and consequently for a better temporal resolution. However, for excitation of small processes (such as dendritic segments or axons), w_0 should be kept of the order of the lateral size of the processes (1–2 μm) so that the gain in temporal resolution with respect to the spiral scanning approach is less evident.

Overall, this technique is well suitable for single-cell excitation or multiple-cell excitation, provided that the number of excited cells is not too elevated or that a short t_{dwell} is needed. A significant gain in temporal resolution can be also achieved by reducing the dwell time, that is, by increasing the excitation power; however, this costs in lateral and axial resolution due to the strong contribution of the out-of-focus light to the evoked responses (Andrasfalvy et al., 2010; Losonczy et al., 2010; Rickgauer and Tank, 2009).

Parallel light shaping with DH requires significant modifications of the optical setup with respect to a conventional 2P microscope and an *ad hoc* software for the calculation of phase profiles.

In terms of light efficiency, in DH simultaneous stimulation of multiple areas implies that the available laser power is divided by the total excitation area, A_{spot}. This, in addition to the power losses at the LC-SLM (30%) and at the TF, grating (15–20%) can limit the maximum number of excitable cells and the penetration depth.

The FOV is limited by the position-dependent diffraction efficiency of the SLM and is given by $FOV=[2\times((\lambda f_{obj}f_1)/(2d_{SLM}f_2))]^2$; this, for a diffraction efficiency of ≥50% (approximate spatial frequency of ~10lp/mm), λ=900nm, and a 40× objective, corresponds to a FOV in 2PE of ~$120\times120\,\mu m^2$.

In comparison to scanning methods, the main advantage of DH is the temporal resolution which is only limited by the dwell time $T_s=t_{dwell}$ and is independent of the number of excited positions.

Holographic light patterning also allows for efficient multiple-scale excitation, going from single or multiple diffraction-limited spots to shapes covering a single subcellular process or a population of sparse neurons, without any adjustments of the optical setup.

For excitation of a single cell, cell process, or multiple sparse cells, DH permits achieving an axial resolution $\propto w_0$ (Dal Maschio et al., 2010; Lutz et al., 2008; Zahid et al., 2010) even without TF, thus allowing for a simpler optical setup and reduced power losses. With DH, 3D light patterning is also possible, which is not achievable with any of the other described approaches (laser scanning or GPC methods).

Advantages and limitations for GPC are very close to those described for DH, with few differences. Similar to DH, parallel light shaping with GPC requires significant modifications of the optical setup, but the software for phase profile calculation, which consists in a simple intensity–binary phase conversion of the original target, is of easy implementation.

In terms of light efficiency, the diffraction efficiency at the LC-SLM can be raised to the theoretical limit of 100%, providing that the total spot surface consists of ¼ of the excitation field: A_{spot}=FOV/4. However, this also sets a limit on the maximum excitation density achievable, given by the ratio between the available excitation power and 1/4 of the excitation field, independently of the excitation spot size. This has limited the excitation field in practical applications to a circle of ~60–100µm diameter (Papagiakoumou et al., 2010).

In GPC, as in DH, the temporal resolution is only limited by the dwell time $T_s=t_{dwell}$ and is independent of the number of excited positions. Multiple-scale excitation is possible, without affecting the temporal resolution or requiring modification of the optical setup.

In contrast with DH, GPC permits the generation of sharp speckle-free excitation patterns. This is particularly interesting for photoactivation of small cellular processes, permitting generation of patterns which precisely match the lateral size, for example, of dendrites. The smooth light distribution achievable with TF-GPC makes it also particularly adapted for fast imaging, for example, with VSD and VSFP.

For experiments requiring sequential excitation with phase modulation, the refresh rate, R_e (60–200Hz), of the LC-SLM has to be taken into account in the definition of temporal resolution. To this end, we can consider a more general expression for T_s, including the sequential projection of n patterns (supposing for simplicity that each pattern comprises the same number of spots) $T_s=N(t_{dwell}+S_t)+1/R_e$, which, for parallel approaches ($N=1$, $S_t=0$), gives $T_s=t_{dwell}+1/R_e$ and, for scanning approaches ($1/R_e=0$), gives $T_s=N(t_{dwell}+S_t)$. Scanning approaches are therefore preferable with respect to parallel approaches when $N(t_{dwell}+S_t)\leq(t_{dwell}+1/R_e)$.

2P optogenetics: Decoupling the imaging from the activation plane

The use of optogenetic photoactivation in combination with functional imaging will permit a full optical control of signal transmission in brain. Ideally, since many brain structures are organized in complex volumes, both imaging and photoactivation should be performed in 3D. Moreover, one should be able to choose the imaging and stimulation planes independently, even when both are combined (as it is often the case) into the same microscope objective. This would allow overcoming current technical limitations, which, for example, restrict the accessible areas for simultaneous imaging and photoactivation to the relatively short portion of a dendrite, which extends in the focal plane of the microscope (Judkewitz et al., 2006; Kwon and Sabatini, 2011) or, in the case of functional mapping of circuit connectivity, to neuronal bodies located on the same plane (Nikolenko et al., 2007). In the following paragraph, we will describe three possible methods permitting to decouple the imaging from the photoactivation plane.

The first approach is based on the use of DH. As described in the section "Parallel approaches" and Fig. 11a, multiple holographic spots can be generated at axial positions different from the objective focal plane by using *the grating and lenses* algorithm or with the GSW algorithm. Arbitrary 2D patterns can also be displaced by adding a global lens effect to the phase hologram used to generate the 2D pattern (Zahid et al., 2010). In both cases, projection to the SLM of different phase holograms, each one introducing a variable lens effect, which compensates for the scanning objective movements, permits to perform 3D imaging while keeping the stimulation plane at a fixed position, as demonstrated a few years ago for 3D optical trapping and imaging (Emiliani et al., 2006; Fig. 11a). Very recently, 3D 2P imaging without moving the objective (Dal Maschio et al., 2011) has been demonstrated by performing lateral scanning of the spot with a conventional 2P laser scanning microscope and axial displacement of the spots holographically with a SLM: it has been shown that the system has little spatial and temporal distortion and its fluorescence imaging performances on cellular structures in the intact mouse brain were demonstrated. This approach, combined with phase-modulated light patterns (TF-DH or TF-GPC), will also permit to achieve a full decoupling of the stimulation and imaging planes.

TF offers yet another alternative for decoupling the excitation plane from the imaging plane. As described above, TF can be considered, in the frequency domain, as pulse shortening ("focusing" in time) during propagation up to the objective focal plane. In complete analogy with the lens effect in DH, addition of a quadratic spectral phase (rather than spatial phase as in DH) results in an axial shift of the focusing plane. Application of an appropriate amount of chirp, $\beta\Omega^2$ (quadratic spectral phase, 2β is the group velocity dispersion (GVD), and $\sqrt{2ln2}\ \Omega$ is the FWHM of the frequency spectrum of the pulse), enables an arbitrary shift of the focal plane, $z_{\text{foc}}=f+\beta\Omega^2 z_{\text{R}}$, as long as the various colors comprising the ultrashort pulse are still spatially overlapping in the shifted plane (corresponding typically to several Rayleigh ranges of the objective lens and depending on the transverse extension of the excitation beam) (Durst et al., 2006; Suchowski et al., 2006; Fig. 11b). This effect has been demonstrated in various realizations of imaging (Du et al., 2009; Durst et al., 2008) and more recently used for remote control of the axial focusing through an optical fiber (Straub et al., 2011).

Finally, a third possibility proposed recently (Anselmi et al., 2011) consists in combining DH with a remote focusing imaging unit (Botcherby et al., 2007). In this system, photostimulation patterns are generated by DH through the principal microscope objective, as described above. 1P fluorescence imaging is carried out by a second microscope, symmetrical to the principal one, which recreates a perfect (aberration-free) 3D remote image of the sample. A mirror, in the remote space, is then used to move this image, in order to select the final imaging plane on the camera, which can be axially displaced with respect to the principal focal plane or tilted by a variable angle (up to $\sim$20°). In this way, 3D imaging can be performed without actually translating the principal microscope objective, that is, independently from photoactivation (Fig. 11c). This technique was first implemented for 1P fluorescence imaging, and excitation light, provided by a LED, was coupled to the principal objective through a conventional epifluorescence pathway (Anselmi et al., 2011). 1P glutamate uncaging was performed by implementing, in the same microscope, a system for DH. Remote focusing is also compatible with 2P imaging (Salter et al.,

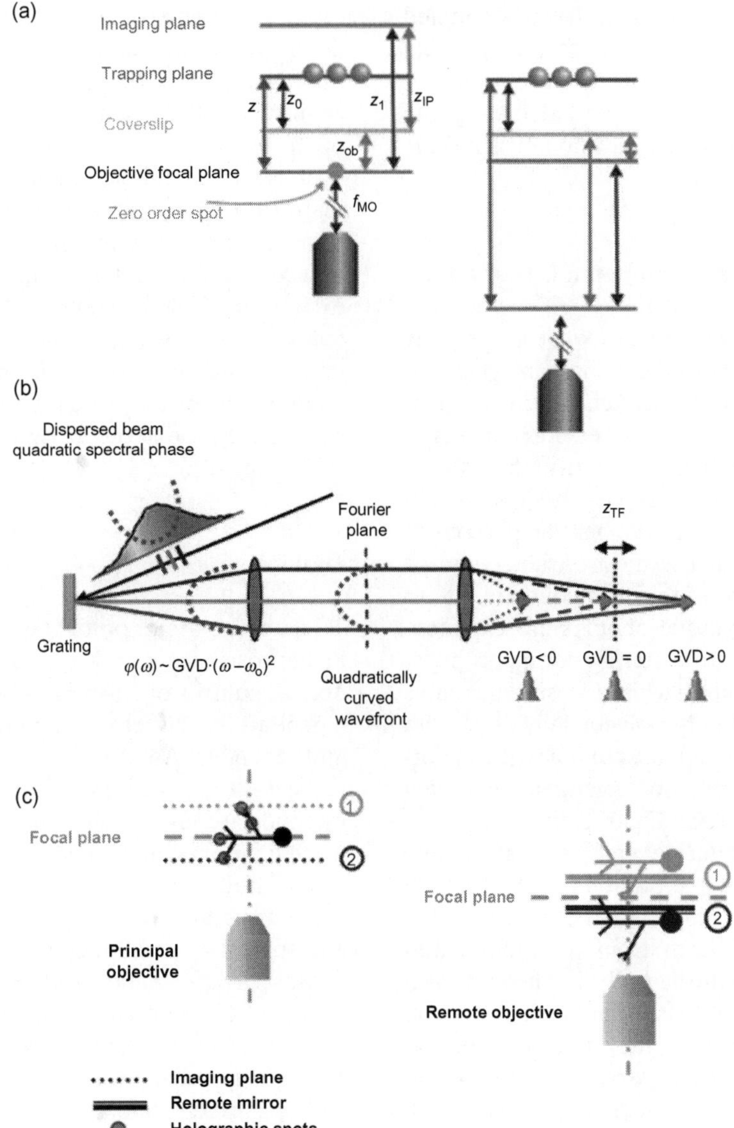

Fig. 11. Decoupling the imaging from the photoactivation plane. (a) Representation of the generation of multiple holographic spots at axial positions different from the objective's focal plane, by projecting different phase holograms, each one introducing a variable lens effect that cancels the defocusing of the objective (adapted from Emiliani et al., 2006). (b) Schematic layout of the experimental setup permitting the displacement of the temporal focus plane position (z_{TF}). Introducing GVD to the original laser beam results in a quadratically curved wavefront at the Fourier plane of the TF telescope, which displaces the temporal focus plane in the axial direction, depending on the amount of the GVD. (c) Schematic layout of a remote focusing setup. Excitation spots (red dots) are placed into the sample through the principal objective (left). A perfect 3D image of the sample is recreated after the remote objective (right). A mirror in the remote space moves this image (right, blue and green) in respect to the focal plane of the remote objective, in order to select the final imaging plane on the camera, which can be axially displaced (left, blue and green dotted lines) from the principal focal plane of the microscope (left, dashed gray line). The position of the principal objective (and so, of the excitation spots) is not affected by remote focusing. (For interpretation of the references to color in this figure legend, the reader is referred to the Web version of this chapter.)

2009). In this case, the imaging laser is coupled directly into the remote unit: lateral scanning can be performed by a pair of galvanometers, while axial scanning can be carried out by the remote mirror associated to galvanometric motors (Botcherby et al., 2010).

2P optogenetics: Deep two-photon activation

Multiphoton excitation provides what is currently one of the best solutions for optical imaging and photoexcitation through scattering tissue. This relies on the fact that the scattered components of the excitation beam typically contribute little to multiphoton processes, as they are both weak and temporally stretched. Thus, the ballistic photons of the excitation beam induce most of the generated signal. The limitations of this technique are due to the fact that these ballistic photons are exponentially attenuated due to scattering during propagation through the sample, while focusing increases the generated signal only polynomially ($1/z$) with propagation. Thus, for too thick samples, the majority of the nonlinear signal no longer generates from the focal plane (Theer and Denk, 2006).

The effect of scattering can be dramatically enhanced for multiple-point or temporally focused excitation, since in both cases the scattered excitation beam from one illuminated spot can interfere with the ballistic photons from another. This can lead to a significant distortion of the excitation shape if the various excitation spots are not enough temporally separated one from the other. Moreover, scattering should lead, due to similar effects, to a decrease in axial resolution of photoactivation. Recent experimental and theoretical investigation of the propagation of temporal-focused low-NA Gaussian beams (Dana and Shoham, 2011), holographic beams, or beams generated with the GPC method through scattering tissue has characterized the deterioration of the axial resolution as a function of propagation distance. It was found that the axial resolution is roughly maintained up to about two scattering lengths and quite rapidly deteriorates thereafter. Interestingly, the excitation shape was found to be extremely robust against scattering within a propagation distance of up to about two scattering lengths, such that even relatively fine features are preserved despite scattering. This latter effect can be understood in terms of the nearly independent speckle patterns induced by the scattering medium to each of the excitation colors comprising the temporally focused pulse, which tend to average out in the multiphoton response. This has enabled to adequately preserve the excitation pattern even upon propagation of about 0.5mm in an acute brain slice (Fig. 12).

Outlook

To date, the use of optogenetics as a research tool in neuroscience grows rapidly. Fast and specific optical control opens a new landscape for studying neuronal systems, by permitting stimulation and imaging of neuronal populations. With respect to classical electrical techniques, optical methods are less invasive, have increased flexibility (e.g., in imaging and stimulating form multiple cells simultaneously), and allow direct access to small structures, such as thin dendritic branches and spines, with submicrometric precision.

Most *in vivo* studies have been performed in 1P widefield illumination, resulting already in important advances toward the understanding of neuronal circuit activity under physiological and pathological conditions, as well as the link between cellular mechanisms and behavior (Gradinaru et al., 2007; Kravitz et al., 2010; Sohal et al., 2009; Tsai et al., 2009). However, the high spatial and temporal specificity provided by 2PE, together with the possibility to penetrate deep into living tissues, would be a fundamental advantage.

Here, we reviewed the latest optical illumination methods used to address the challenging issue of 2PE of optogenetic tools. In particular, the low conductivity of optogenetic channels was overcome by

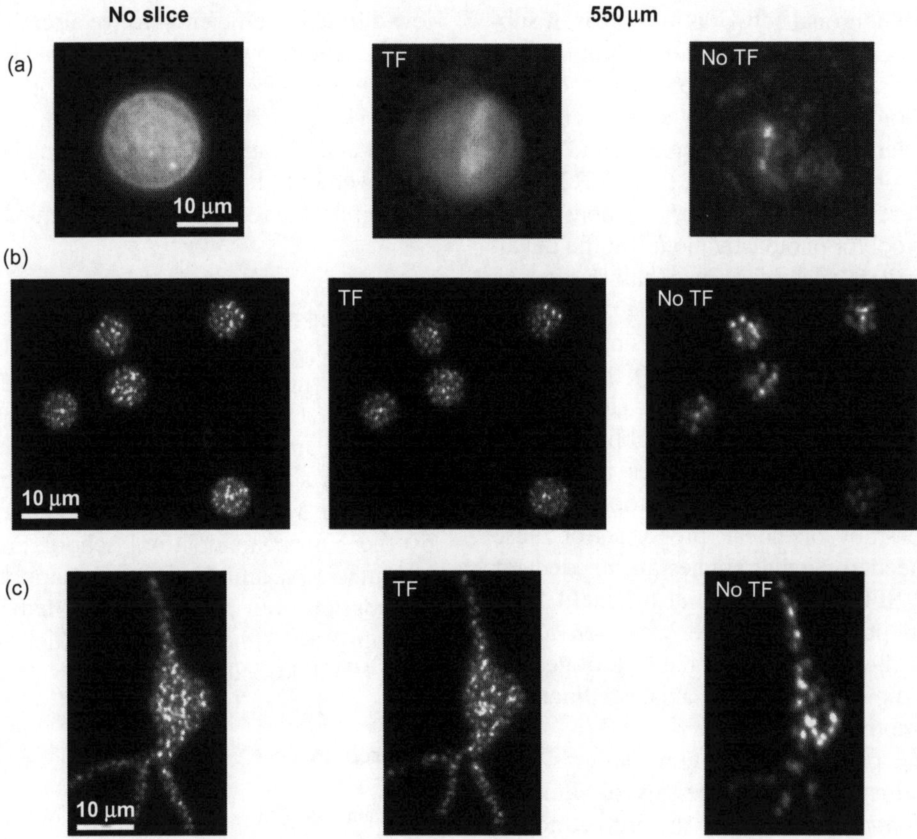

Fig. 12. In depth two-photon excitation. 2P-excited fluorescence images on a thin fluorescent film of: (a) a 15-μm diameter spot generated with GPC, (b) a configuration of multiple spots of 7-μm diameter and (c) a holographic pattern mimicking the neuron with its small processes, without scattering (no slice; left) and after propagation through acute cortical brain slices of 550 μm with TF (middle) and without TF in the optical setup (right). 2P fluorescence was excited at λ=950nm, with a 60x, 0.9 NA, objective lens.

scanning a low-NA laser spot through the target excitation area (Andrasfalvy et al., 2010; Rickgauer and Tank, 2009), or by using phase modulation techniques (TF-DH or TF-GPC), which shape the excitation beam to generate axially resolved extended light patterns (Papagiakoumou et al., 2010). Despite these recent works, the greatest part of 2P photoactivation studies has been carried out, so far, by photolysis of caged neurotransmitters, since photolysis is more easily induced by 2P diffraction-limited spots (Branco et al., 2010; Carter and Sabatini, 2004; Losonczy et al., 2008; Nikolenko et al., 2007). In particular, recently, simultaneous 2P photolysis and Ca^{2+} imaging were demonstrated by combining a holographic setup for glutamate uncaging and a galvo-steered for imaging system or vice versa (Dal Maschio et al., 2010). A sufficiently high SNR was achieved in both cases, even though the number of frames per second recorded was limited by the necessity to efficiently excite the fluorophore.

Similar setups could be used for combined functional imaging and photoactivation of optogenetic tools, thus advancing our understanding on the functioning of neuronal networks at cellular or subcellular level. However, information is still missing on the 2P cross sections of optogenetic tools. So far, to our knowledge, only 2P cross section of ChR2 has been measured (Rickgauer and Tank, 2009), owing probably to the fact that ChR2 is the first discovered and the most widely used optogenetic tool for photoactivation. But the development of other types of opsins (halorhodopsin, *NpHR*; Volvox-carteri channelrhodopsins, *VChR1-2*; ultrafast channelrhodopsin, *ChETA* etc.) (Gradinaru et al., 2007; Lin, 2011), that can target different cell populations and are activated in different wavelength regions, thus enabling parallel studies of different networks, moves really fast. Thus, for further development of 2P optogenetics, an extensive study of the 2P properties of these proteins is needed. Similar studies are needed for VSDs and VSFPs, which could also benefit from the use of depth-resolved large excitation areas. Especially in the case of VSDs, the lack of genetic specificity makes the need for axial confinement even more important.

Concerning parallel illumination methods, further improvements may be necessary to alternate multiple excitation patterns into the preparation at a rate >100 Hz, in order to mimic rapidly changing neuronal activity patterns. For both GPC and DH, the phase profiles addressed to the SLM can be calculated in advance, in order to reduce as much as possible the computation time. In this case, the refresh rate of the SLM becomes the limiting factor. Typical nematic LC matrices that are used for phase modulation in applications like DH (entire interval 0–2π of phase range) are limited to a refresh rate of 60 Hz, but new nematic LCs operating at 200 Hz are now appearing on the market.

Another interesting possibility to be explored would be the use of ferroelectric liquid crystal matrices, binary SLMs (phase delay either 0 or π), which can operate at rates >1 kHz. This would be straightforward with GPC, since this technique only requires a binary phase modulation, contrary to DH. However, ferroelectric LC-SLMs suffer from low-diffraction efficiency, as we already mentioned, which is far below the theoretical maximum of 40% (Golan et al., 2009), making, thus, these devices a poor choice for 2P excitation, where available power at the sample is often a limiting factor. Future improvements, however, might render this technology applicable to 2P patterned illumination.

Acknowledgments

D. O. acknowledges support by the European Research Council starting investigator grant SINSLIM and the Crown center of photonics. E. P. was supported by the "Fondation pour la Recherche Médicale" (FRM). F. A. was supported by the European doctoral school Frontières du Vivant. F. A. and V. E. were supported by the "Fondation pour la Recherche Médicale" (FRM équipe) and by the Human Frontier Science Program (RGP0013/2010).

References

Akemann, W., Mutoh, H., Perron, A., Rossier, J., & Knopfel, T. (2010). Imaging brain electric signals with genetically targeted voltage-sensitive fluorescent proteins. *Nature Methods*, *7*, 643–649.

Andrasfalvy, B. K., Zemelman, B. V., Tang, J., & Vaziri, A. (2010). Two-photon single-cell optogenetic control of neuronal activity by sculpted light. *Proceedings of the National Academy of Sciences of the United States of America*, *107*, 11981–11986.

Anselmi, F., Ventalon, C., Begue, A., Ogden, D., & Emiliani, V. (2011). 3D imaging and photostimulation by remote focusing and holographic light patterning. *Proceedings of the National Academy of Sciences of the United States of America*, *108*, 19504–19509.

Aravanis, A. M., Wang, L. P., Zhang, F., Meltzer, L. A., Mogri, M. Z., Schneider, M. B., et al. (2007). An optical neural interface: In vivo control of rodent motor cortex with integrated fiberoptic and optogenetic technology. *Journal of Neural Engineering*, *4*, S143–S156.

Arenkiel, B. R., Peca, J., Davison, I. G., Feliciano, C., Deisseroth, K., Augustine, G. J., et al. (2007). In vivo light-induced activation of neural circuitry in transgenic mice expressing channelrhodopsin-2. *Neuron*, *54*, 205–218.

Botcherby, E. J., Juskaitis, R., Booth, M. J., & Wilson, T. (2007). Aberration-free optical refocusing in high numerical aperture microscopy. *Optics Letters*, *32*, 2007–2009.

Botcherby, E. J., Smith, C. W., Booth, M. J., Juskaitis, R., & Wilson, T. (2010). Arbitrary-scan imaging for two-photon microscopy. *Proceedings of SPIE*, *756917*, 1–8.

Branco, T., Clark, B. A., & Hausser, M. (2010). Dendritic discrimination of temporal input sequences in cortical neurons. *Science*, *329*, 1671–1675.

Cardin, J. A., Carlen, M., Meletis, K., Knoblich, U., Zhang, F., Deisseroth, K., et al. (2009). Driving fast-spiking cells induces gamma rhythm and controls sensory responses. *Nature*, *459*, 663–667.

Carter, A. G., & Sabatini, B. L. (2004). State-dependent calcium signaling in dendritic spines of striatal medium spiny neurons. *Neuron*, *44*, 483–493.

Chanda, B., Blunck, R., Faria, L. C., Schweizer, F. E., Mody, I., & Bezanilla, F. (2005). A hybrid approach to measuring electrical activity in genetically specified neurons. *Nature Neuroscience*, *8*, 1619–1626.

Curtis, J. E., Koss, B. A., & Grier, D. G. (2002). Dynamic holographic optical tweezers. *Optics Communications*, *207*, 169.

Dal Maschio, M., De Stasi, A. M., Benfenati, F., & Fellin, T. (2011). Three-dimensional in vivo scanning microscopy with inertia-free focus control. *Optics Letters*, *36*, 3503–3505.

Dal Maschio, M., Difato, F., Beltramo, R., Blau, A., Benfenati, F., & Fellin, T. (2010). Simultaneous two-photon imaging and photo-stimulation with structured light illumination. *Optics Express*, *18*, 18720–18731.

Dana, H., & Shoham, S. (2011). Numerical evaluation of temporal focusing characteristics in transparent and scattering media. *Optics Express*, *19*, 4937–4948.

Daria, V. R., Stricker, C., Bowman, R., Redman, S., & Bachor, H. A. (2009). Arbitrary multisite two-photon excitation in four dimensions. *Applied Physics Letters*, *95*, 093701.

Dhawale, A. K., Hagiwara, A., Bhalla, U. S., Murthy, V. N., & Albeanu, D. F. (2010). Non-redundant odor coding by sister mitral cells revealed by light addressable glomeruli in the mouse. *Nature Neuroscience*, *13*, 1404–1412.

Di Leonardo, R., Ianni, F., & Ruocco, G. (2007). Computer generation of optimal holograms for optical trap arrays. *Optics Express*, *15*, 1913–1922.

Du, R., Bi, K., Zeng, S., Li, D., Xue, S., & Luo, Q. (2009). Analysis of fast axial scanning scheme using temporal focusing with acousto-optic deflectors. *Journal of Modern Optics*, *56*, 81–84.

Durst, M. E., Zhu, G., & Xu, C. (2006). Simultaneous spatial and temporal focusing for axial scanning. *Optics Express*, *14*, 12243–12254.

Durst, M. E., Zhu, G., & Xu, C. (2008). Simultaneous spatial and temporal focusing in nonlinear microscopy. *Optics Communications*, *281*, 1796–1805.

Emiliani, V., Cojoc, D., Ferrari, E., Garbin, V., Durieux, C., Di Fabrizio, E., et al. (2006). Wavefront engineering microscopy to study 3D mechanotransduction in living cells. *Proceedings of SPIE*, *6195*, 61950J.

Farah, N., Reutsky, I., & Shoham, S. (2007). Patterned optical activation of retinal ganglion cells. *Conference Proceedings: Annual International Conference of the IEEE Engineering in Medicine and Biology Society*, *2007*, 6369–6371.

Feldbauer, K., Zimmermann, D., Pintschovius, V., Spitz, J., Bamann, C., & Bamberg, E. (2009). Channelrhodopsin-2 is a leaky proton pump. *Proceedings of the National Academy of Sciences of the United States of America*, *106*, 12317–12322.

Fisher, J. A., Barchi, J. R., Welle, C. G., Kim, G. H., Kosterin, P., Obaid, A. L., et al. (2008). Two-photon excitation of potentiometric probes enables optical recording of action potentials from mammalian nerve terminals in situ. *Journal of Neurophysiology*, *99*, 1545–1553.

Fisher, J. A., Salzberg, B. M., & Yodh, A. G. (2005). Near infrared two-photon excitation cross-sections of voltage-sensitive dyes. *Journal of Neuroscience Methods*, *148*, 94–102.

Fricke, M., & Nielsen, T. (2005). Two-dimensional imaging without scanning by multifocal multiphoton microscopy. *Applied Optics*, *44*, 2984–2988.

Gerchberg, R. W., & Saxton, W. O. (1972). A practical algorithm for the determination of the phase from image and diffraction pictures. *Optik*, *35*, 237–246.

Glückstad, J. (1996). Phase contrast image synthesis. *Optics Communications*, *130*, 225.

Glückstad, J., & Mogensen, P. C. (2001). Optimal phase contrast in common-path interferometry. *Applied Optics*, *40*, 268–282.

Golan, L., Reutsky, I., Farah, N., & Shoham, S. (2009). Design and characteristics of holographic neural photo-stimulation systems. *Journal of Neural Engineering*, *6*, 66004.

Golan, L., & Shoham, S. (2009). Speckle elimination using shift-averaging in high-rate holographic projection. *Optics Express*, *17*, 1330–1339.

Goodman, J. W. (2005). *Introduction to Fourier optics* (3rd ed.). Roberts & Company.

Gradinaru, V., Mogri, M., Thompson, K. R., Henderson, J. M., & Deisseroth, K. (2009). Optical deconstruction of Parkinsonian neural circuitry. *Science*, *324*, 354–359.

Gradinaru, V., Thompson, K. R., Zhang, F., Mogri, M., Kay, K., Schneider, M. B., et al. (2007). Targeting and readout strategies for fast optical neural control in vitro and in vivo. *Journal of Neuroscience*, *27*, 14231–14238.

Grossman, N., Poher, V., Grubb, M. S., Kennedy, G. T., Nikolic, K., McGovern, B., et al. (2010). Multi-site optical excitation using ChR2 and micro-LED array. *Journal of Neural Engineering*, *7*, 16004.

Judkewitz, B., Roth, A., & Hausser, M. (2006). Dendritic enlightenment: Using patterned two-photon uncaging to reveal the secrets of the brain's smallest dendrites. *Neuron, 50*, 180–183.

Kramer, R. H., Fortin, D. L., & Trauner, D. (2009). New photochemical tools for controlling neuronal activity. *Current Opinion in Neurobiology, 19*, 544–552.

Kravitz, A. V., Freeze, B. S., Parker, P. R., Kay, K., Thwin, M. T., Deisseroth, K., et al. (2010). Regulation of Parkinsonian motor behaviours by optogenetic control of basal ganglia circuitry. *Nature, 466*, 622–626.

Kuhn, B., Denk, W., & Bruno, R. M. (2008). In vivo two-photon voltage-sensitive dye imaging reveals top-down control of cortical layers 1 and 2 during wakefulness. *Proceedings of the National Academy of Sciences of the United States of America, 105*, 7588–7593.

Kwon, H. B., & Sabatini, B. L. (2011). Glutamate induces de novo growth of functional spines in developing cortex. *Nature, 474*, 100–104.

Leach, J., Wulff, K., Sinclair, G., Jordan, P., Courtial, J., Thomson, L., et al. (2006). Interactive approach to optical tweezers control. *Applied Optics, 45*, 897.

Liesener, J., Reicherter, M., Haist, T., & Tiziani, H. J. (2000). Multi-functional optical tweezers using computer-generated holograms. *Optics Communications, 185*, 77–82.

Lin, J. Y. (2011). A user's guide to channelrhodopsin variants: Features, limitations and future developments. *Experimental Physiology, 96*, 19–25.

Losonczy, A., Makara, J. K., & Magee, J. C. (2008). Compartmentalized dendritic plasticity and input feature storage in neurons. *Nature, 452*, 436–441.

Losonczy, A., Zemelman, B. V., Vaziri, A., & Magee, J. C. (2010). Network mechanisms of theta related neuronal activity in hippocampal CA1 pyramidal neurons. *Nature Neuroscience, 13*, 967–972.

Lutz, C., Otis, T. S., de Sars, V., Charpak, S., DiGregorio, D. A., & Emiliani, V. (2008). Holographic photolysis of caged neurotransmitters. *Nature Methods, 5*, 821–827.

Mutoh, H., Perron, A., Akemann, W., Iwamoto, Y., & Knopfel, T. (2011). Optogenetic monitoring of membrane potentials. *Experimental Physiology, 96*, 13–18.

Nikolenko, V., Peterka, D. S., & Yuste, R. (2010). A portable laser photostimulation and imaging microscope. *Journal of Neural Engineering, 7*, 045001.

Nikolenko, V., Poskanzer, K. E., & Yuste, R. (2007). Two-photon photostimulation and imaging of neural circuits. *Nature Methods, 4*, 943–950.

Nikolenko, V., Watson, B. O., Araya, R., Woodruff, A., Peterka, D. S., & Yuste, R. (2008). SLM microscopy: Scanless two-photon imaging and photostimulation with spatial light modulators. *Frontiers in Neural Circuits, 2*, 1–14.

Oron, D., & Silberberg, Y. (2005). Harmonic generation with temporally focused ultrashort pulses. *Journal of the Optical Society of America B: Optical Physics, 22*, 2660–2663.

Oron, D., Tal, E., & Silberberg, Y. (2005). Scanningless depth-resolved microscopy. *Optics Express, 13*, 1468–1476.

Palima, D., & Glückstad, J. (2008). Multi-wavelength spatial light shaping using generalized phase contrast. *Optics Express, 16*, 1331–1342.

Papagiakoumou, E., Anselmi, F., Bègue, A., de Sars, V., Gluckstad, J., Isacoff, E. Y., et al. (2010). Scanless two-photon excitation of channelrhodopsin-2. *Nature Methods, 7*, 848–854.

Papagiakoumou, E., de Sars, V., Emiliani, V., & Oron, D. (2009). Temporal focusing with spatially modulated excitation. *Optics Express, 17*, 5391–5401.

Papagiakoumou, E., de Sars, V., Oron, D., & Emiliani, V. (2008). Patterned two-photon illumination by spatiotemporal shaping of ultrashort pulses. *Optics Express, 16*, 22039–22047.

Peterka, D. S., Takahashi, H., & Yuste, R. (2011). Imaging voltage in neurons. *Neuron, 69*, 9–21.

Petreanu, L., Huber, D., Sobczyk, A., & Svoboda, K. (2007). Channelrhodopsin-2-assisted circuit mapping of long-range callosal projections. *Nature Neuroscience, 10*, 663–668.

Petreanu, L., Mao, T., Sternson, S. M., & Svoboda, K. (2009). The subcellular organization of neocortical excitatory connections. *Nature, 457*, 1142–1145.

Reicherter, M., Haist, T., Wagemann, E. U., & Tiziani, H. J. (1999). Optical particle trapping with computer-generated holograms written on a liquid-crystal display. *Optics Letters, 24*, 608–610.

Rhodes, W. T. (Ed.), (2009). *Generalized phase contrast. Application in optics and photonics*. Atlanta: Springer.

Rickgauer, J. P., & Tank, D. W. (2009). Two-photon excitation of channelrhodopsin-2 at saturation. *Proceedings of the National Academy of Sciences of the United States of America, 106*, 15025–15030.

Rodrigo, P. J., Daria, V. R., & Glückstad, J. (2004). Real-time three-dimensional optical micromanipulation of multiple particles and living cells. *Optics Letters, 29*, 2270–2272.

Salter, P. S., Carbone, G., Botcherby, E. J., Wilson, T., Elston, S. J., & Raynes, E. P. (2009). Liquid crystal director dynamics imaged using two-photon fluorescence microscopy with remote focusing. *Physical Review Letters, 103*, 257803.

Sohal, V. S., Zhang, F., Yizhar, O., & Deisseroth, K. (2009). Parvalbumin neurons and gamma rhythms enhance cortical circuit performance. *Nature, 459*, 698–702.

Straub, A., Durst, M. E., & Xu, C. (2011). High speed multi-photon axial scanning through an optical fiber in a remotely scanned temporal focusing setup. *Biomedical Optics Express, 2*, 80–88.

Suchowski, H., Oron, D., & Silberberg, Y. (2006). Generation of a dark nonlinear focus by spatio-temporal coherent control. *Optics Communications, 264*, 482–487.

Tal, E., Oron, D., & Silberberg, Y. (2005). Improved depth resolution in video-rate line-scanning multiphoton microscopy using temporal focusing. *Optics Letters, 30*, 1686–1688.

Theer, P., & Denk, W. (2006). On the fundamental imaging-depth limit in two-photon microscopy. *Journal of the Optical*

Society of America A: Optics, Image Science, and Vision, *23*, 3139–3149.

Tsai, H. C., Zhang, F., Adamantidis, A., Stuber, G. D., Bonci, A., de Lecea, L., et al. (2009). Phasic firing in dopaminergic neurons is sufficient for behavioral conditioning. *Science*, *324*, 1080–1084.

Vaziri, A., & Emiliani, V. (2011). *Current Opinion in Neurobiology*, [Epub ahead of print] PMID: 22209216.

Wang, H., Peca, J., Matsuzaki, M., Matsuzaki, K., Noguchi, J., Qiu, L., et al. (2007). High-speed mapping of synaptic connectivity using photostimulation in Channelrhodopsin-2 transgenic mice. *Proceedings of the National Academy of Sciences of the United States of America*, *104*, 8143–8148.

Witten, I. B., Lin, S. C., Brodsky, M., Prakash, R., Diester, I., Anikeeva, P., et al. (2010). Cholinergic interneurons control local circuit activity and cocaine conditioning. *Science*, *330*, 1677–1681.

Wyart, C., Del Bene, F., Warp, E., Scott, E. K., Trauner, D., Baier, H., et al. (2009). Optogenetic dissection of a behavioural module in the vertebrate spinal cord. *Nature*, *461*, 407–410.

Yang, S., Papagiakoumou, E., Guillon, M., de Sars, V., Tang, C. M., & Emiliani, V. (2011). Three-dimensional holographic photostimulation of the dendritic arbor. *Journal of Neural Engineering*, *8*, 046002.

Zahid, M., Velez-Fort, M., Papagiakoumou, E., Ventalon, C., Angulo, M. C., & Emiliani, V. (2010). Holographic photolysis for multiple cell stimulation in mouse hippocampal slices. *PLoS One*, *5*, e9431.

Zernike, F. (1955). How I discovered phase contrast. *Science*, *121*, 345–349.

Zhu, G., van Howe, J., Durst, M., Zipfel, W., & Xu, C. (2005). Simultaneous spatial and temporal focusing of femtosecond pulses. *Optics Express*, *13*, 2153–2159.

T. Knöpfel and E. Boyden (Eds.)
Progress in Brain Research, Vol. 196
ISSN: 0079-6123

CHAPTER 8

Zebrafish as an appealing model for optogenetic studies

Joshua Simmich[†], Eric Staykov[†] and Ethan Scott*

School of Biomedical Sciences, The University of Queensland, St. Lucia, Queensland, Australia

Abstract: Optogenetics, the use of light-based protein tools, has begun to revolutionize biological research. The approach has proven especially useful in the nervous system, where light has been used both to detect and to manipulate activity in targeted neurons. Optogenetic tools have been deployed in systems ranging from cultured cells to primates, with each offering a particular combination of advantages and drawbacks. In this chapter, we provide an overview of optogenetics in zebrafish. Two of the greatest attributes of the zebrafish model system are external fertilization and transparency in early life stages. Combined, these allow researchers to observe the internal structures of developing zebrafish embryos and larvae without dissections or other interference. This transparency, combined with the animals' small size, simple husbandry, and similarity to mammals in many structures and processes, has made zebrafish a particularly popular model system in developmental biology. The easy optical access also dovetails with optogenetic tools, allowing their use in intact, developing, and behaving animals. This means that optogenetic studies in embryonic and larval zebrafish can be carried out in a high-throughput fashion with relatively simple equipment. As a consequence, zebrafish have been an important proving ground for optogenetic tools and approaches and have already yielded important new knowledge about the neural circuits underlying behavior. Here, we provide a general introduction to zebrafish as a model system for optogenetics. Through descriptions and analyses of important optogenetic studies that have been done in zebrafish, we highlight the advantages and liabilities that the system brings to optogenetic experiments.

Keywords: behavior; neural circuits; optogenetics; zebrafish.

*Corresponding author.
Tel.: +61-7-3346-9471; Fax: +61-7-3365-1766
E-mail: ethan.scott@uq.edu.au

[†]These authors contributed equally.

DOI: 10.1016/B978-0-444-59426-6.00008-2

Introduction

Optogenetics involves the application of light to observe and manipulate cell activity using genetically encoded tools. A good model organism for optogenetics should be easily accessible to light, amenable to transgenesis and, because optogenetics often focuses on neurons, possess brains and behaviors that are interesting to study. As zebrafish satisfy all of these requirements, they have been an important platform for early optogenetic experiments and hold enormous potential for future research into brain function.

The zebrafish (*Danio rerio*), a member of the carp family, is a small fish native to slow-moving freshwaters of tropical South Asia (Engeszer et al., 2007; Spence et al., 2008). Zebrafish was established as a model system in the 1980s (Streisinger et al., 1981) and grew in popularity largely because it is an externally developing vertebrate that is transparent in its early life stages. These properties have proven especially helpful for developmental biologists who are able to observe internal tissues directly, starting at fertilization. With the advent of forward genetic screens (e.g., Mullins et al., 1994) and the use of morpholinos for gene disruption (Nasevicius and Ekker, 2000), the properties inherent to zebrafish have aided studies of gene function. Finally, the establishment of transgenesis techniques in zebrafish allowed exogenous proteins, especially tissue-specific markers, to be expressed. Transgenesis is also a prerequisite for optogenetics, given the need to express genetically encoded tools.

Chief among the advantages zebrafish offer for optogenetics is that they are, in their embryonic and larval forms, transparent. The brain of a larval zebrafish at 5 days post fertilization (dpf) has a thickness less than 500 μm, making almost all neurons accessible to confocal or multiphoton microscopy. Although skin pigmentation can obscure internal structures, this pigmentation can be prevented by the addition of phenylthiourea to the water or by utilizing mutants that never develop such pigment (Lister et al., 1999; White et al., 2008).

Further, the zebrafish brain and its associated behaviors develop quickly. By 7 dpf, when the larva is roughly 4 mm long (Parichy et al., 2009), its brain consists of ~78,000 neurons (Hill et al., 2003). Despite this simplicity, larvae at this stage are capable of a wide range of behaviors. Assays for visual behaviors (reviewed by Baier and Scott, 2009; Portugues and Engert, 2009) and for control of the trunk during swimming and escape responses (reviewed by McLean and Fetcho, 2011) are the best established, and many of the initial forays into optogenetics in zebrafish have used these behaviors. The zebrafish nervous system shares the basic architecture with other vertebrates, especially in structures such as somatosensory neurons, olfactory bulb, retina, cerebellum, and spinal cord. While the circuits in these regions are generally simpler than their mammalian counterparts, their similar overall structure and connectivity allows many discoveries in zebrafish to be generalized to other vertebrates, including mammals (Burne et al., 2011). In some cases, generally involving interneurons, differences in the circuits across evolution preclude this generalization (Fetcho, 2007).

Zebrafish therefore offer an appealing combination of the genetic accessibility and transparency of small invertebrate model organisms, with the organs, neural structures, and complex behaviors of other vertebrates. These advantages, however, are accompanied by certain limitations. Zebrafish lack a distinct cerebral cortex, hippocampus, and amygdala and do not exhibit all of the behaviors of mammalian model organisms. For this reason, studies of the zebrafish nervous system are easily generalized for some circuits (e.g., retinal and spinal) but have less bearing on mammalian circuits in other regions (especially in the forebrain). Larval zebrafish also lack more nuanced and complex behaviors, including social behaviors. While adult fish show many of these behaviors (Norton and Bally-Cuif, 2010), they are no longer transparent and are less amenable to high-throughput analyses. Another possible complication comes from the fact that so many of the popular behavioral assays in zebrafish are based on vision. With the intense light that is needed to drive optogenetic tools, there is the risk that larvae will be blinded, making it difficult to apply visual stimuli to the larvae at the same time that the optogenetic tools are being used. The fact that visual stimuli have been successfully applied during calcium (Ca^{2+}) imaging (Aizenberg and Schuman, 2011; Del Bene et al., 2010; Niell and Smith, 2005; Sumbre et al., 2008), however, hints that this problem will be surmountable.

In this review, we provide an overview of zebrafish as a platform for optogenetic studies. We will first outline the transgenic techniques and genetic methods that can be used to target optogenetic tools to tissues and cell types of interest. We will then dedicate sections to recent optogenetic studies that have described neural anatomy or observed patterns of activity in the larval brain. Finally, we will focus on a few incisive studies that have combined different optogenetic strategies to elucidate the functions of specific neural circuits in the zebrafish nervous system. These studies are summarized and represented graphically in Fig. 1.

Getting genes into zebrafish

Since optogenetics relies on the expression of exogenous proteins, it also relies on the ability of researchers to deliver optogenetic tools into their model organisms. Zebrafish eggs are externally fertilized, making zygotes easily accessible for transgenic manipulations. Transgenes can be introduced to zebrafish simply by injecting a plasmid containing a transgene into the early zebrafish embryo, and indeed, this was the

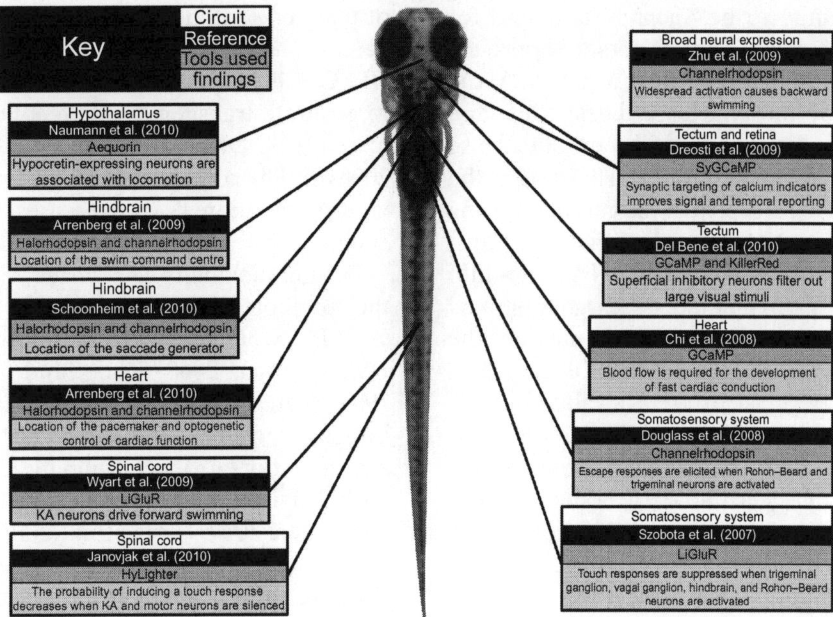

Fig. 1. Summary of optogenetic studies in zebrafish. In addition to proving the utility of optogenetic tools in zebrafish, several of the studies presented in this review have made contributions to our knowledge of developmental and neuroscientific processes. This figure highlights some of these studies, providing the structure studied, the optogenetic tools used, and a simplified description of the findings. For more details on these reports, see the main text.

method by which the first transgenic zebrafish were created (Stuart et al., 1988). Though a number of zebrafish lines expressing GFP in various tissues were created by this method (Amsterdam et al., 1999; Higashijima et al., 1997; Long et al., 1997), integration into the genome is rare. This low efficiency meant that simple plasmid injection had already been superseded by the time optogenetic tools arrived on the zebrafish scene.

Several techniques have since been used to increase the efficiency of transgenesis in zebrafish. Transgenesis mediated by retroviruses (Lin et al., 1994) can achieve 100% efficiency, but engineering retroviruses has traditionally been labor intense as compared to traditional subcloning. This has made retroviruses popular for making many lines for the same transgene (Amsterdam et al., 1999; Ellingsen et al., 2005; Gaiano et al., 1996), but less so for making lines for a variety of different genetic constructs. Recent advances, however, have made retroviruses easier to engineer, and this approach will likely continue to be an important one for generating transgenic fish. Another approach involves generating a DNA construct with the cargo flanked by meganuclease digest sites, and coinjecting this construct with meganuclease enzyme (Thermes et al., 2002). This raises the number of linearized copies of the cargo, boosting transgenesis rates. A third method involves transposase-mediated insertion of the DNA cargo into the genome. With this method, the cargo is flanked by target sites for a transposase enzyme, and the construct is coinjected with mRNA for the transposase (Davidson et al., 2003; Kawakami et al., 2000; Miskey et al., 2005). The enzyme, which has been translated from the mRNA in the zygote, transposes the transgene into the genome. Because of a high rate of germline insertion, efficient stable transgenesis in the F_1 generation, and the relative ease of generating DNA constructs for this approach (Kwan et al., 2007), transposase-mediated transgenesis has been adopted by most zebrafish labs, with the Tol2 transposase being the most commonly used (reviewed by Suster et al., 2009a).

In most cases, it is important that optogenetic tools be expressed in the tissue or cell types of interest with as little off-target background expression as possible. This is normally achieved by pairing the transgene with a tissue-specific promoter or enhancer. In the simplest case, this involves creating a construct consisting of the gene for the optogenetic tool downstream of the promoter or enhancer element of a gene that is known to be specific to the desired cell type(s). However, the location of the regulatory elements responsible for the tissue specificity is not always obvious and is often remote from the gene itself (Navratilova et al., 2009). Traditional constructs for transgenesis incorporate promoters of just a few kilobases and, as a consequence, are at risk of including only a subset of the enhancers and suppressors that control the specificity of expression. Surrounding an open reading frame with large regions of DNA flanking an endogenous gene better assures that the expression pattern of that gene will be accurately replicated in transgenic animals. Bacterial artificial chromosomes (BACs) allow precisely this and have been used to generate transgenic zebrafish both by injecting naked BAC DNA (Jessen et al., 1999; Yang et al., 2006) and by coinjecting DNA along with the previously mentioned transposases (Suster et al., 2009b).

Further flexibility can be achieved by linking the promoter/enhancer and the transgene via Gal4/UAS (Asakawa and Kawakami, 2009; Halpern et al., 2008; Scheer and Campos-Ortega, 1999; Scott, 2009) or Cre/LoxP (Hans et al., 2009; Langenau et al., 2005). Gal4 has been used in zebrafish for enhancer trapping (Asakawa et al., 2008; Davison et al., 2007; Distel et al., 2009; Scott and Baier, 2009; Scott et al., 2007), generating a great number of varied expression patterns. Because these combinatorial systems allow separate control over the gene that is to be expressed (with the UAS or flanked by LoxP sites) and the pattern in which it will be expressed (with the Gal4 or Cre), they provide the flexibility to target a host of different transgenes to numerous patterns with relatively few transgenic lines. A different sort of flexibility is afforded by the Tet system (Zhu et al., 2009), which allows conditional expression (regulated by bath application of doxycycline) of transgenes in specific expression patterns (based on the enhancers surrounding the transgene).

Imaging and tracing neurons and circuits

Although optogenetics usually involves the observation or manipulation of neural activity, there are many light-mediated genetic tools for describing the anatomy or connectivity of neuronal circuits. Particularly, when deployed in zebrafish, these tools allow individual or groups of neurons to be traced in living animals, or observed without interruption as development proceeds.

One of the first optogenetic tools to be used in this way was the photoconvertible protein Kaede. Kaede is a tetrameric fluorescent protein derived from coral and exhibits green fluorescence unless permanently converted by UV irradiation to exhibit red fluorescence (Ando et al., 2002). This useful tool for labeling neurons was used in the zebrafish by Sato et al. (2006) to label specific neurons in the trigeminal sensory ganglion and tectum and individual Rohon–Beard neurons in the spinal cord in live 4dpf zebrafish larvae. The authors were able to examine the developmental changes in the dendritic and axonal morphology of those neurons for days after labeling. Kaede has also proven useful for tracing the axon trajectories of targeted cells, as red Kaede diffuses into axons after photoconversion in the somae of targeted neurons (Scott, 2009; Scott et al., 2007).

Another fluorescent reporter protein, Dronpa, was used by Aramaki and Hatta (2006) to image single hindbrain neurons in live zebrafish. Dronpa is a monomeric fluorescent protein that, like Kaede, was derived from coral (Ando et al., 2007; Habuchi et al., 2005). Unlike Kaede, however, Dronpa can be switched from a green fluorescent state to a nonfluorescent state by irradiation at 490nm and switched back to fluorescent by irradiation of 400nm light. In an approach that they termed "optical dissection," they erased fluorescence from embryos expressing Dronpa in the hindbrain, and then serially illuminated and extinguished fluorescence in individual neurons, each time imaging their structure. This permitted them to describe each neuron's structure in detail and to reconstruct the structure as a whole at the single cell level. As with Kaede, Dronpa allows for the anterograde or retrograde diffusion of converted protein and can thus label the axons of irradiated somae or the somae of irradiated axons (Aramaki and Hatta, 2006). Finally, Dronpa's convertibility can be used to subtract the background fluorescence from a Dronpa signal (Marriott et al., 2008). This is because the background remains stable through repeated switching of Dronpa, and the constant signal can be digitally identified and subtracted from the overall image.

Photoconvertible proteins can be used not only to define and image circuits but also to link behaviors with neuronal development. This is exemplified in two studies using Kaede to describe the development of spinal circuitry controlling tail movements in larval zebrafish. Kimura et al. (2006) irradiated Kaede-expressing spinal neurons at a chosen timepoint and then used the red and green fluorescence at a later timepoint to determine whether a given cell was born before or after the irradiation, respectively. Their rationale was that any cell with photoconverted red Kaede must have differentiated (and therefore expressed Kaede) before the time of photoconversion. These early-born neurons contained both red Kaede that had been present at the time of the photoconversion and green Kaede that had been produced subsequently. Cells expressing green unconverted Kaede alone were assumed to have differentiated after the photoconversion. McLean and Fetcho (2009) used the same approach, but with more timepoints. They photoconverted Kaede, expressed in postmitotic spinal cord neurons of embryonic zebrafish larvae, at intervals 1 day apart and then imaged the larvae at 4dpf. Through this simple use of a photoconvertible protein, combined with electrophysiology and behavioral imaging, they confirmed the hypothesis that neurons responsible for the fast, large-amplitude movements emerged first and were more dorsal, with the younger neurons driving slower and finer movements and positioned more ventrally.

Monitoring neural activity

Genetically encoded activity indicators

One conceptually simple approach for defining behavioral circuits is to see which neurons are active as a behavior unfolds. Physiological changes in active neurons, such as voltage spikes, increase in intracellular Ca^{2+}, and the fusion of neurotransmitter-bearing vesicles provide means by which to observe activity. As detailed in this section, optogenetic tools have been developed to detect each of these cell biological events, and several studies have already used these to link neural activity with behavior. The transparency afforded by zebrafish facilitates the use of these indicators since it allows their observation in intact, behaving animals.

For decades, electrophysiology has been, and in some cases remains, the best method of measuring neural activity. Optogenetic tools offer distinct and complementary properties for observing activity and have begun to answer questions that are difficult to address electrophysiologically. While electrophysiological approaches typically observe one or a few cells at a time, optogenetic tools can provide information on the activity of many genetically targeted cells spread over large areas. Electrophysiology is also invasive, requiring an electrode to be inserted, whereas optogenetic tools, especially in transparent model systems, generally cause minimal interference with the normal functioning of an animal or neuronal circuit.

Another set of alternatives for observing activity are Ca^{2+}- and voltage-sensitive dyes. Once taken up by neurons of interest, signals from these dyes report on the targeted neurons' activity. Dyes have the advantage of strong signals as compared to genetically encoded indicators (Knöpfel et al., 2006), although this gap is being closed as improved proteins are engineered. Protein indicators, however, have the advantage that they can be genetically targeted. This allows them to be expressed broadly (spatially), but specifically (genetically). In other words, they can be used to report on the activity of a particular type of neuron, even if it is densely packed with other neurons or broadly spaced within the brain.

Ca^{2+}-sensitive proteins

The Ca^{2+} concentration in a neuron increases during neural activity, especially in dendrites responding to excitatory synaptic transmission, and in the presynaptic terminal prior to neurotransmitter release. Ca^{2+} imaging with either synthetic or genetically encoded Ca^{2+} indicators therefore provides a method of visualizing the activity of neurons.

Cameleon is one such genetically encoded Ca^{2+} indicator and makes use of a Förster resonance energy transfer (FRET)-based fluorescent signal. In its original form, cameleon comprised the Ca^{2+}-binding peptide calmodulin (CaM), enhanced blue fluorescent protein (eBFP), and enhanced GFP (eGFP; Miyawaki et al., 1997, 1999), although the fluorophores used have changed in subsequent forms. Ca^{2+} binding to the CaM induces a conformational change in the protein, bringing the pair of fluorophores closer to one another. This closer proximity leads to energy transfer from eBFP to eGFP, which in turn boosts the green signal while dimming the blue signal. Therefore, the ratio of blue to green signals is an indicator of Ca^{2+} concentration and, therefore, neural activity. The use of FRET has the added advantage that it ameliorates problems caused by movements in the sample. Since both fluorophores are affected equally by movement artifacts, the ratio remains a consistent indicator even if there is movement in the sample (Miyawaki, 2005). Despite strong signals *in vitro*, early versions of cameleon exhibited relatively modest FRET signals in neurons. This problem has more recently been addressed by a combination of improved cameleons (Mank et al., 2008; Nagai et al., 2004; Palmer et al., 2004, 2006; Wallace et al., 2008), and statistical optimization of the raw signals (Fan et al., 2007).

Higashijima et al. (2003) led the way in conclusively establishing cameleon as an indicator of neuronal activity in zebrafish, even during behavior. As an initial confirmation, they used electrophysiology to evoke a yellow cameleon 2.1 signal in individual neurons in paralyzed zebrafish, thus confirming that cameleon could indeed detect Ca^{2+} transients in live cells. They created transgenic fish expressing yellow cameleon 2.1 either in all neurons or specifically in Rohon–Beard neurons, which are touch sensitive. As is typical of optogenetic experiments, they embedded live larvae in agar and, using a confocal microscope, imaged Rohon–Beard neurons and motor neurons. By touching the larva on the head, they evoked escape responses and, despite movement of the larvae, observed coincident FRET signals from the cameleon, representing transient increases in Ca^{2+} in the neurons studied.

Optical reporters of neuronal activity are very useful for mapping brain regions with specific topographic arrangements. Yaksi et al. (2007) used yellow cameleon 2.1 to map the olfactory bulb in explants from adult zebrafish during their responses to odors. Mapping both the mitral cells and interneurons, they found that different odorants produced short-lived foci of activity in the two cell types. Interneurons showed a long-lasting slightly chemotopic map, while mitral cells exhibited a strong but short-lived period of focussed activity, which became more sparse within a few hundred milliseconds of the initial response.

Because Ca^{2+} is a second messenger utilized in many cellular processes, the use of genetically encoded Ca^{2+} indicators is not restricted to imaging of neural activity. Making good use of the external development and transparency afforded by the zebrafish model organism, a study by Tsuruwaka et al. (2007) imaged intracellular Ca^{2+} concentration during gastrulation using yellow cameleon 2.12. They found that the Ca^{2+} concentration is highly dynamic during early development, and that the areas with highest Ca^{2+} change as development proceeds. The authors propose that these regulated changes in Ca^{2+} concentration are a part of essential developmental signaling pathways, including Wnt and bone morphogenic protein. In comparison to a synthetic Ca^{2+} dye, in this case Fluo-3, the signal from yellow cameleon was much clearer and extended further into development, and this was an apparent consequence of sequestration of the dye in the yolk.

Another Ca^{2+} indicator used in zebrafish, GCaMP, is based on a Ca^{2+}-dependent mutant of GFP (Nakai et al., 2001). Though GCaMP is a fusion between CaM and a fluorescent protein, it uses a single GFP molecule. *In vitro* characterization of GCaMP shows that it increases in fluorescence ~4.5-fold in the presence of 1 mM Ca^{2+}, although it should be noted that this is a nonphysiological concentration. *In vivo*, GCaMP has been used to visualize cardiac conductance and the dynamics of contractions in the developing heart (Chi et al., 2008). Recently, GCaMP-HS (hyper sensitive), a more sensitive mutant form of GCaMP, has been introduced and used to observe activity in spinal neurons (Muto et al., 2011). In a particularly incisive study, Del Bene et al. (2010) used GCaMP, in combination with the photosensitizer KillerRed (KR), to probe visual filtering in the optic tectum. As that study involved both observing and manipulating neurons (by ablation), it will be discussed in depth in the next section.

Ca^{2+} indicators can also be targeted to specific cellular locations, for instance to synapses in order to measure synaptic activity *in vivo*. By creating a fusion protein between synaptophysin and GCaMP2, Dreosti et al. (2009) demonstrated that presynaptically targeted SyGaMP2, as they termed their fusion protein, can detect the brief

Ca^{2+} transient that occurs very close to the voltage-gated Ca^{2+} channels in the presynaptic terminal. Using live transgenic zebrafish larvae, they observed the activity of synapses within the optic tectum and graded synaptic activity in ribbon synapses of retinal bipolar cells.

Optogenetic observations in zebrafish usually require that larvae be immobilized under a microscope, and this limits the behaviors that can be studied. One tool, the bioluminescent Ca^{2+} reporter Aequorin, allows for activity to be detected in free-swimming larvae. As proof of this technique, Naumann et al. (2010) generated transgenic zebrafish expressing an Aequorin–GFP fusion protein in all neurons. Aequorin is an enzyme that, on binding Ca^{2+}, catalyzes the oxidation of coelenterazine and, by doing so, emits a blue photon. The emitted blue photons only reach detectable intensities above the basal Ca^{2+} concentration and in contrast to fluorescent reporters, luminescent proteins do not require light excitation. Though the intensity of the emitted light increases when Aequorin is fused to GFP, the signal is modest and requires the emitted photons from many neurons to be pooled in order for the signal to be detectable. This means that temporal, but not spatial, information can be gathered on the activity of the targeted neurons. This, in turn, means that Aequorin must be strictly targeted to the neurons of interest, since noise from off-target cells is indistinguishable from the desired signal.

Other optogenetic approaches for detecting neural activity

As detailed above, a majority of optogenetic observations of neural activity in zebrafish have been performed with various sensors of Ca^{2+}. An alternative is to observe changes in membrane voltage directly using voltage-sensitive fluorescent proteins (VSFPs). These proteins consist of the voltage-sensing domain (modern VSFPs use the voltage-sensing domain of a voltage-sensitive phosphatase from *Ciona intestinalis*), fused to a single unaltered or circularly permuted fluorescent protein, or a FRET pair (Dimitrov et al., 2007, Gautam et al., 2009; Lundby et al., 2008, 2010; Perron et al., 2009a,b; Tsutsui et al., 2008). Changes in voltage lead to conformational changes in the VSD, resulting in a change in FRET efficiency or intensity of the single fluorescent protein (Dimitrov et al., 2007; Lundby et al., 2008). Like cameleon, most VSFPs are ratiometric FRET-based reporters and so provide the same advantages. VSFPs have the advantage that they report on physiological (voltage) changes that are faster than the corresponding changes in Ca^{2+}. VSFPs therefore have the potential for higher temporal resolution in analyzing neural activity than do Ca^{2+} indicators. Recently, Tsutsui et al. (2010) mapped the voltage dynamics across a beating zebrafish heart using Mermaid, a VSFP. This, along with concurrent work in the mouse somatosensory cortex (Akemann et al., 2010), demonstrated the utility of genetically encoded reporters of voltage *in vivo* and also showed the significant benefits afforded by optogenetic methods when an experiment involves observing an entire organ at once.

Future studies will likely use voltage-sensing optogenetic tools, as these more rapidly and directly report on neuronal activity. In addition, optogenetic tools exist for visualizing the release of neurotransmitters at synapses (Miesenbock et al., 1998; Sankaranarayanan et al., 2000; Yuste et al., 2000), though no published studies have used these reporters in zebrafish. There still exists considerable promise to use these, and other optogenetic tools, to image activity in genetically defined neurons in live and behaving zebrafish.

Controlling neurons to describe circuits

Overview of optogenetic tools for manipulation of zebrafish neurons

Indicators of Ca^{2+} concentration, voltage, or vesicle release can reveal patterns of activity in targeted neurons as behaviors unfold. Determining the necessity or sufficiency of those circuits for a behavior, however, requires the ability to block or drive activity in the neurons of interest. A number of tools have been used in zebrafish to meet this need. Selective manipulations have been achieved using *Natronomonas pharaonis* halorhodopsin (NpHR), channelrhodopsin-2 (ChR2), light-sensitive ionotropic glutamate receptor (LiGluR), the modified iGluR HyLighter, and KR. NpHR is a membrane pump that transports chloride ions into cells when illuminated with yellow light (Schobert and Lanyi, 1982), thus hyperpolarizing and silencing them (Gradinaru et al., 2008). ChR2 is a channel that allows sodium influx when illuminated with blue light (Li et al., 2005; Nagel et al., 2003). This depolarizes the cell, triggering action potentials. LiGluR is a glutamate receptor that has its ligand attached to a photoswitchable tether called MAG (Volgraf et al., 2006). Ultraviolet light shifts the tether to its *cis* form, bringing the ligand into the channel's binding pocket. With the channel open, ions, primarily Na^+, flow in and depolarize the cell, facilitating action potentials. Green light reverses the tether to its *trans* state, thus removing the ligand and closing the channel. HyLighter is the functional counterpart of LiGluR in that it transports potassium ions out of neurons in response to ultraviolet light, effectively silencing them (Janovjak et al., 2010). Green light closes the channel, reversing the effect. KR is a red fluorescent protein that produces reactive oxygen species as it goes through its excitation/emission cycles. Irradiation of KR with green light can flood the targeted cells with radical oxygen species, leading eventually to apoptosis (Bulina et al., 2006). Stated simply, NpHR and HyLighter have the capacity to silence neurons with yellow or ultraviolet light, respectively; ChR2 and LiGluR can drive activity in neurons when pulsed with blue or ultraviolet light, respectively; and KR can induce apoptosis in cells irradiated with green light.

Each of the aforementioned tools has advantages and disadvantages. While NpHR has been shown to silence neurons effectively in the zebrafish, extended periods of inhibition cause "rebound" firing when the inhibition is lifted, presumably because of homeostatic adjustments that the neurons have made during the period of silencing (Arrenberg et al., 2009). The use of HyLighter does not seem to produce this effect (Janovjak et al., 2010). The degree of depolarization is greater in neurons expressing LiGluR than in neurons expressing ChR2, and it is also faster to close LiGluR with light than it is to wait for ChR2 to close following photostimulation (Szobota et al., 2007). This means that action potentials can be induced more frequently with LiGluR. However, the operation of HyLighter or LiGluR requires the use of two light wavelengths. This is problematic, especially when other optogenetic tools with overlapping wavelengths are expressed in the same cells. Further, LiGluR and HyLighter cannot be used in conjunction because they will be activated and deactivated by the same wavelengths, which is not the case with ChR2 and NpHR. An additional obstacle to using LiGluR or HyLighter is the need for MAG, which is difficult and expensive to synthesize, and has limited tissue penetration. KR provides the advantage of killing neurons outright, thus providing an unambiguous removal of their function. However, KR is slow, taking roughly 10min to destroy cultured neurons (Bulina et al., 2006), is irreversible, and runs the risk of contaminating the ablated area with cellular debris.

Each of these tools has utility for circuit analysis, and most have already proven useful in zebrafish. As detailed above, each has its limitations as well. Several research groups are dedicated to the improvement of these proteins, with the goals of improving expression levels, light sensitivity, and temporal dynamics, and improved forms of the tools reported above have already been developed (Berndt et al., 2011; Chow et al., 2010; Gradinaru et al., 2010; Gunaydin et al., 2010). As these and further optimized forms find their way into zebrafish, many of the current experimental limitations of these approaches should be overcome.

Optogenetic discoveries in zebrafish circuits

One of the simplest approaches for defining the behavioral function of a circuit or neuron is to trigger it and look for a resulting response from the animal. This, in principle, reveals the sufficiency of that circuit for the behavior that is observed. As we will outline in this section, this approach can be applied broadly, or to circuits that are defined genetically (using a promoter for a particular cell type or types), or spatially (by illuminating a targeted region of the nervous system). Less targeted expression allows for the broad sampling of circuits for roles in a particular behavior of interest, while highly targeted expression permits specific circuits to be scrutinized.

In an example of broad expression, Zhu et al. (2009) used viral delivery and the Tet system to drive expression of ChR2 broadly in the nervous system of zebrafish larvae. The application of blue light to these animals resulted in bouts of backward swimming. The authors suggest that the backward swimming is not a reversal of the motor program of forward swimming because it was slower and did not include large-amplitude tail movements. Since larvae do not normally swim backward, it is possible that the behavior resulted from the activation of an unnatural combination of motor centers. It is also possible that a particular command center is responsible for the behavior, but that the breadth of the manipulation makes it difficult to identify. One approach for finding such a center would be systematically to illuminate different parts of the central nervous system (CNS), searching for a location that consistently elicits the behavior.

This is essentially what Arrenberg et al. (2009) recently did to identify a swim command center in the larval hindbrain. In this case, they used NpHR to find areas necessary for forward swimming. NpHR was expressed in most CNS neurons through the use of the *Gal4*^{s1101t} and *UAS:NpHR-mCherry* transgenic lines. Unsurprisingly, during whole-body illumination, the larvae stopped moving. However, when the illumination stopped, the larvae performed forward swims. The authors showed that this "rebound" swimming was linked to a burst of activity in NpHR-expressing neurons immediately following the end of illumination. They then used rebound swimming to investigate the region involved in initiating forward swims, also known as the swim command center. Systematically, regions of the brain were scanned with an optical fiber. When the hindbrain was found to be most effective in eliciting rebound swimming, a thinner fiber optic cable was used to determine the area within the hindbrain responsible for forward swimming. In each case, Kaede was used to photoconvert the affected neurons. Using this technique, the authors zeroed in on a region rostral to the *commissura infima Halleri*. They further used the temporal dynamics of rebound swimming to gauge the neural dynamics of the swim command center. Rebound swimming, they found, happened roughly 300ms after the light was switched off. If the light was quickly turned back on, with a lack of illumination lasting less than about 200ms, no swimming occurred, indicating that the command center had not initiated a swim signal. For periods between 200 and 300ms, the swim became stronger, and for periods of rebound >300ms, the swim response plateaued. Since the temporal dynamics of NpHR offset and spinal circuits are both very rapid, the authors interpreted their results to mean that the command center is responsible for most of the delay. They theorize that it needs time to build a critical mass of activity that is capable of driving the downstream motor circuitry. These experiments highlight the utility of optogenetic tools for both spatial and temporal control of circuits and provide examples of how this control can yield biological insights.

Schoonheim et al. (2010) used a similar approach to locate and control the circuit that initiates saccades in zebrafish. Saccades are fast eye movements that mediate gaze changes, and occur spontaneously, or can be induced as part of the optokinetic response (OKR; Easter and Nicola, 1997). In the OKR, the eyes sweep smoothly to follow a horizontally moving visual stimulus, occasionally saccading back to a starting point in advance of another sweep. These are similar to the eye movements of a passenger looking out of a train window at the passing scenery. In mammals, saccades are initiated by a circuit called the burst generator (Strassman et al., 1986a,b). To locate the region of the zebrafish brain that initiates saccades, the experimenters expressed NpHR and ChR2 in nearly every neuron of the CNS using the *Gal4^{s1101t}* line (Scott et al., 2007). Using yellow light from an optic fiber to activate NpHR, they silenced regions of neurons and found that saccades stopped when an area in rhombomere 5 was photostimulated. They concluded that this region was necessary for generating saccades. When they silenced the left side of the area, they found that saccades to the left in both eyes were suppressed, and vice versa with the right side. The neurons that composed the identified region in rhombomere 5 were then excited by illumination with blue light. Saccades, albeit slower than normal, were induced both during free viewing and OKR experiments. Further, the saccades were synchronized to the pulses of light. Using the photoconvertible protein, Kaede, they labeled the region red with ultraviolet light, thus marking the cells involved in producing the saccades. This study has paved the way for a more detailed optogenetic analysis of the saccade-producing region in zebrafish. Future optogenetic studies could identify and characterize the cells composing this region of rhombomere 5, identifying the microcircuits that are necessary for saccades to occur.

Another example of spatially restricted optogenetic manipulations has helped to identify the electrical requirements for generating a heartbeat (Arrenberg et al., 2010). In this study, the researchers expressed NpHR and ChR2 in all cardiomyocytes of 3 dpf larvae, again using *Gal4^{s1101t}*. When they illuminated the entire heart with orange light to activate NpHR, contraction ceased. In the absence of the light, normal cardiac function resumed. To determine the region of the pacemaker, they varied the site of photosilencing in developing larvae. Throughout development, the pacemaker region progressively became smaller. The locus at 5 dpf was an area in the dorsal right quadrant of the sinoatrial ring. They found the pacemaker generally consisted of 10–30 cells, but in some larvae, the silencing of as few as three cardiomyocytes caused cardiac arrest. Further, the experimenters were able to control the heart by photoactivating the sinoatrial ring with ChR2 and induce contractions at 2.7–4.7 Hz. In essence, it was found that there are only a small number of pacemaker cells in the zebrafish heart and disease-like states can be induced and reversed using light-mediated tools. This experiment highlights how the dual restriction (genetic and spatial) allowed by optogenetic tools permits the controlled manipulation of a very small number of targeted cells.

The utility of tight genetic control of expression was recently reinforced by Douglass et al. (2008). In this study, they used the *islet-1* promoter (Higashijima et al., 2000) to drive ChR2 transiently in a subset of Rohon–Beard and trigeminal neurons, both of which perceive touch stimuli. When these neurons were photostimulated in 24-h postfertilization embryos, escape responses were elicited. It took the larvae 30 ms after light stimulation to respond, which is similar to the latency for a mechanical touch response (Bhatt et al., 2007; Ritter et al., 2001). The direction of the tail in the responses was usually consistent for each fish, which the authors believe was due to imbalances in the distribution of ChR2-positive neurons along the lateral plane. By adding spatial restriction to this tight genetic restriction, they could photoactivate small numbers of, or even individual, sensory neurons. They found that a single photoactivated trigeminal neuron can elicit escape behavior. This suggests that somatosensation in larval zebrafish places a premium on sensitivity, since a single-unit signal can elicit a full response. This characteristic is consistent with the low spontaneous firing frequencies that they found in trigeminal neurons using electrophysiology.

Szobota et al. (2007) also examined the touch response after photostimulating neurons. Rather than eliciting the touch response, they were able to suppress it when they used LiGluR to photostimulate select neurons. LiGluR was expressed in trigeminal ganglion, vagal ganglion, hindbrain, and Rohon–Beard neurons by combining *UAS: LiGluR* with *et101.2:Gal4-VP16* (Köster and Fraser, 2001; Scott and Baier, 2009). After illumination with ultraviolet light for 15min, most of the light-sensitive fish did not undergo a tactile response when prodded, but still had the ability to swim spontaneously. When the ion channel was closed, the fish were able to perform the response normally. This result seems counterintuitive, but the authors suggest that illumination of sensory neurons on both sides of the larvae may have made them unable to detect the direction of the stimulus and therefore direction to turn away from. Alternatively, the larvae may not have been able to detect the mechanical stimulus over the increased activity of the somatosensory system, or habituation may have occurred in the circuits due to the extended period of stimulation.

In a conceptually similar but more nuanced set of experiments, Wyart et al. (2009) expressed LiGluR in different subsets of spinal neurons as a means of analyzing their roles in forward swimming and startle response. They focused on one particular type of spinal neuron, the Kolmer–Agduhr (KA) cell, which had long been described anatomically but had no known function. This was achieved by driving *UAS:LiGluR* with the *Gal4*^{s1020t} line (Scott et al., 2007), which is expressed in KA cells and motor neurons. When activated with ultraviolet light, most of the larvae performed motions that were kinematically similar to routine slow forward swims. To determine whether motor neurons were responsible, the same experiment was repeated in a *Gal4*^{s1003t} line (expressed only in KA cells) and again in *Gal*^{s1041t} and *Hb9:Gal4* (motor neuron specific). The *Gal4*^{s1003t} larvae showed the same response as *Gal4*^{s1020t}, whereas *Gal*^{s1041t} and *Hb9:Gal4* larvae displayed no response. This shows that motor neurons were not responsible for the induced slow swim. When the hindbrain was ablated, they found that slow swims still occurred, consistent with their theory that the KA cells directly influence the central pattern generator instead of sending signals to the hindbrain. Using *UAS:tetanus toxin light chain*, they genetically silenced KA cells and saw a decrease in spontaneous free swimming. When tetanus toxin was driven by *Gal4*^{s1020t}, the larvae were paralyzed (probably because the motor neurons were silenced), and in *Gal4*^{s1003t}, they were not paralyzed but the number of times they swam spontaneously was significantly reduced. This is evidence that KA cells are necessary for normal levels of spontaneous locomotion. This approach (termed "intersectional optogenetics" by the authors) of combining opposing manipulation tools (LiGluR and tetanus toxin) with distinct but overlapping expression patterns (different Gal4 lines, in this case) is appealing for its ability to zero in on circuits that are both necessary and sufficient for particular behaviors (summarized in Fig. 2).

Perhaps, the most biologically significant optogenetic zebrafish study to date is the work done by Del Bene et al. (2010) in the optic tectum, a midbrain visual structure involved in high acuity vision (Nevin et al., 2010). By using GCaMP and KR, the authors were able to clarify the mechanism by which the zebrafish optic tectum filters out large visual stimuli. In the first part of the study, GCaMP1.6 was expressed in retinal axons that deliver stimuli to the tectum. Immobilized larvae were exposed to a variety of visual stimuli, provided by an adjacent LCD screen. It was found that the activity in the retinal axons did not vary significantly in response to stimuli ranging from 2° to 32° in width. GCaMP was then expressed in periventricular neurons in the superficial (retinorecipient) layers and the deeper (higher order processing) layers of the tectum. In response to large stimuli, the superficial layers had significantly more activity than the deeper layers. However, the small stimuli evoked the same amount of activity in superficial and

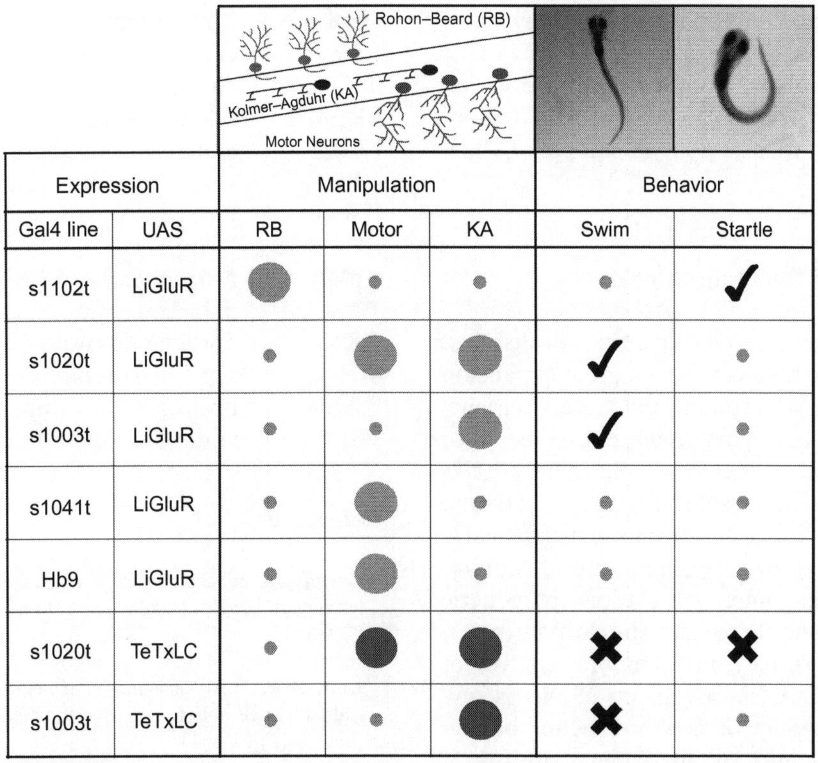

Rohon–Beard (RB)
Kolmer–Agduhr (KA)
Motor Neurons

Expression		Manipulation			Behavior	
Gal4 line	UAS	RB	Motor	KA	Swim	Startle
s1102t	LiGluR	●	•	•	•	✓
s1020t	LiGluR	•	●	●	✓	•
s1003t	LiGluR	•	•	●	✓	•
s1041t	LiGluR	•	●	•	•	•
Hb9	LiGluR	•	●	•	•	•
s1020t	TeTxLC	•	●	●	✖	✖
s1003t	TeTxLC	•	•	●	✖	•

Fig. 2. Intersectional optogenetics, applied to the zebrafish spinal cord. This figure summarizes the approach used by Wyart et al. (2009) to assign a function to KA neurons in the zebrafish spinal cord. The "Expression" section shows the transgenes used to express optogenetic proteins, with Gal4 dictating the expression pattern and UAS determining the tool to be expressed. $Gal4^{s1103t}$ drives expression in Rohon–Beard neurons. $Gal4^{s1020t}$ expresses Gal4 in KA neurons and motor neurons, while $Gal4^{s1003t}$ has expression only in KA neurons. $Gal4^{s1041t}$ and *Hb9* drive expression only in motor neurons. These targeted neurons are activated or silenced by *UAS:LiGluR* or *UAS:Tetanus Toxin*, respectively. For each experiment, the "Manipulation" section is paired with the effect on each of two behaviors: forward swimming and startle response. By looking at different combinations of expression patterns and the behavioral effects of the manipulations, the authors were able to establish the KA neurons' necessity and sufficiency for driving forward swimming behavior. (For color version of this figure, the reader is referred to the Web version of this chapter.)

deep layers. This indicates that large stimuli are being filtered out, and the authors used the $GABA_A$ inhibitor bicuculline to prevent this filtering, demonstrating that inhibitory GABA activity was responsible for the effect. GCaMP was then expressed in a population of superficial interneurons in the top layer of the tectum using *UAS:GCampl.6* and $Gal4^{s1156t}$ (Scott and Baier, 2009). These neurons were found to be GABAergic, have no activity in response to small visual stimuli, and increased Ca^{2+} concentration in response to large stimuli. These superficial inhibitory neurons (SINs), therefore, fit the profile of cells that could be mediating size selectivity. They then tested the necessity of SINs for visual filtering by two means. First, they destroyed the SINs using KR and used GCaMP imaging to confirm that all layers of the tectum responded to large visual stimuli. Second, they expressed tetanus toxin in the SINs, and this led to a reduced ability for the affected larvae to capture paramecia, consistent with a failure to resolve small visual targets. This study shows how combining varied genetic targeting, optogenetic triggering and ablation, optogenetic observation of activity, and behavior can convincingly demonstrate the roles of single cells in complex behavioral circuits.

Conclusions and future directions

Since the first optogenetic studies in zebrafish just a few years ago, there has been a great expansion in the number of protein tools used, circuits targeted, and behaviors analyzed. We have described several of the more important studies to use optogenetics in zebrafish, along with the biological relevance of the results. This is clearly just the beginning of an exciting line of inquiry. With the ongoing improvement of optogenetic tools, studies along these lines should improve in terms of the sensitivity and temporal control of their manipulations. Improvements in microscopy and the development of new and better behavioral assays will also aid in linking patterns of neural activity to the behaviors that they drive. Presently, the weakest facet of this system is the targeting of optogenetic tools to the circuits and cell types of interest. In a majority of the studies done so far, whole regions of the brain have been observed or manipulated, and few studies have been able to look, in a convincing fashion, at individual cell types in isolation. This speaks to the complexity of regulatory elements in vertebrate genomes, and the fact that few genes are expressed exclusively in single cell types. In the coming years, the ability to drive gene expression selectively in neurons of interest will be one of the keys to fulfilling the potential of the zebrafish model system for optogenetics.

Abbreviations

BAC	bacterial artificial chromosome
BFP	blue fluorescent protein
CaM	calmodulin
ChR2	channelrhodopsin-2
dpf	days postfertilization
FRET	Förster resonance energy transfer
GABA	γ-aminobutyric acid
GFP	green fluorescent protein
KA	Kolmer–Agduhr
KR	KillerRed
NpHR	*Natronomonas pharaonis* halorhodopsin
OKR	optokinetic response
SIN	superficial inhibitory neuron
UAS	upstream activating sequence
VSFP	voltage-sensitive fluorescent protein

References

Aizenberg, M., & Schuman, E. M. (2011). Cerebellar-dependent learning in larval zebrafish. *The Journal of Neuroscience, 31*, 8708–8712.

Akemann, W., Mutoh, H., Perron, A., Rossier, J., & Knöpfel, T. (2010). Imaging brain electric signals with genetically targeted voltage-sensitive fluorescent proteins. *Nature Methods, 7*, 643–649.

Amsterdam, A., Burgess, S., Golling, G., Chen, W., Sun, Z., Townsend, K., et al. (1999). A large-scale insertional mutagenesis screen in zebrafish. *Genes & Development, 13*, 2713–2724.

Ando, R., Flors, C., Mizuno, H., Hofkens, J., & Miyawaki, A. (2007). Highlighted generation of fluorescence signals using simultaneous two-color irradiation on Dronpa mutants. *Biophysical Journal, 92*, L97–L99.

Ando, R., Hama, H., Yamamoto-Hino, M., Mizuno, H., & Miyawaki, A. (2002). An optical marker based on the UV-induced green-to-red photoconversion of a fluorescent protein. *Proceedings of the National Academy of Sciences of the United States of America, 99*, 12651–12656.

Aramaki, S., & Hatta, K. (2006). Visualizing neurons one-by-one *in vivo*: Optical dissection and reconstruction of neural networks with reversible fluorescent proteins. *Developmental Dynamics, 235*, 2192–2199.

Arrenberg, A. B., Del Bene, F., & Baier, H. (2009). Optical control of zebrafish behavior with halorhodopsin. *Proceedings of the National Academy of Sciences of the United States of America, 106*, 17968–17973.

Arrenberg, A. B., Stainier, D. Y., Baier, H., & Huisken, J. (2010). Optogenetic control of cardiac function. *Science, 330*, 971–974.

Asakawa, K., & Kawakami, K. (2009). The Tol2-mediated Gal4-UAS method for gene and enhancer trapping in zebrafish. *Methods, 49*, 275–281.

Asakawa, K., Suster, M. L., Mizusawa, K., Nagayoshi, S., Kotani, T., Urasaki, A., et al. (2008). Genetic dissection of neural circuits by Tol2 transposon-mediated Gal4 gene and enhancer trapping in zebrafish. *Proceedings of the National Academy of Sciences of the United States of America, 105*, 1255–1260.

Baier, H., & Scott, E. K. (2009). Genetic and optical targeting of neural circuits and behavior—Zebrafish in the spotlight. *Current Opinion in Neurobiology, 19*, 553–560.

Berndt, A., Schoenenberger, P., Mattis, J., Tye, K. M., Deisseroth, K., Hegemann, P., et al. (2011). High-efficiency channelrhodopsins for fast neuronal stimulation at low light levels. *Proceedings of the National Academy of Sciences of the United States of America, 108*, 7595–7600.

Bhatt, D. H., Mclean, D. L., Hale, M. E., & Fetcho, J. R. (2007). Grading movement strength by changes in firing intensity versus recruitment of spinal interneurons. *Neuron, 53*, 91–102.

Bulina, M. E., Chudakov, D. M., Britanova, O. V., Yanushevich, Y. G., Staroverov, D. B., Chepurnykh, T. V., et al. (2006). A genetically encoded photosensitizer. *Nature Biotechnology, 24*, 95–99.

Burne, T., Scott, E., Van Swinderen, B., Hilliard, M., Reinhard, J., Claudianos, C., et al. (2011). Big ideas for small brains: What can psychiatry learn from worms, flies, bees and fish? *Molecular Psychiatry, 16*, 7–16.

Chi, N. C., Shaw, R. M., Jungblut, B., Huisken, J., Ferrer, T., Arnaout, R., et al. (2008). Genetic and physiologic dissection of the vertebrate cardiac conduction system. *PLoS Biology, 6*, 1006–1019.

Chow, B. Y., Han, X., Dobry, A. S., Qian, X., Chuong, A. S., Li, M., et al. (2010). High-performance genetically targetable optical neural silencing by light-driven proton pumps. *Nature, 463*, 98–102.

Davidson, A. E., Balciunas, D., Mohn, D., Shaffer, J., Hermanson, S., Sivasubbu, S., et al. (2003). Efficient gene delivery and gene expression in zebrafish using the Sleeping Beauty transposon. *Developmental Biology, 263*, 191–202.

Davison, J. M., Akitake, C. M., Goll, M. G., Rhee, J. M., Gosse, N., Baier, H., et al. (2007). Transactivation from Gal4-VP16 transgenic insertions for tissue-specific cell labeling and ablation in zebrafish. *Developmental Biology, 304*, 811–824.

Del Bene, F., Wyart, C., Robles, E., Tran, A., Looger, L., Scott, E. K., et al. (2010). Filtering of visual information in the tectum by an identified neural circuit. *Science, 330*, 669–673.

Dimitrov, D., He, Y., Mutoh, H., Baker, B. J., Cohen, L., Akemann, W., et al. (2007). Engineering and characterization of an enhanced fluorescent protein voltage sensor. *PLoS One, 2*, e440.

Distel, M., Wullimann, M. F., & Köster, R. W. (2009). Optimized Gal4 genetics for permanent gene expression mapping in zebrafish. *Proceedings of the National Academy of Sciences of the United States of America, 106*, 13365–13370.

Douglass, A. D., Kraves, S., Deisseroth, K., Schier, A. F., & Engert, F. (2008). Escape behavior elicited by single, channelrhodopsin-2-evoked spikes in zebrafish somatosensory neurons. *Current Biology, 18*, 1133–1137.

Dreosti, E., Odermatt, B., Dorostkar, M. M., & Lagnado, L. (2009). A genetically encoded reporter of synaptic activity *in vivo*. *Nature Methods, 6*, 883–889.

Easter, S. S., Jr., & Nicola, G. N. (1997). The development of eye movements in the zebrafish (Danio rerio). *Developmental Psychobiology, 31*, 267–276.

Ellingsen, S., Laplante, M. A., Konig, M., Kikuta, H., Furmanek, T., Hoivik, E. A., et al. (2005). Large-scale enhancer detection in the zebrafish genome. *Development, 132*, 3799–3811.

Engeszer, R. E., Patterson, L. B., Rao, A. A., & Parichy, D. M. (2007). Zebrafish in the wild: A review of natural history and new notes from the field. *Zebrafish, 4*, 21–40.

Fan, X., Majumder, A., Reagin, S. S., Porter, E. L., Sornborger, A. T., Keith, C. H., et al. (2007). New statistical methods enhance imaging of cameleon fluorescence resonance energy transfer in cultured zebrafish spinal neurons. *Journal of Biomedical Optics, 12*, 034017.

Fetcho, J. R. (2007). The utility of zebrafish for studies of the comparative biology of motor systems. *Journal of Experimental Zoology Part B: Molecular and Developmental Evolution, 308*, 550–562.

Gaiano, N., Amsterdam, A., Kawakami, K., Allende, M., Becker, T., & Hopkins, N. (1996). Insertional mutagenesis and rapid cloning of essential genes in zebrafish. *Nature, 383*, 829–832.

Gautam, S. G., Perron, A., Mutoh, H., & Knopfel, T. (2009). Exploration of fluorescent protein voltage probes based on circularly permuted fluorescent proteins. *Frontiers in Neuroengineering, 2*, 14.

Gradinaru, V., Thompson, K. R., & Deisseroth, K. (2008). eNpHR: A Natronomonas halorhodopsin enhanced for optogenetic applications. *Brain Cell Biology, 36*, 129–139.

Gradinaru, V., Zhang, F., Ramakrishnan, C., Mattis, J., Prakash, R., Diester, I., et al. (2010). Molecular and cellular approaches for diversifying and extending optogenetics. *Cell, 141*, 154–165.

Gunaydin, L. A., Yizhar, O., Berndt, A., Sohal, V. S., Deisseroth, K., & Hegemann, P. (2010). Ultrafast optogenetic control. *Nature Neuroscience, 13*, 387–392.

Habuchi, S., Ando, R., Dedecker, P., Verheijen, W., Mizuno, H., Miyawaki, A., et al. (2005). Reversible single-molecule photoswitching in the GFP-like fluorescent protein Dronpa. *Proceedings of the National Academy of Sciences of the United States of America, 102*, 9511–9516.

Halpern, M. E., Rhee, J., Goll, M. G., Akitake, C. M., Parsons, M., & Leach, S. D. (2008). Gal4/UAS transgenic tools and their application to zebrafish. *Zebrafish, 5*, 97–110.

Hans, S., Kaslin, J., Freudenreich, D., & Brand, M. (2009). Temporally-controlled site-specific recombination in zebrafish. *PLoS One*, *4*, e4640.

Higashijima, S., Hotta, Y., & Okamoto, H. (2000). Visualization of cranial motor neurons in live transgenic zebrafish expressing green fluorescent protein under the control of the islet-1 promoter/enhancer. *The Journal of Neuroscience*, *20*, 206–218.

Higashijima, S. I., Masino, M. A., Mandel, G., & Fetcho, J. R. (2003). Imaging neuronal activity during zebrafish behavior with a genetically encoded calcium indicator. *Journal of Neurophysiology*, *90*, 3986–3997.

Higashijima, S. I., Okamoto, H., Ueno, N., Hotta, Y., & Eguchi, G. (1997). High-frequency generation of transgenic zebrafish which reliably express GFP in whole muscles or the whole body by using promoters of zebrafish origin. *Developmental Biology*, *192*, 289–299.

Hill, A., Howard, C. V., Strahle, U., & Cossins, A. (2003). Neurodevelopmental defects in zebrafish (Danio rerio) at environmentally relevant dioxin (TCDD) concentrations. *Toxicological Sciences*, *76*, 392–399.

Janovjak, H., Szobota, S., Wyart, C., Trauner, D., & Isacoff, E. Y. (2010). A light-gated, potassium-selective glutamate receptor for the optical inhibition of neuronal firing. *Nature Neuroscience*, *13*, 1027–1032.

Jessen, J. R., Willett, C. E., & Lin, S. (1999). Artificial chromosome transgenesis reveals long-distance negative regulation of rag1 in zebrafish. *Nature Genetics*, *23*, 15–16.

Kawakami, K., Shima, A., & Kawakami, N. (2000). Identification of a functional transposase of the Tol2 element, an Ac-like element from the Japanese medaka fish, and its transposition in the zebrafish germ lineage. *Proceedings of the National Academy of Sciences of the United States of America*, *97*, 11403–11408.

Kimura, Y., Okamura, Y., & Higashijima, S. (2006). alx, a zebrafish homolog of Chx10, marks ipsilateral descending excitatory interneurons that participate in the regulation of spinal locomotor circuits. *The Journal of Neuroscience*, *26*, 5684–5697.

Knöpfel, T., Diez-Garcia, J., & Akemann, W. (2006). Optical probing of neuronal circuit dynamics: Genetically encoded versus classical fluorescent sensors. *Trends in Neurosciences*, *29*, 160–166.

Köster, R. W., & Fraser, S. E. (2001). Tracing transgene expression in living zebrafish embryos. *Developmental Biology*, *233*, 329–346.

Kwan, K. M., Fujimoto, E., Grabher, C., Mangum, B. D., Hardy, M. E., Campbell, D. S., et al. (2007). The Tol2kit: A multisite gateway-based construction kit for Tol2 transposon transgenesis constructs. *Developmental Dynamics*, *236*, 3088–3099.

Langenau, D. M., Feng, H., Berghmans, S., Kanki, J. P., Kutok, J. L., & Look, A. T. (2005). Cre/lox-regulated transgenic zebrafish model with conditional myc-induced T cell acute lymphoblastic leukemia. *Proceedings of the National Academy of Sciences of the United States of America*, *102*, 6068–6073.

Li, X., Gutierrez, D. V., Hanson, M. G., Han, J., Mark, M. D., Chiel, H., et al. (2005). Fast noninvasive activation and inhibition of neural and network activity by vertebrate rhodopsin and green algae channelrhodopsin. *Proceedings of the National Academy of Sciences of the United States of America*, *102*, 17816–17821.

Lin, S., Gaiano, N., Culp, P., Burns, J. C., Friedmann, T., Yee, J. K., et al. (1994). Integration and germ-line transmission of a pseudotyped retroviral vector in zebrafish. *Science*, *265*, 666–669.

Lister, J. A., Robertson, C. P., Lepage, T., Johnson, S. L., & Raible, D. W. (1999). nacre encodes a zebrafish microphthalmia-related protein that regulates neural-crest-derived pigment cell fate. *Development*, *126*, 3757–3767.

Long, Q., Meng, A., Wang, M., Jessen, J. R., Farrell, M. J., & Lin, S. (1997). GATA-1 expression pattern can be recapitulated in living transgenic zebrafish using GFP reporter gene. *Development*, *124*, 4105–4111.

Lundby, A., Akemann, W., & Knöpfel, T. (2010). Biophysical characterization of the fluorescent protein voltage probe VSFP2.3 based on the voltage-sensing domain of Ci-VSP. *European Biophysics Journal*, *39*, 1625–1635.

Lundby, A., Mutoh, H., Dimitrov, D., Akemann, W., & Knöpfel, T. (2008). Engineering of a genetically encodable fluorescent voltage sensor exploiting fast Ci-VSP voltage-sensing movements. *PLoS One*, *3*, e2514.

Mank, M., Santos, A. F., Direnberger, S., Mrsic-Flogel, T. D., Hofer, S. B., Stein, V., et al. (2008). A genetically encoded calcium indicator for chronic *in vivo* two-photon imaging. *Nature Methods*, *5*, 805–811.

Marriott, G., Mao, S., Sakata, T., Ran, J., Jackson, D. K., Petchprayoon, C., et al. (2008). Optical lock-in detection imaging microscopy for contrast-enhanced imaging in living cells. *Proceedings of the National Academy of Sciences of the United States of America*, *105*, 17789–17794.

Mclean, D. L., & Fetcho, J. R. (2009). Spinal interneurons differentiate sequentially from those driving the fastest swimming movements in larval zebrafish to those driving the slowest ones. *The Journal of Neuroscience*, *29*, 13566–13577.

Mclean, D. L., & Fetcho, J. R. (2011). Movement, technology and discovery in the zebrafish. *Current Opinion in Neurobiology*, *21*, 110–115.

Miesenbock, G., De Angelis, D. A., & Rothman, J. E. (1998). Visualizing secretion and synaptic transmission with pH-sensitive green fluorescent proteins. *Nature*, *394*, 192–195.

Miskey, C., Izsvaãåk, Z., Kawakami, K., & Ivics, Z. (2005). DNA transposons in vertebrate functional genomics. *Cellular and Molecular Life Sciences*, *62*, 629–641.

Miyawaki, A. (2005). Innovations in the imaging of brain functions using fluorescent proteins. *Neuron*, *48*, 189–199.

Miyawaki, A., Griesbeck, O., Heim, R., & Tsien, R. Y. (1999). Dynamic and quantitative Ca^{2+} measurements using improved cameleons. *Proceedings of the National Academy of Sciences of the United States of America*, *96*, 2135–2140.

Miyawaki, A., Llopis, J., Heim, R., Mccaffery, J. M., Adams, J. A., Ikura, M., et al. (1997). Fluorescent indicators for Ca^{2+} based on green fluorescent proteins and calmodulin. *Nature*, *388*, 882–887.

Mullins, M. C., Hammerschmidt, M., Haffter, P., & Nusslein-Volhard, C. (1994). Large-scale mutagenesis in the zebrafish: In search of genes controlling development in a vertebrate. *Current Biology*, *4*, 189–202.

Muto, A., Ohkura, M., Kotani, T., Higashijima, S., Nakai, J., & Kawakami, K. (2011). Genetic visualization with an improved GCaMP calcium indicator reveals spatiotemporal activation of the spinal motor neurons in zebrafish. *Proceedings of the National Academy of Sciences of the United States of America*, *108*, 5425–5430.

Nagai, T., Yamada, S., Tominaga, T., Ichikawa, M., & Miyawaki, A. (2004). Expanded dynamic range of fluorescent indicators for Ca(2+) by circularly permuted yellow fluorescent proteins. *Proceedings of the National Academy of Sciences of the United States of America*, *101*, 10554–10559.

Nagel, G., Szellas, T., Huhn, W., Kateriya, S., Adeishvili, N., Berthold, P., et al. (2003). Channelrhodopsin-2, a directly light-gated cation-selective membrane channel. *Proceedings of the National Academy of Sciences of the United States of America*, *100*, 13940–13945.

Nakai, J., Ohkura, M., & Imoto, K. (2001). A high signal-to-noise Ca^{2+} probe composed of a single green fluorescent protein. *Nature Biotechnology*, *19*, 137–141.

Nasevicius, A., & Ekker, S. C. (2000). Effective targeted gene 'knockdown' in zebrafish. *Nature Genetics*, *26*, 216–220.

Naumann, E. A., Kampff, A. R., Prober, D. A., Schier, A. F., & Engert, F. (2010). Monitoring neural activity with bioluminescence during natural behavior. *Nature Neuroscience*, *13*, 513–522.

Navratilova, P., Fredman, D., Hawkins, T. A., Turner, K., Lenhard, B., & Becker, T. S. (2009). Systematic human/zebrafish comparative identification of *cis*-regulatory activity around vertebrate developmental transcription factor genes. *Developmental Biology*, *327*, 526–540.

Nevin, L. M., Robles, E., Baier, H., & Scott, E. K. (2010). Focusing on optic tectum circuitry through the lens of genetics. *BMC Biology*, *8*, 126.

Niell, C. M., & Smith, S. J. (2005). Functional imaging reveals rapid development of visual response properties in the zebrafish tectum. *Neuron*, *45*, 941–951.

Norton, W., & Bally-Cuif, L. (2010). Adult zebrafish as a model organism for behavioural genetics. *BMC Neuroscience*, *11*, 90.

Palmer, A. E., Giacomello, M., Kortemme, T., Hires, S. A., Lev-Ram, V., Baker, D., et al. (2006). Ca^{2+} indicators based on computationally redesigned calmodulin–peptide pairs. *Chemistry & Biology*, *13*, 521–530.

Palmer, A. E., Jin, C., Reed, J. C., & Tsien, R. Y. (2004). Bcl-2-mediated alterations in endoplasmic reticulum Ca^{2+} analyzed with an improved genetically encoded fluorescent sensor. *Proceedings of the National Academy of Sciences of the United States of America*, *101*, 17404–17409.

Parichy, D. M., Elizondo, M. R., Mills, M. G., Gordon, T. N., & Engeszer, R. E. (2009). Normal table of postembryonic zebrafish development: Staging by externally visible anatomy of the living fish. *Developmental Dynamics*, *238*, 2975–3015.

Perron, A., Mutoh, H., Akemann, W., Gautam, S. G., Dimitrov, D., Iwamoto, Y., et al. (2009a). Second and third generation voltage-sensitive fluorescent proteins for monitoring membrane potential. *Frontiers in Molecular Neuroscience*, *2*, 5.

Perron, A., Mutoh, H., Launey, T., & Knopfel, T. (2009b). Red-shifted voltage-sensitive fluorescent proteins. *Chemical Biology*, *16*, 1268–1277.

Portugues, R., & Engert, F. (2009). The neural basis of visual behaviors in the larval zebrafish. *Current Opinion in Neurobiology*, *19*, 644–647.

Ritter, D. A., Bhatt, D. H., & Fetcho, J. R. (2001). *In vivo* imaging of zebrafish reveals differences in the spinal networks for escape and swimming movements. *The Journal of Neuroscience*, *21*, 8956–8965.

Sankaranarayanan, S., De Angelis, D., Rothman, J. E., & Ryan, T. A. (2000). The use of pHluorins for optical measurements of presynaptic activity. *Biophysical Journal*, *79*, 2199–2208.

Sato, T., Takahoko, M., & Okamoto, H. (2006). HuC:Kaede, a useful tool to label neural morphologies in networks *in vivo*. *Genesis*, *44*, 136–142.

Scheer, N., & Campos-Ortega, J. A. (1999). Use of the Gal4-UAS technique for targeted gene expression in the zebrafish. *Mechanisms of Development*, *80*, 153–158.

Schobert, B., & Lanyi, J. K. (1982). Halorhodopsin is a light-driven chloride pump. *The Journal of Biological Chemistry*, *257*, 10306–10313.

Schoonheim, P. J., Arrenberg, A. B., Del Bene, F., & Baier, H. (2010). Optogenetic localization and genetic perturbation of saccade-generating neurons in zebrafish. *The Journal of Neuroscience*, *30*, 7111–7120.

Scott, E. K. (2009). The Gal4/UAS toolbox in zebrafish: New approaches for defining behavioral circuits. *Journal of Neurochemistry*, *110*, 441–456.

Scott, E. K., & Baier, H. (2009). The cellular architecture of the larval zebrafish tectum, as revealed by gal4 enhancer trap lines. *Frontiers in Neural Circuits*, *3*, 13.

Scott, E. K., Mason, L., Arrenberg, A. B., Ziv, L., Gosse, N. J., Xiao, T., et al. (2007). Targeting neural circuitry in zebrafish using GAL4 enhancer trapping. *Nature Methods*, *4*, 323–326.

Spence, R., Gerlach, G., Lawrence, C., & Smith, C. (2008). The behaviour and ecology of the zebrafish, Danio rerio. *Biological Reviews*, *83*, 13–34.

Strassman, A., Highstein, S. M., & Mccrea, R. A. (1986a). Anatomy and physiology of saccadic burst neurons in the alert squirrel monkey. I. Excitatory burst neurons. *The Journal of Comparative Neurology*, *249*, 337–357.

Strassman, A., Highstein, S. M., & Mccrea, R. A. (1986b). Anatomy and physiology of saccadic burst neurons in the alert squirrel monkey. II. Inhibitory burst neurons. *The Journal of Comparative Neurology*, *249*, 358–380.

Streisinger, G., Walker, C., & Dower, N. (1981). Production of clones of homozygous diploid zebra fish (Brachydanio rerio). *Nature*, *291*, 293–296.

Stuart, G. W., Mcmurray, J. V., & Westerfield, M. (1988). Replication, integration and stable germ-line transmission of foreign sequences injected into early zebrafish embryos. *Development*, *103*, 403–412.

Sumbre, G., Muto, A., Baier, H., & Poo, M. M. (2008). Entrained rhythmic activities of neuronal ensembles as perceptual memory of time interval. *Nature*, *456*, 102–106.

Suster, M. L., Kikuta, H., Urasaki, A., Asakawa, K., & Kawakami, K. (2009a). Transgenesis in zebrafish with the tol2 transposon system. *Methods in Molecular Biology (Clifton, N.J.)*, *561*, 41–63.

Suster, M. L., Sumiyama, K., & Kawakami, K. (2009b). Transposon-mediated BAC transgenesis in zebrafish and mice. *BMC Genomics*, *10*, 477.

Szobota, S., Gorostiza, P., Del Bene, F., Wyart, C., Fortin, D. L., Kolstad, K. D., et al. (2007). Remote control of neuronal activity with a light-gated glutamate receptor. *Neuron*, *54*, 535–545.

Thermes, V., Grabher, C., Ristoratore, F., Bourrat, F., Choulika, A., Wittbrodt, J., et al. (2002). I-SceI meganuclease mediates highly efficient transgenesis in fish. *Mechanisms of Development*, *118*, 91–98.

Tsuruwaka, Y., Konishi, T., Miyawaki, A., & Takagi, M. (2007). Real-time monitoring of dynamic intracellular Ca^2 movement during early embryogenesis through expression of yellow cameleon. *Zebrafish*, *4*, 253–260.

Tsutsui, H., Higashijima, S.-I., Miyawaki, A., & Okamura, Y. (2010). Visualizing voltage dynamics in zebrafish heart. *The Journal of Physiology*, *588*, 2017–2021.

Tsutsui, H., Karasawa, S., Okamura, Y., & Miyawaki, A. (2008). Improving membrane voltage measurements using FRET with new fluorescent proteins. *Nature Methods*, *5*, 683–685.

Volgraf, M., Gorostiza, P., Numano, R., Kramer, R. H., Isacoff, E. Y., & Trauner, D. (2006). Allosteric control of an ionotropic glutamate receptor with an optical switch. *Nature Chemical Biology*, *2*, 47–52.

Wallace, D. J., Zum Alten Borgloh, S. M., Astori, S., Yang, Y., Bausen, M., Kügler, S., et al. (2008). Single-spike detection *in vitro* and *in vivo* with a genetic Ca^{2+} sensor. *Nature Methods*, *5*, 797–804.

White, R. M., Sessa, A., Burke, C., Bowman, T., Leblanc, J., Ceol, C., et al. (2008). Transparent adult zebrafish as a tool for *in vivo* transplantation analysis. *Cell Stem Cell*, *2*, 183–189.

Wyart, C., Bene, F. D., Warp, E., Scott, E. K., Trauner, D., Baier, H., et al. (2009). Optogenetic dissection of a behavioural module in the vertebrate spinal cord. *Nature*, *461*, 407–410.

Yaksi, E., Judkewitz, B., & Friedrich, R. W. (2007). Topological reorganization of odor representations in the olfactory bulb. *PLoS Biology*, *5*, 1453–1473.

Yang, Z., Jiang, H., Chachainasakul, T., Gong, S., Yang, X. W., Heintz, N., et al. (2006). Modified bacterial artificial chromosomes for zebrafish transgenesis. *Methods*, *39*, 183–188.

Yuste, R., Miller, R. B., Holthoff, K., Zhang, S., & Miesenbock, G. (2000). Synapto-pHluorins: Chimeras between pH-sensitive mutants of green fluorescent protein and synaptic vesicle membrane proteins as reporters of neurotransmitter release. *Methods in Enzymology*, *327*, 522–546.

Zhu, P., Narita, Y., Bundschuh, S. T., Fajardo, O., Scharer, Y. P., Chattopadhyaya, B., et al. (2009). Optogenetic dissection of neuronal circuits in zebrafish using viral gene transfer and the Tet system. *Frontiers in Neural Circuits*, *3*, 21.

T. Knöpfel and E. Boyden (Eds.)
Progress in Brain Research, Vol. 196
ISSN: 0079-6123

CHAPTER 9

Genetic targeting of specific neuronal cell types in the cerebral cortex

Alan Urban* and Jean Rossier

Laboratoire de Neurobiologie et Diversité Cellulaire, Centre National de la Recherche Scientifique, Unité Mixte de Recherche 7637, Ecole Supérieure de Physique et de Chimie Industrielles, Paris, France

Abstract: Understanding the structure and function of cortical circuits requires the identification of and control over specific cell types in the cortex. To address these obstacles, recent optogenetic approaches have been developed. The capacity to activate, silence, or monitor specific cell types by combining genetics, virology, and optics will decipher the role of specific groups of neurons within circuits with a spatiotemporal resolution that overcomes standard approaches. In this review, the various strategies for selective genetic targeting of a defined neuronal population are discussed as well as the pros and cons of the use of transgenic animals and recombinant viral vectors for the expression of transgenes in a specific set of neurons.

Keywords: optogenetics; opsins; transgenesis; recombinase; AAV; neuronal network.

Introduction

Of all brain structures, the cortex, which represents over 80% of the volume of the human brain, is what makes us human. Paradoxically, the basic local architecture of the neocortex is quite similar in all mammals, from mouse to man, comprising cortical columns composed of several neuron types organized in units (Hubel and Wiesel, 1977). It is estimated that the human cortex contains two millions of these cortical modules, each of which composed of 10,000–70,000 neurons. These local cortical modules are organized in a framework with a six-layered architecture, in which neurons with distinct functional properties are distributed in discrete layers.

In order to understand how the brain works, it is necessary to experimentally modulate the activity of neuronal circuits in a highly specific and temporally controlled manner. Electrical or pharmacological stimulation of neuronal networks were extensively used in the past; today, the recent developments of optogenetics greatly extend our ability to decipher neural circuits.

*Corresponding author.
Tel.: +33-140-795-182; Fax: +33-140-794-757
E-mail: alan.urban@espci.fr

DOI: 10.1016/B978-0-444-59426-6.00009-4

Optogenetic tools are divided into two families referred as optogenetic actuators and optogenetic sensors: actuators are suitable for controlling specific neuronal populations and are described in Chapters 2 and 3; sensors are suitable for monitoring neuronal networks. Optogenetic sensors are described in details in Chapter 5 of this book.

Briefly, optogenetic actuators are based on opsins that are seven-transmembrane proteins containing the light-isomerizable chromophore all-*trans*-retinal. When illuminated, these proteins can rapidly translocate ions across the membranes of the cells in which they are expressed.

Several features have made actuators of prime importance for studying the cortex. First, light activation has a higher spatiotemporal resolution than electrical or pharmacological brain stimulation. Second, it has been demonstrated that actuators could be used in living cells with minimal toxicity (Zhang et al., 2007). Third, they can be fused to fluorescent proteins without loss of activity, so it allows visualization of correct expression in cells. Last but not least, actuators are proteins, so their expression can be selectively restricted to certain cell types and/or at specific locations. As a result of these unique properties, actuators and sensors have been used in an increasing number of studies to control neuronal networks in a large number of species both *in vitro* and *in vivo*.

One of the biggest challenges for optogenetics is to genetically modify a chosen population of cells for expressing the light sensitive proteins. In this review, we will first explain how to identify molecular markers (enhancer/promoter) specific to a given cortical cell type. Then, we will overview the various methods to deliver genes coding for opsins by the development of transgenic animal lines or direct transfection of neurons. We will go ahead by highlighting the available strategies to express more selectively and more efficiently the opsins in the targeted cells using regulation of the transcription or recombinase-based conditional systems. Finally, we will provide a summary of the technical progress in viral mediated gene delivery, with an emphasis on new strategies to label defined cell populations with high specificity.

Targeting cell types by controlling gene expression

As it is possible to characterize populations of neuron that express specific molecular markers, it is possible to use *cis*-regulatory elements controlling the expression of these markers to genetically target a desired protein. Here, we present the general concepts to regulate gene expression and how transgenesis of optogenetic actuator/sensors could be used to control and monitor specific neuronal populations in the brain.

Neuronal cell types: From phenotype to genotype

The cortex is constituted of several different cell types such as neurons, astrocytes, microglia, oligodendroglia, and epithelial cells which are mixed in a highly complex three-dimensional structure. Here, we will focus on neurons. To understand how the brain processes information, we must understand the structure of its neural circuits, defined as functional entities of interconnected neurons that influence each other. On a local level, the function of a neuron derives from its morphological, electrophysiological molecular properties, and its embryonic origin. Even if these parameters are implemented by rules that are only partially understood, it is reasonable to assume that a phenotypic network architecture gradually emerges from the execution of the various developmental instructions encoded in the genome. In this chapter, we will present molecular parameters commonly used to classify neurons and demonstrate that a defined cell population can be characterized by a specific pattern of gene expression.

As our understanding of the vast diversity of neurons in the cortex develops, it has become clear that one or more molecular markers should be sufficient to define a specific neural populations. In fact, for both pyramidal cells and interneurons, global expression profiling can help grouping neurons into classes that reflect meaningful biological properties (Cauli et al., 1997; Karagiannis et al., 2009; Subkhankulova et al., 2010). Although the full transcriptional map at single cell resolution is not yet achievable, a lot of progress has been made in identifying a number of the relevant genes based on single cell RT-mPCR (Lambolez et al., 1992). If excitatory cells express only a limited set of specific molecular markers (Kubota et al., 1994), different types of inhibitory interneurons might be distinguished by a single marker or by a combination of markers (molecular profile) including transcription factors (TFs), neurotransmitters, neuropeptides, calcium binding protein, receptors (iono/metabotropic), structural proteins, cell-surface markers, ion channels, connexins, transporters, and more (Ascoli et al., 2008).

Today, there is no single morphological, electrophysiological, or developmental parameter that is able to specifically describe a neural subtype. The regulation of gene expression within a specific cell type is spatially and temporally defined. We thus assume that molecular markers whose expression are controlled by specific promoters are suitable to group classes of neurons. In order to restrict the expression of optogenetic trangenes to a class of neurons, the transgenes are incorporated under the control of specific promoters activated only in a class of neurons. In conclusion, optogenetics are based on expression of genetically encoded actuators and are becoming a powerful tool for deciphering neural circuits as described in the following part of this review.

Cis-*regulatory elements (promoter/enhancer)*

The factors controlling gene expression are complex because the ability to produce biologically active proteins comes under regulation at DNA, RNA, and protein levels. Nevertheless, the control of transcriptional initiation is one of the most important regulation mode. For a gene to produce a protein, it requires a promoter. A promoter is a section of DNA in front of the gene that functions to recruit the cellular machinery that will initiate the multistep process of protein production. Enhancer is a short region of DNA that can modulate transcription of genes. They consist of several composite elements and/or individual binding sites for TFs but their position in respect to transcription start are different. A promoter can roughly be divided in two parts: a proximal part, referred to as the core, and a distal part. The proximal part is believed to be the regulatory part of the gene that promotes recognition of transcriptional start sites by RNA polymerase and that is responsible for the basal level of transcription (Berk, 1999; Nikolov and Burley, 1997). It is mediated by elements, such as the TATA and Initiator boxes through the binding of the TATA box-binding protein, and other general TFs specific for RNA polymerase II (Featherstone, 2002). TFs are molecules involved in regulating gene expression. They are usually proteins, although they can also consist of short, noncoding RNA. TFs are also usually found working in groups or complexes, forming multiple interactions that allow for varying degrees of control over rates of transcription. In eukaryotes, genes are usually in a default “off” state, so TFs serve mainly to turn gene expression on. The distal part of the promoter includes elements that regulate the spatiotemporal expression (Fessele et al., 2002; Tjian and Maniatis, 1994). In addition to the proximal and distal parts, regulatory regions that contain enhancer and/or repressors elements have also been described (Bagga et al., 2000; Barton et al., 1997). They modulate the level of transcription depending on the type of tissue, developmental stage, stage of the cell cycle, induction by hormones, or other molecular signals. Positions of enhancers can vary significantly within the gene as they could be located just before the promoter or at a distance of several thousand base pair in the 5′ region, within introns, or in 3′ regions (Kadonaga, 2004). Lastly, eukaryotic genomes can be organized into domains of transcriptional activity or transcriptional silencing, encompassing one or more genes (Oki and Kamakaka, 2002).

Even if the regulatory motifs of some specific genes have been investigated in detail, there are not yet clear and unequivocal descriptions of genomic segments that contain all elements required to activate transcription. Nevertheless, to increase scientific community access to general information about promoters and their functions, regulations, there are many specialized databases such as the Eukaryotic Promoter Database (EPD) containing description about promoters, as they are defined by an experimentally proven transcription start site and their tissue specificity (Perier et al., 1999).

Genetic modification of host genome by transgenesis approaches

Germline transgenesis

As discussed in the previous chapter, the *cis*-regulatory elements can be hundreds of kilobases pairs long and are thus difficult to manipulate both *in vitro* and *in vivo*. However, many strategies have been extensively developed in mouse because of the powerful genetic tools that are available to manipulate the mouse genome and to integrate large DNA fragments into the host genome. Actually, two different strategies could be applied to generate a transgenic mouse termed random or positional transgenesis in relationship with the control site of transgene integration. A transgenic construct requires three main components: (1) a promoter to drive expression of the transgene, (2) a transgenic open reading frame encoding gene to be expressed, and (3) a polyadenylation signal to terminate transcription. Bacterial artificial chromosome (BAC)-mediated transgenic fulfill these criteria as BAC vectors can accommodate large genomic fragment up to 700kb that may contain several contiguous genes with their *cis*-regulatory elements (Shizuya et al., 1992). BACs offer several advantages because they are easy to manipulate *in vitro* and their integration is generally stable (Marra et al., 1997) with a linear relationship between the copy number and the level of expression of the integrated gene (Chandler et al., 2007). In addition, BACs can be modified by homologous recombination in *Escherichia coli*, to introduce desirable mutations including insertions, deletions, and point mutations (Gong et al., 2002).

Random transgenesis

Random transgenesis (Fig. 1a) is the most rapid and effective method to generate transgenic mice. The main technical step is the direct microinjection of BAC DNA into the pronucleus of fertilized mouse eggs, followed by transfer of the injected oocytes to pseudogestant foster mothers. Pups that arise from a transfer have the foreign DNA stably and randomly integrated into the genome and can then germline transmit the integrated transgene to their offspring to establish a transgenic mouse line. BAC DNA constructs are large enough to contain all the regulatory elements necessary to confer accurate transgene expression *in vivo*. The major advantage of using BAC transgenesis is to overcome positional effects (Heintz, 2001), meaning that the integration site has little or no effect on the expression of the transgene. Random transgenesis is an effective and efficient method to produce genetically engineered mouse models as demonstrated in GENSAT project that makes available to researchers with a collection of around 600 mouse lines expressing GFP under a particular promoter (Heintz, 2004). Since the transgene integration is a random event, each transgenic line is unique because the transgene is integrated at a distinct chromosomal location and with a definite number of copies. Rather than recapitulating the activity of each promoter/enhancer, expression of the target gene is often limited to a subset of cells in which *cis*-regulatory elements are active. Thus, different transgenic lines made with the same transgene selectively label different highly restricted subsets of neurons. For example, random insertion of the CamKIIa promoter, which is

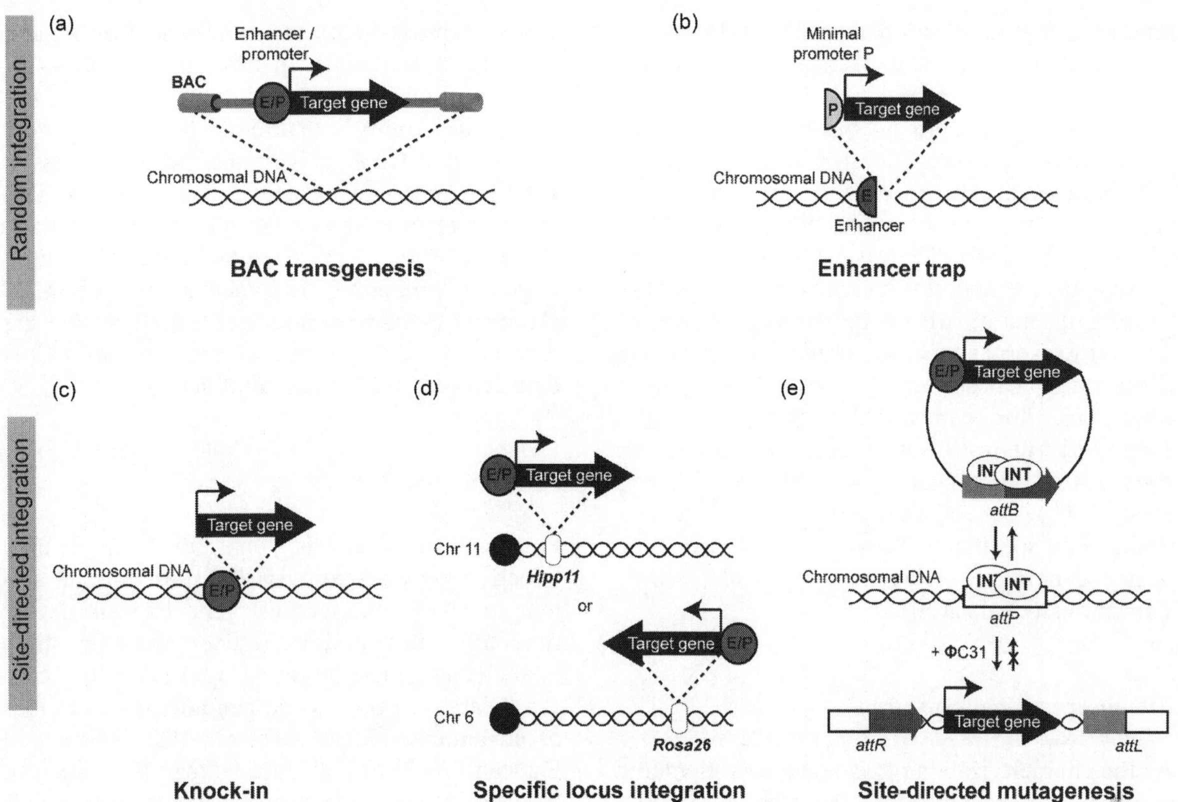

Fig. 1. Methods for targeting gene expression. (a) Bacterial artificial chromosomes (BACs) are large-insert DNA clones that can accommodate up to 200kb of genomic DNA, and are likely to contain all the regulatory elements E/P necessary to confer accurate transgene expression *in vivo*. (b) The enhancer trap technique uses a reporter gene fused to a minimal promoter, typically containing the TATA box and transcription start site. The minimal promoter is located on the reporter gene so that it can only be activated by nearby enhancer sequences near the chromosome insertion site. (c) Gene knock-in refers to a genetic engineering method that involves the insertion of a protein coding cDNA sequence at a particular locus in the genome. (d) Transgenes targeted by homologous recombination to the Rosa26/Hipp11 locus are stably and efficiently expressed in undifferentiated cells as well as the differentiated cell types generated from mouse ES cells. (e) Plasmid derived from bacteriophage ΦC31 inserts its target gene into that of its host via the integrase (INT) enzyme, which catalyzes recombination between a phage attachment site (attP) and a bacterial attachment site (attB) present in the chromosome. Integrase requires no accessory factors and has a high efficiency of recombination. (For color version of this figure, the reader is referred to the Web version of this chapter.)

normally expressed in most excitatory forebrain neurons can be restricted to specific cell types of the striatum and hippocampus (Kellendonk et al., 2006; Nakazawa et al., 2002; Tsien et al., 1996). In fact, in many cases transgenes are integrated as multiple copies from one to several hundred that usually form head-to-tail repeats. Because of these limitations (unknown location in the genome and copy number), it is often necessary to perform phenotypic studies on several transgenic lines that have been generated. Even if BAC transgenic animals contain a high number of transgene's copies, the amount of protein in each cell is dependent on the strength of the promoter. Because most of the promoters are relatively weak, this strategy could be nonsuitable for optogenetic tools as they require a high expression level. Nevertheless, a number of mouse strains expressing actuators have been generated (VChR1, eNphr3.0 for more details see http://jaxmice.jax.org/). One recent success is the generation of healthy ChR2-eYFP transgenic mouse with expression throughout the brain using the Thy1 promoter (Arenkiel et al., 2007; Zhao et al., 2008). Thy1 is an immunoglobulin superfamily member that is expressed in projection neurons in many parts of the nervous system, as well as by several nonneuronal cell types, including thymocytes (Gordon et al., 1987). Under these conditions, ChR2 can be used to map synaptically connected neurons in slices and even in anesthetized mice (Arenkiel et al., 2007; Wang et al., 2007). Several transgenic mice expressing opsins with specific neuronal targeting have now been generated (Hagglund et al., 2010; Thyagarajan et al., 2010).

Enhancer and repressor trap

As the characterization of *cis*-regulatory elements of each gene is difficult to perform, alternative strategies based on random integration have been developed. An example is the enhancer trap (Fig. 1b) approach that uses positional effects dependent on transgene insertion site in the host genome. Short promoter segments from marker genes can restrict expression to the population of cells expressing the marker and positional effects can further restrict expression within that population. The power of this strategy was revealed following detailed analysis of lines in which highly restricted subsets of neurons were labeled by fluorescent proteins under the control of a short segment of the Thy1 promoter (Feng et al., 2000) or the Gad promoters (Chattopadhyaya et al., 2004; Lopez-Bendito et al., 2004; Oliva et al., 2000). This method has been also successful in flies (Bellen et al., 1989) and zebrafish (Davison et al., 2007; Nagayoshi et al., 2008; Scott and Baier, 2009).

On the contrary to the enhancer trap, another strategy called repressor trap could be used to inhibit expression of the transgene (for review, see Luo et al., 2008). In this case, the structure of the chromatin close to the integration site can affect the ability of transcriptional regulatory proteins and RNA polymerases to find access to specific genes and to activate transcription. Two primary mechanisms exist that alter chromatin structure and as a consequence affect alterations in gene expression. These mechanisms are methylation of cytosine residues in the DNA that are found in the dinucleotide, CG referred as a CpG dinucleotide and histone modification.

Positional transgenesis

Positional transgenesis (Fig. 1c) often termed knock-in gene targeting is the most faithful system to mimic endogenous gene expression. As randomly integrated transgenes are susceptible to silencing genes, targeting transgenes to a chosen location in the mouse genome has a number of advantages (Bronson et al., 1996; Misra and Duncan, 2002). Firstly, the integration site can be chosen to allow insertion of the transgene into a region of chromatin favorable for expression and that avoids an undesirable insertional mutagenesis. Additionally, only a single copy is introduced which avoids problems associated with large multicopy arrays. This strategy requires embryonic stem (ES) cells that are harvested from the inner cell mass of mouse blastocysts. They can be grown in culture and retain their full potential to produce all the cells of the mature animal, including its gametes. The introduction of the gene into ES cells is a multiple step process. Targeting constructs are introduced into the ES cells by electroporation. Then, the cell undergoes homologous recombination at the shared sequence, during which two crossover events replace the WT gene with the targeting construct. Correctly targeted cells inherit an antibiotic resistance gene and are able to grow in the presence of that antibiotic. The presence of the desired mutation in the ES cell DNA is then directly confirmed by PCR or genomic southern blot. ES cells with one copy of the desired mutation (heterozygous) are expanded, injected into blastocysts harvested from mice with black coat color, and implanted into pseudogestant females. Chimeric offspring in part deriving from the ES cells (dominant agouti coat color) and in part from the donor blastocysts (black coat color), are identified by patchy agouti/black coat color. Subsequent crosses will be made until the desired genotype is obtained. ES cells can thus be used as a vehicle to obtain transgenic mice from germline ES cell-mouse chimeras. One additional advantage of positional transgenesis is the ability to prescreen ES cell clones for expression (Bronson et al., 1996). This strategy has also its limitations including the limited size of genomic DNA that can be inserted. In addition, DNA fragments smaller than 20kb often result in positional effects including lack of transgene expression, expression restricted to only a subset of cells or extinction of transgene expression in successive generations. Moreover, availability of ES cells is limited for a number of mouse strains (i.e., 129SV, C57/bL6J). Positional transgenesis is usually more time consuming and costly compared to random mutagenesis because of technical manipulations required for insertion of drug selection cassettes (Testa et al., 2003), ES cell culture, generation of ES cell-mouse chimeras, and germline breeding. Nevertheless, this strategy has been successfully applied to express ChR2 in GABAergic neurons (Katzel et al., 2011).

Integration in a specific locus

At best, targeting the site of integration by homologous recombination in ES cells would virtually solve any chromosomal position effects since the transgenic construct would then be controlled by all regulatory sequences present in the chosen endogenous locus. Alternatively, particular locations in the host genome can be selected according to their capacity to allow adequate expression patterns of experimental transgenes (Wallace et al., 2000).

ROSA26 locus This strategy has been successfully applied in mice since the discovery of the Gt (ROSA)26Sor locus (ROSA26). The ROSA26 locus (Fig. 1d) was first isolated in a gene-trap mutagenesis screening performed in mouse ES cells (Friedrich and Soriano, 1991). The ubiquitous expression of ROSA26 in embryonic and adult tissues, together with the high frequency of gene-targeting events observed at this locus in murine ES cells has led to the establishment of several ROSA26 knock-in lines. These represent a variety of transgenes including reporters (Soriano, 1999), site-specific recombinases and recently optogenetic actuators (hChR2(H134R)::tdTomato and hChR2(H134R)::YFP from H. Zeng, Allen Institute for Brain Science).

Hipp11 locus As transgenic strategies are in constant evolution, the new Hipp11 (H11) locus (Fig. 1d) will probably replace ROSA26 in the future because homozygous insertions into this locus are not predicted to disrupt any endogenous genes and the resulting mice are completely healthy and fertile (Hippenmeyer et al., 2010).

HPRT locus Recent studies have also shown that it is possible to take advantage of both random and positional transgenesis method to insert BAC as a single copy at a specific genomic location such as hypoxanthine phosphoribosyltransferase (HPRT) locus and thus express transgene with the appropriate tissue and cell-specific pattern (Heaney et al., 2004; Miyazaki et al., 2005). However as the HPRT locus is on the X chromosome it will be randomly inactivated in female mice, which may not be ideal for every experiment.

attB site Recently, it has been demonstrate that an intact single-copy transgene can be inserted into predetermined chromosomal locus containing attB site with high efficiency and faithfully transmitted through generations by ΦC31 integrase approach (Fig. 1e). This system allows production of transgenic mice with the advantages of random transgenesis via pronuclear injection and precision of positional transgenesis (Tasic et al., 2011).

Somatic gene transfer into the brain by DNA electroporation

As mentioned before, the specific factors that exert control of gene expression include the strength of promoter elements, the presence of enhancer sequences, and the interaction between multiple activator and inhibitor proteins. All these elements are encoded in DNA and could therefore be used to target optogenetic tools into defined group of neuron. Electroporation has become a common laboratory technique for enhancing the efficiency of DNA delivery into cells. Application of low voltage rectangular pulses after local injection of DNA temporarily increases in the transmembrane potential difference, which provokes cell membrane permeabilization and facilitates DNA uptake. Developed initially on chick embryo (Itasaki et al., 1999), *in vivo* electroporation has now been successfully applied in several model systems for the delivery of genetic materials (Isaka and Imai, 2007).

If performed in embryonic mouse, *in utero in vivo* electroporation provides a powerful tool for the manipulation of neurons in the cortex and allows for the targeting of specific neuronal layers. Technically, DNA is injected in the lateral ventricles of the developing embryo and electroporated into neuronal progenitors lining the walls of the lateral ventricle. DNA electroporation in mouse was initially restricted to pyramidal cells (Saito and Nakatsuji, 2001). Additional technical developments made it possible to specifically target gene expression to interneurons by *in utero* electroporation directed to the ganglionic eminences (Borrell et al., 2005). With stereotaxic apparatus facilitated microinjection, *in vivo* electroporation could also target a defined small area in the adult brain (Wei et al., 2003). Optogenetic tools have been successfully electroporated into mouse embryos (Adesnik and Scanziani, 2010; Gradinaru et al., 2007; Lewis et al., 2009 Petreanu et al., 2007).

Recently, electroporation has been refined to allow delivery on a single neuron scale both *in vitro* (Mertz et al., 2002; Teruel et al., 1999), in slices (Haas et al., 2002) and *in vivo* (Judkewitz et al., 2009). Single neuron electroporation presents many advantages such as the transfection of multiple genes using plasmids at the same time or the labeling of very few neurons at precise locations within the brain. As an example of such strategy for optogenetics, the expression of ChR2 has been successfully used to activate sparse pyramidal neurons in barrel cortex (Huber et al., 2008).

Even if electroporation requires very specific technical skills, this strategy provides high expression levels that are needed when using bacterial opsins and thus is optimal for development of new optogenetic constructs as it requires less time (2–5 weeks instead of several months) than transgenic approaches. Moreover, the size of the DNA used for *in utero* electroporation is not limited which allows for a very specific targeting of cortical neurons. Another advantage is that opsins are expressed early, which makes possible electrophysiological studies in brain slices at young age. Nevertheless, there are a number of limitations including the restricted spatial expression of optogenetic probes (especially in mouse where labeled cells are mainly pyramidal cells of the cortex) and the variability of expression level among experiments.

Genetic strategies for refined gene targeting

Over the past decades, an expanding repertoire of genetic tools has greatly facilitated the visualization and manipulation of cell populations in model organisms. We will describe here the general concept of driving the expression of a target transgene by using two (binary expression) or more (combinatorial expression) promoters that could be used following the Boolean logic gates AND (expression is only possible by the intersection of two expression patterns, i.e., in a population of cells expressing A and B) or NOT (expression is exclusively restricted in one pattern).

Binary expression strategies

The most versatile strategies for cell-type-specific targeting are binary systems, which use the natural *cis*-regulatory elements of endogenous genes to drive the expression of a primary effector, usually a TF (Fig. 2a) or recombinase (Fig. 2b). This primary effector then activates expression of the transgene encoding a secondary effector that allows the expression of the transgene of interest.

Control of transcriptional activity

Ectopic expression has proved to be an excellent technique for analyzing gene function in *Drosophila* and other model organisms as demonstrated by the GAL4 system allowing the expression of any given open reading frame. The GAL4 system was built on the characterization of transcriptional regulation in yeast. GAL4 is an archetypal eukaryotic TF isolated as an activator of the genes responsible for galactose metabolism in *Saccharomyces cerevisiae* (Hashimoto et al., 1983). The target sequence of GAL4 was defined as a 17-mer, four copies of which are found in the upstream activation sequence (UAS) of the galactose metabolism genes, Gal10 and Gal1 (Bram et al., 1986; Giniger et al., 1985; Webster et al., 1988). Further, the activity of GAL4 is repressed by a physical interaction with the GAL80 protein, which is repressed when galactose is the only carbon source (Lue et al., 1987; Wu et al., 1996). The high level of conservation in the eukaryotic transcriptional machinery means that GAL4 can activate transcription in other species, as distantly related as humans and plants (Kakidani and Ptashne, 1988; Ma et al., 1988; Webster et al., 1988). The generation of GAL4 transgenes under the control of short promoter/enhancer fragments offer a powerful strategy for labeling more restricted subsets of cells. By using an artificial promoter that contains a tandem array of GAL4-binding sites and a transcriptional start site, expression of a target gene can be controlled by the expression of GAL4 (Fig. 2a). The advantage of this system is that simply expressing GAL4 in a different tissue can change the expression pattern of the target gene. This system has multiple applications, especially when a large collection of different tissue-specific GAL4 expressing transgenes is available. Thus, besides the use as a flexible system to express genes in different tissues, it can also be used to generate conditional overexpression lines, for targeted ablation of cells by expression of toxins and for tissue-specific RNAi. GAL4/UAS has been successfully used in combination with the enhancer trap screens in flies what led to characterization of almost 7000 GAL4 lines (Hayashi et al., 2002). This system has also been used in mouse (Ornitz et al., 1991) and in zebrafish (Scott et al., 2007).

Another binary expression system available in fruit fly is based on the repressor LexA that is a regulator of the SOS response to DNA damage in *E. coli* (Walker, 1984). LexA is a two-domain protein including a DNA-binding domain (DBD) and a dimerization domain. LexA binds as a dimer with varying affinities to single or multiple copies of gene-specific LexA DNA-binding motifs (Lex Aop) found upstream of its target genes. Fusing the C-terminal activation domain (AD) derived from various eukaryotic TFs to LexA allows it to drive *in vivo* the transcription of reporter transgenes in *Drosophila* whose promoters contain Lex Aop motifs (Lai and Lee, 2006; Szuts and Bienz, 2000).

Recently, a new repressible binary expression system based on the regulatory genes from the *Neurospora* quinic acid gene cluster has been reported (Potter et al., 2010). This Q system offers many applications for labeling more restricted subsets including combinatorial logic gates with the GAL4 system as described in the section "Somatic gene transfer into the brain by DNA electroporation".

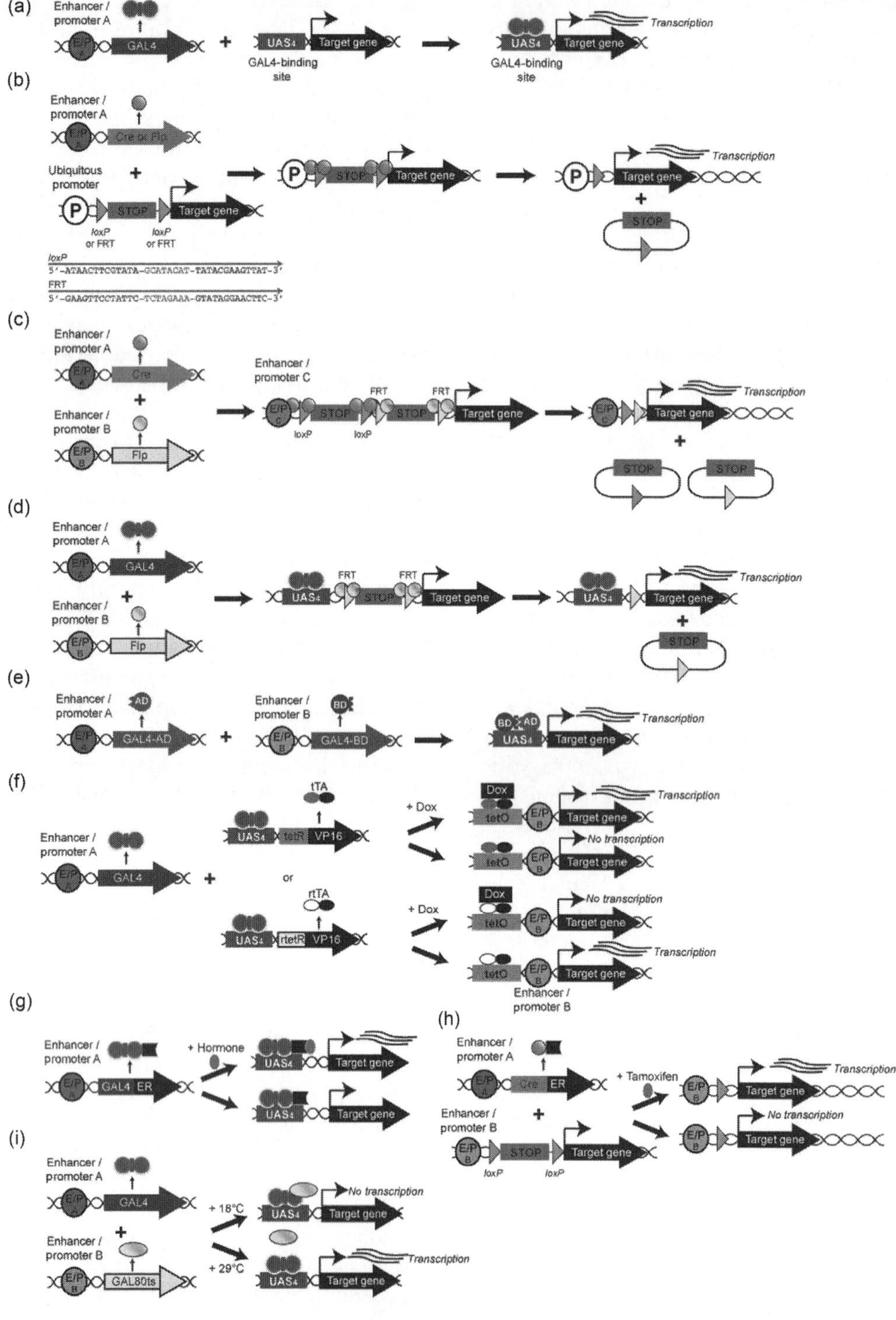
(a)
Enhancer / promoter A
E/P A
GAL4
UAS4
Target gene
GAL4-binding site
Transcription
(b)
Enhancer / promoter A
Cre or Flp
Ubiquitous promoter
P
STOP
Target gene
loxP or FRT
loxP or FRT
Transcription
loxP
5'-ATAACTTCGTATA-GCATACAT-TATACGAAGTTAT-3'
FRT
5'-GAAGTTCCTATTC-TCTAGAAA-GTATAGGAACTTC-3'
(c)
Enhancer / promoter A
Cre
Enhancer / promoter B
Flp
Enhancer / promoter C
FRT
FRT
loxP
loxP
STOP
Target gene
Transcription
(d)
Enhancer / promoter A
GAL4
Enhancer / promoter B
Flp
UAS4
FRT
FRT
STOP
Target gene
Transcription
(e)
Enhancer / promoter A
AD
GAL4-AD
Enhancer / promoter B
BD
GAL4-BD
UAS4
Target gene
Transcription
(f)
Enhancer / promoter A
GAL4
tTA
UAS4
tetR
VP16
or
rtTA
rtetR
+ Dox
Dox
tetO
Target gene
Transcription
No transcription
Enhancer / promoter B
(g)
Enhancer / promoter A
GAL4
ER
+ Hormone
UAS4
Target gene
(h)
Enhancer / promoter A
Cre
ER
+ Tamoxifen
Enhancer / promoter B
STOP
Target gene
loxP
loxP
Transcription
No transcription
(i)
Enhancer / promoter A
GAL4
Enhancer / promoter B
GAL80ts
+ 18°C
+ 29°C
UAS4
Target gene
No transcription
Transcription

Site-specific recombination approach

Cre/lox The site-specific recombination system is also a binary system. The system begins with the *cre* gene, short for cyclization recombination, which encodes a 38kDa recombinase protein from bacteriophage P1 that mediates intramolecular (excessive or inversional) and intermolecular (integrative) site-specific recombination between loxP sites (for review, see Sauer, 1993). These sites are known as loxP (locus of X-over P1) sequences, which are 34bp long (two 13bp inverted repeats separated by an 8bp asymmetric spacer region) and are recognition sites for the Cre to recombine the DNA surrounding them. One molecule of Cre binds per inverted repeat or two Cre molecules line up at one loxP site. The recombination occurs in the asymmetric spacer region. Those eight bases are also responsible for the directionality of the site. Two loxP sequences in opposite orientation to each other invert the intervening piece of DNA; two sites in direct orientation dictate excision of the intervening DNA between the sites leaving one loxP site behind. The Cre/lox system is a bipartite system in which one transgenic line, the driver, expresses Cre recombinase in a known temporal and spatial pattern and a second transgenic line, the reporter, contains a Cre recombinase dependent transgene that is under the control of an ubiquitous promoter. This reporter line contains a transcription stop flanked with two loxP sites in the same orientation even termed as "floxed stop" or LSL (lox-stop-lox) sequence. The stop sequence is a short sequence with several transcriptional stop codons that will prevent the gene from producing a protein. When Cre is present in the cells of this organism, it catalyzes recombination between the loxP sites, thereby deleting the stop sequence and allowing the expression of the target gene (Fig. 2b). Therefore, if the Cre gene is bound to a promoter that allows Cre production only in neuronal cells, the target gene will be specifically expressed in those cells.

Since the Cre/lox system has been extensively used over the past 15 years, there are now numerous transgenic animals, plants and bacterial stocks that already contain the *cre* gene driven by ubiquitous or tissue-specific promoters. For example, there is an extensive collection of Cre driver mice from the NIH Neuroscience Blueprint that were initially designed for the tissue and cell-type-specific perturbation of gene function in the nervous system (e.g., http://credrivermice.org; http://www.

Fig. 2. Genetic strategies for refined gene expression. (a) The GAL4 protein is present only where the promoter A (E/PA) is active. GAL4 binds to a sequence called the upstream activating sequences (UAS) element and is induces translation of the gene linked to UAS. (b) Cre/loxP or Flp/FRT leads to expression in cells where the promoter A is active and therefore removes the transcription stop to allow expression of the target gene under the control of a constitutive promoter P. LoxP and FRT consist of a 34bp DNA sequence containing an asymmetric 8bp sequence (red) in between two sets of palindromic, 13bp sequences flanking it. (c) A combination of Cre/loxP and Flp/FRT recombination systems allows for the target gene to be expressed only when promoters A and B are active in the same cell. (d) As for (c), a combination of GAL4/UAS and Flp/FRT increases cell specificity for the target gene which is only expressed when promoters A and B are active. (e) Split-GAL4 system, which independently targets the Gal4 DNA-binding domain (Gal4DBD) and a cognate transcription activation domain (AD) using two different promoters A and B, drives transgene expression in a restricted fashion: only cells in which both promoters are active at the same time express the two heterodimerizing transcription factor domains to reconstitute transcriptional activity. This system is also available for Cre recombinase. (f) Tetracycline inducible GAL4 system encompasses two complementary control circuits, described as tTA-dependant (Tet-Off) or rtTA-dependant (Tet-On) expression system. Expression of the target gene is only possible in cells where promoter A is active and if doxycycline (Dox), a tetracycline analog is present (Tet-On) or absent (Tet-Off). (g) and (h) Inducible version of GAL4-ER/UAS and Cre-ER/loxP recombinase systems in which addition of tamoxifen will cause dose-dependent activation of the target gene. ER: estrogen receptor. (i) Temporal control of the GAL4/UAS system is based on a temperature-sensitive GAL80ts able to repress expression of the target gene at 18°C but not at 29°C. (For interpretation of the references to color in this figure legend, the reader is referred to the Web version of this chapter.)

gensat.org; Gong et al., 2002, 2007; Heintz, 2004). Cre driver lines in combination with "floxed stop" transgenic lines are an invaluable system that has the advantage of working in almost any type of cell. This strategy will probably be used in the future to express ChR2 in a specific cell types, using the JAX Rosa-CAG-LSL-ChR2(H134R)-EYFP-WPRE transgenic strain from H. Zeng.

Nevertheless, there are some disadvantages of the Cre/lox system. First, as mentioned in the section "Neuronal cell types: From phenotype to genotype" it could be difficult to identify a promoter that is perfectly specific for a class of cells (Sauer, 1993). Second, the Cre recombination is an irreversible process that can produce "non-physiological" response when the driver promoter is transiently activated. This limitation has been overcome by to the development of inducible Cre (see section "Temporal control of transgene expression"). Last but not least, the establishment of transgenic systems with inserted genes requires a significant amount of time and money.

Flippase/FRT Another similar recombination strategy came from the yeast in the form of Flippase/FRT analogous to the Cre/Lox recombination system. The 2μm plasmid of *S. cerevisiae* codes for a site-specific recombinase, the Flippase recombination enzyme (Flp), that catalyzes efficient recombination across two 599-bp inverted repeats (referred as Flippase Recognition Target (FRT) sites) of the plasmid DNA both *in vivo* (Broach et al., 1982) and *in vitro* (Vetter et al., 1983). Flp/FRT has been used to control gene expression by "FLP-out": a recombinase-catalyzed intramolecular excision of spacer DNA including a transcriptional stop between tandemly oriented FRT sites. As for Cre/lox system, the gene downstream of the spacer is not transcribed until activation of the FLP recombinase and subsequent FLP-out (Golic and Lindquist, 1989; Struhl and Basler, 1993). After the FRT-containing cassette is excised by the FLP recombinase, the downstream gene is brought into proximity to the ubiquitous and constitutively active actin promoter and is therefore expressed (Fig. 2b). This system and related systems have proven quite powerful and flexible in model organisms including *Drosophila* (Struhl and Basler, 1993) and mouse (Dymecki, 1996).

Resolvase/invertase More recently, a screen of recombination systems derived from the resolvase/invertase family for site-specific recombinase activity in the fission yeast *Schizosaccharomyces pombe* has highlighted seven new recombination systems (Thomson and Ow, 2006) that could be used independently or in combination with other binary expression systems to increase specificity of transgene expression.

Combinatorial expression strategies

Despite the success of binary expression strategies, the small number of genes with a highly restricted expression might limit its efficiency. Even the most restricted marker genes are likely to be expressed by multiple cell types. As an example, it has been demonstrated that the calcium-binding protein parvalbumin that is a specific marker for fast spiking interneurons is also expressed in some of layer V pyramidal cells in barrel cortex (Mcmullen et al., 1994; Tanahira et al., 2009). To restrict with a higher precision expression of a transgene, is it possible to use combinatorial approaches. In these strategies, expression of a genetic marker is dependent on the presence of two factors, each of which is restricted to a subpopulation of neurons. Therefore, expression is more specific because it only occurs in cells that are in the intersection of the two populations.

Intersectional strategy

Cell types are typically defined by expression of a unique combination of genes, rather than a single gene. Intersectional methods are important to selectively access specific cell types with a higher resolution. To achieve this goal, one possibility is to modify previously described binary strategies (i.e.,

Cre/loxP system) by replacing the ubiquitous promoter with a second tissue-specific promoter, such that the transgene of interest can only be expressed in cells in which both promoters are active.

Dual binary system

A useful way to access specific cell types with precision depends on combinatorial restriction of transgene expression by independently targeting the recombinases Cre and Flp using distinct promoters (Awatramani et al., 2003). In this dual recombinase system, expression of the transgene of interest is made contingent upon the excision not of one but of two stop cassettes placed between the transgene and a broadly active promoter. Each stop cassette is flanked by target sites for only one of the two recombinases so that only in cells expressing both Cre and Flp will both cassettes be removed, allowing transcription of the transgene of interest (Fig. 2c).

One approach to refining spatial regulation of the GAL4 system is to combine it with the FLP-out technique (Struhl and Basler, 1993). In this strategy, a terminator cassette flanked by FRT sites is inserted between the UAS promoter and the gene to be expressed, rendering the transgene silent. Activating the transgene requires the expression of the Flp recombinase to remove the terminator cassette (Fig. 2d). The use of a heat-shock inducible hsFLP recombinase affords temporal control of the onset of transgene expression. A reverse strategy can give a similar result by placing a FRT-flanked terminator cassette in front of the GAL4-coding sequences (Ito et al., 1997).

Another approach to regulate temporal expression of a transgene is to use a subtractive gene strategy. This strategy was first introduced as an essential component of the MARCM system for Mosaic Analysis with a Repressible Cell Marker, which combines Flp/FRT and Gal4/Gal80 to couple transgene expression with mitotic recombination (Lee and Luo, 1999). The GAL80 protein binds to the C-terminus of GAL4 and blocks transcription by interfering with the recruitment of other components of the transcriptional machinery. The use of GAL80 repressor offers extensive possibility to express a transgene in a restricted population of cells by using the logic gate in population expressing A NOT B.

Split protein system

This strategy rests on components that are inactive alone, but when combined reconstitute a desired function as for protein complementation, in which two inactive fragments of a protein associate to reconstitute function. By independently targeting the expression of the two nonfunctional parts of a split protein, reconstitution of function can be achieved at the intersection, and only at the intersection, of the expression patterns of the two promoters used to target them. The typical example is the "split-GAL4" method that takes advantage of the modular nature of the GAL4 TF (Luan and White, 2007). In this technique, the separate DBD and AD of GAL4 are fused to a heterodimerizing leucine zipper motif and each fusion protein is expressed separately. Only when they are present in the same cell can the leucine zippers direct heterodimerization, resulting in the formation of a functional activator (Fig. 2e).

A similar strategy has been used to design a "split-Cre" system based on the complementation of Cre protein fragments. Here, the Cre recombinase was divided in two halves and both Cre fragments were fused to the constitutively dimerizing coiled-coil leucine zipper domain of the yeast transcriptional activator GCN4 to force the association of split-Cre fragments, thereby enhancing Cre activity by functional complementation (Hirrlinger et al., 2009b).

Temporal control of transgene expression

While global expression of transgenes can be used as an effective tool for studying the consequences of gene activation or inactivation, in many instances it is desirable to control both spatial and temporal aspects of transgene expression. As described above, spatial resolution is achieved by the use of tissue- or cell-specific promoters but timing of transgene expression can be controlled by addition or withdrawal of a small molecule able to either induce or inhibit gene expression.

A widely used tool in both mouse and fly is the tetracycline-dependant expression of transgene. This system is based on the transcriptional activity of the tetracycline-transactivator (tTA). This TF was modified by the incorporation of a fusion between the *E. coli* tetracycline repressor and the strong transcriptional AD of the herpes simplex virus VP16 (TetR-VP16). TetR-VP16 is expressed under the control of a tissue-specific promoter and can promote expression of genes bearing tetO operator sequences. Addition of the cell-permeable ligand tetracycline or its derivatives (e.g., doxycycline), which binds to and prevents TetR-VP16 from binding tetO (Gossen and Bujard, 1992), can be used to repress expression at specific times. Two versions of the tetracycline systems exist: Tet-On, in which the addition of the drug results in an active reverse-tetR (rtTA) causing transgene activation from the tet operator and Tet-Off, in which addition of the drug inactivates tTA, and in turn, expression from the tet operator is switched off (Fig. 2f). To take advantage of the number of established tissue-specific GAL4 lines, both the Tet-On and the Tet-Off expression systems have been linked to the GAL4/UAS system (Stebbins et al., 2001). To do this, the tetracycline transactivators were placed under the control of the UAS. Therefore, expression of rTA can be regulated by crossing the UAS-rTA (or an optimized version of rtTA with a high level of transgene induction called rtTAs-M2-altTA) transgenics to a given GAL4 driver strain (Stebbins et al., 2001).

Another similar approach to regulate temporal expression is to use hormone inducible variants of GAL4. Two systems are available, GAL4-estrogen receptor (GAL4-ER) (Han et al., 2000) and a second called GeneSwitch, which is a fusion of GAL4-progesterone receptor and the AD of p65 (Osterwalder et al., 2001; Fig. 2g). In the GAL4-ER, the receptor is engineered such that exogenous administration of the appropriate ligand results in the fusion protein relocating into the nucleus to activate transcription upon the administration of tamoxifen, an estrogen analog. Similarly, Cre recombinase fused to the estrogen receptor (Cre-ER), and expressed in specific tissues, can be temporally activated by the addition of tamoxifen (Feil et al., 1996; (Fig. 2h). Drawbacks include the relatively slow response in gene expression following the cessation of therapy and the difficulties of delivering an oil-soluble ligand. It led to the development of a second-generation mutant Cre-ER(T2), which gives a four-fold increase in the efficiency of recombination induced by 4-hydroxy-tamoxifen in cultured cells (Indra et al., 1999). A refinement of this strategy called split-CreERT2 has been recently developed to allow spatially and temporally precise genetic access to cell populations defined by the simultaneous activity of two promoters (Hirrlinger et al., 2009a).

Proteins whose functions are regulated by temperature give an alternative strategy to small molecules for temporal control. For example, in *Drosophila*, a heat-shock-promoter driver Flp recombinase (hsFlp) has been developed to induce gene expression in a temporally controlled manner. Another example is the development of the temporal and regional gene expression targeting (TARGET) technique (Mcguire et al., 2003). In this approach, a temperature-sensitive variant of the GAL80 protein (GAL80ts) is expressed ubiquitously under the control of the tubulin 1α promoter. GAL80 repression of GAL4 is activated by a simple temperature shift (18–35°C), giving a precise temporal control of the onset of expression. Interestingly, the TARGET system is fully compatible with the vast array of GAL4 lines already established (Fig. 2i).

Viral mediated gene delivery

Classical methods of gene transfer, such as transfection, have many limitations for gene delivery. First, they are only applicable to cell cultures *in vitro* and then are often limited to dividing cells. The development of viral vectors has circumvented these limitations. Replacing genes necessary for viral replication with an expression cassette containing the genes of interest transforms the viruses into safe vectors able to deliver genetic material to various tissues including the brain. Viral vectors can also be combined with transgenic animals and/or genetic strategies described in the section "Genetic modification of host genome by transgenesis approaches". In this section, we will describe viral vectors currently used for gene transfer into the brain and recent developments in viral technology to improve targeting, transcriptional regulation and transgene expression.

Viral vector description

Viral expression systems combine many advantages including fast and easy implementation, high expression of transgenes in infected cells, and efficiency in a lot of species such as primates where transgenesis is difficult to perform. Here, we will present briefly several kinds of viruses, including retrovirus, adenovirus, adeno-associated virus (AAV), and herpes simplex virus (HSV) that are used in gene transfer into the brain (Table 1).

Retrovirus/lentivirus vectors

Retroviruses are a class of enveloped single-stranded RNA virus. Following infection, the viral genome is reverse transcribed into double-stranded DNA, which integrates into the host genome and is expressed as proteins. The viral genome is about 7–10kb, composed of three gene regions termed gag (coding for viral protease and integrase), pol (coding for reverse transcriptase), and env (coding for the viral envelope glycoprotein). At each end of the genome are long terminal repeats (LTRs), which include promoter/enhancer regions and sequences involved in integration. The genome also has a packaging signal (Ψ) and RNA splice sites in the env gene. Retroviral vectors are most frequently based upon the Moloney murine leukemia virus (Mo-MLV), an amphotrophic virus, capable of infecting both mouse cells (via the cationic amino acid transporter CAT-1; Weiss and Tailor, 1995) and human cells (via the transmembrane phosphate transporter RAM-1; Weiss and Tailor, 1995).

In the recombinant retroviral vector, the viral genes are replaced with the transgene of interest and expressed on plasmids in the packaging cell line. Because the nonessential genes lack the packaging sequence (Ψ) they are not included in the virion particle. To prevent recombination resulting in replication-competent retroviruses, all regions of homology with the vector backbone should be removed and the nonessential genes should be expressed by at least two transcriptional units (Markowitz et al., 1988). Even so, replication-competent retroviruses do occur at a low frequency. With this system, it is possible to produce viral titers of 10^5 to 10^7 colony forming units/ml.

The disadvantages of retroviral vectors include the random insertion into the host genome, which could possibly cause oncogene activation and the limited insert capacity (around 8kb). Moreover, a requirement for retroviral integration and expression of viral genes is that the target cells should be dividing. This limits its use to proliferating cells *in vivo*. For example, when treating cancers *in vivo*, tumor cells are preferentially targeted (Roth et al., 1996).

Lentiviruses belong to the general category of retroviruses that are mostly based on the human immunodeficiency virus type 1 (HIV-1). Once inside cells, the RNA genome of the lentivirus

Table 1. Properties of viral vectors used for gene delivery in the brain

Viral vector	Retrovirus (RV)/ lentivirus (LV)	Adenovirus (ADV)	Adeno-associated virus (AAV)	Herpes simplex virus (HSV) amplicon
Description	Spheric 80–100 nm, enveloped	Icosahedric 60–90 nm, non-enveloped	Icosahedric 25 nm, non-enveloped	Icosahedric 120–300 nm, enveloped
Genetic material	Single-stranded RNA	Double-stranded DNA	Single-stranded DNA	Double-stranded DNA
Packaging capacity	6–8 kb	Less than 10 kb	4.7 kb	up to 150 kb
Speed of expression	3–5 weeks	3–5 weeks	1–2 weeks	Hours to days
Integration into the host genome	Yes	No, episomal	Yes	No, episomal
Duration of expression	Years	Weeks	Years	Days
Infected cells	Dividing cell only (RV) dividing and quiescent cells (LV)	Dividing and quiescent cells	Dividing and quiescent cells	Dividing and quiescent cells
Natural tropism	Helper T cells, macrophage, monocytes, intestinal epithelia, brain	Broad	Different serotypes with various tropism	Mucosal epithelia, skin, cornea
Tropism through pseudotyping	Broad tropism when pseudotyped with VSV-G	Could be reduced when pseudotyped with fibers from different subgroup	Transcapsidation display differential efficiency and cell tropism	Broad tropism when pseudotyped with VSV-G
Viral production	Low titers	Growth to high titers	Can be concentrated to high titers	Can be concentrated to high titers
Advantages	High level of expression of foreign genes	High level of expression of foreign genes	Helper virus-free stocks possible nonpathogenic	Large insert capacity
Limitations	Potent human pathogen	High immunogenicity genetic manipulation is unwieldy	Limited insert capacity	Occasional cytotoxicity

is reverse transcribed by reverse transcriptase into a double-stranded DNA provirus that is incorporated into a pre-integration nucleoprotein complex able to pass through the pores of intact nuclear membranes. Therefore lentiviral vectors have all the advantages of Mo-MLV-based vectors, alongside the ability to infect both dividing and nondividing cells (Naldini et al., 1996). HIV vectors can accommodate fairly large gene inserts and can provide long-term expression through chromosomal integration. Lenti-vectors have a capacity of ~10kb for genetic material and sufficient amounts of high-concentration vector (10^8 to 10^9infection units/ml) can easily be produced. Stability of expression of lentivirus vectors in the brain is their greatest advantage. Long-term expression of the transgenes has been observed in rat neurons for at least 6 months following intracerebral injection of lentiviral vectors, with no sign of tissue pathology or immune response (Blomer et al., 1997). Nevertheless, a major limitation of lentiviruses is the limited genetic payload length that is often incompatible with full size enhancer/promoter needed for strong and cell-type-specific expression.

Adenovirus vectors

Adenoviruses are large (60–90nm diameter), non-enveloped, linear double-stranded DNA viruses that are usually associated with mild human infections including upper respiratory tract infections, keratoconjunctivitis and gastroenteritis. The adenovirus genome is 36kb in length and contains inverted terminal repeat (ITR) sequences at both ends containing the *cis*-acting DNA sequences that define the origin of DNA replication. The gene transcription of this virus can be divided into two phases of gene expression: early genes (E) and late gene (L), expressed before and after the onset of viral DNA, respectively. The first mRNA/protein to be made around 1h after infection is E1A. This protein is a transacting transcriptional regulatory factor that is necessary for transcriptional activation of early genes. The protein is also capable of activating transcription from a variety of other viral and cellular promoters and shows no sequence specificity, indicating that it functions by modifying the cellular environment.

The uptake of the adenovirus particle is a two stage process involving an initial interaction of the fiber knob protein with a range of cellular receptors, which include the MHC class I molecule and the high affinity cell-surface receptor called the Coxsackie and adenovirus receptor (CAR) (Bergelson et al., 1997). The capsid penton base protein then binds to the αvβ3 and αvβ5 integrin family of cell-surface heterodimers allowing internalization via receptor-mediated endocytosis (Wickham et al., 1993).

The first generation of recombinant adenovirus (rAds) is derived from the human adenovirus serotypes 5 and 2 and their replication was made defective-through deletion of the E1 gene regions. Adenoviruses are commonly used for gene transfer, as they can be generated at high titers and efficiently infect and express their genes in a variety of cell types including both dividing and quiescent cells. Other advantages of this vector include ease of manipulation (Graham and Prevec, 1991) and large insert capacity up to 35 kb of foreign DNA (Schiedner et al., 1998).

This vector has some drawbacks that may prevent its future use. First, most adenoviral vectors in their current form are episomal thus they do not integrate into the host DNA and therefore only cause a transient transgene expression. Second, because most mammals have been exposed to natural adenovirus infections, immunologic responses may hamper gene transfer efficacy. In addition, because the adenoviral genes express hundreds of proteins adenoviruses stimulate the immune system and trigger inflammatory responses. Newer second and third generation of rAd vectors that are deficient or defective in the E2, E3, or E4 gene regions are less immunogenic than the first generation and can be propagated on trans-complementary cell lines (Brough et al., 1996).

Vector systems have been developed in which most or all adenovirus proteins coding sequences are removed. From the viewpoint of safety, size of transgene and absence of vector gene expression, these so-called "gutless" adenovirus vectors contains only the ITRs and packaging signal of the wild-type virus. The gutless vectors require virtually all adenovirus gene functions to be provided for vector propagation and, since this cannot yet be achieved using a packaging cell line, they must be provided using a helper cell (Parks et al., 1996). Interestingly, gutless adenoviruses can be produced with a high titer (Parks et al., 1996) and give rise to long-term expression compared with first-generation adenoviruses (Morsy et al., 1998).

Adeno-associated vectors

AAV is a small (25nm diameter), non-enveloped, and single-stranded DNA parvovirus. The upstream open reading frame encodes four replication (rep) proteins that allow AAV rep-proteins to package AAV ITR-flanked transgenes into nearly all serotype virions. As a dependovirus, AAV requires Adenovirus or HSV as a helper virus to complete its lytic life (Conway et al., 1997). In the absence of the helper virus, wild-type AAV establishes latency by integration with the assistance of Rep-proteins through the interaction of the ITR with chromosome 19 (Berns and Giraud, 1995). The first stage in the AAV life cycle is the binding of the particle to a host cell. To bind cell receptors, AAVs use heparan sulfate proteoglycan (HSP) structures on the cell surface. Once bound to these receptors, they utilize coreceptors on the cell surface, for example, $\alpha v\beta 5$ integrin and fibroblast growth factor receptor 1 (FGFR1), which aid in internalization via receptor-mediated endocytosis followed by endosomal sorting and trafficking into the nucleus (Ding et al., 2005).

Until recently, the majority of the research conducted using AAV-based vectors employed serotype 2. Vectors based on AAV2 have been the most studied and are currently used in clinical trials for some diseases (Hildinger and Auricchio, 2004). To date at least 10 additional serotypes of AAV have been identified, the majority of which have been isolated as contaminants of adenoviral cultures. Many *in vivo* studies have clearly demonstrated that the various AAV serotypes display different tissue or cell tropisms (Zincarelli et al., 2008).

Recombinant AAV (rAAV) vectors are constructed by cotransfection of two plasmids. The first one contains the transcription unit of interest flanked by the ITRs and the second contains the rep and cap ORFs. In order to propagate rAAVs, infection with a helper virus (classically an adenovirus) is required. Although this technique enables the production of rAAVs with high titers (10^{10}infectious particles/ml) (Samulski et al., 1989), it requires extensive purification steps involving heating and cesium chloride gradient purification to remove adenoviral contamination. Here again, a main drawback is the packaging capacity for the transgene (4.7kb) that is too limited to generate vectors with a restricted expression in a specific cell type. Nevertheless, AAV has gained attention because of its safety, lack of immunogenic viral proteins and efficient transgene expression in a very broad host range (Samulski et al., 1989). AAV is able to infect a large number of both dividing and nondividing cells including neurons (Miao et al., 2000). AAV is not pathogenic and is not associated with disease, even though it has a broad range of infectivity. Moreover, transgene expression from rAAV vectors has been shown to continue for long periods of time, including up to 15 months in the CNS (Lo et al., 1999).

Herpes simplex virus vectors

HSV is an enveloped, doubled-stranded DNA virus with a genome of around 150kb encoding at least 80 proteins including many enzymes and surface glycoproteins. HSV is a member of the Herpesviridae family. It possesses an icosahedral capsid and is considered to be relatively large for a virus, with virions ranging from 120nm to 300 nm in size. HSV is a neurotropic DNA virus with a wide host range due to binding of viral envelope glycoproteins (gB and gC) to the extracellular HSP molecules (Wudunn and Spear, 1989). Internalization of the virus requires FGFR1 and envelope glycoprotein gD that binds specifically to a receptor called the Herpes Virus Entry Mediator (HVEM) receptor and provides a strong, fixed attachment to the host cell (Kaner et al., 1990).

Three different classes of vectors can be derived from HSV: replication-competent attenuated vectors, replication-incompetent recombinant vectors, and defective helper-dependent vectors known as amplicons (Neve et al., 2005). Amplicons are HSV-1 particles identical to wild-type HSV-1 from the structural, immunological, and host-range points of view, but which carry a concatemeric form of a DNA plasmid, named the amplicon plasmid, instead of the viral genome. An amplicon plasmid is a standard *E. coli* plasmid carrying an origin of DNA replication and a cleavage/packaging signal (pac or "a" site) from HSV-1, in addition to the transgenic sequences of interest. The major interest of amplicons as gene transfer tools stems from the fact that they carry no virus genes and consequently do not induce synthesis of virus proteins. Therefore, these vectors are fully nontoxic for the infected cells and nonpathogenic for the inoculated organisms. A second and major advantage is that most of the 150kbp capacity of the HSV-1 particle can be used to accommodate very large pieces of foreign DNA such as BAC clones allowing integration of full size enhancer/promoter for cell-specific expression. Herpes viruses are currently used as gene transfer vectors due to their specific advantages over other viral vectors. Among the unique features of HSV-derived vectors is their ability to invade and establish lifelong nontoxic latent infections in neurons (Carpenter and Stevens, 1996).

Targeted gene delivery

Viral vectors present many advantages for specific cell type delivery of transgene compared to transgenesis and DNA electroporation. In fact, recent developments have focused on the improvement of cell-type specificity in brain by using modified viral tropism, specific promoters, and intersections of viral infection with genetically modified organisms.

Modification of viral tropism

At the cellular level, a virus undergoes five major steps prior to achieving gene expression: (1) binding or attachment to cellular surface receptors, (2) endocytosis, (3) trafficking to the nucleus, (4) uncoating of the virus to release the genome, and (5) conversion of the genome to double-stranded DNA as a template for transcription in the nucleus. In this cascade of events, the first step, often termed viral tropism, defines the specificity of a virus for a particular host-cell type. Therefore by replacing the envelope or capsid proteins from a virus with that of another virus, the host range can be extended, in a technique known as pseudotyping.

Retrovirus/lentivirus The retroviral envelope interacts with a specific cellular protein to determine the target cell range. Altering the env gene or its product has proved a successful means of manipulating the cell range. Approaches have included direct modifications of the binding site between the envelope protein and the cellular receptor; however, these approaches tend to interfere with subsequent internalization of the viral particle (Harris and Lemoine, 1996). Another strategy to target-specific cell types is to genetically insert an antibody bridge between the envelope glycoprotein and specific cellular receptors (Etienne-Julan et al., 1992). Nevertheless, the retroviral vector constructed from the murine leukemia virus can only express transgenes in cells undergoing mitosis, indicating its inability as a delivery vehicle for neuronal expression.

Recombinant lentiviruses are often modified with the G glycoproteins (VSV-G) coming from the vesicular stomatitis virus, a prototypic member of the genus Vesiculovirus of the family Rhabdoviridae. VSV-G glycoproteins enable viral entry, mediate virus attachment to the host cell and fusion of the viral envelope with the endosomal membrane. VSV-G glycoproteins bind to ubiquitous phospholipid components of the plasma membrane but not to a specific cell-surface receptor, such viruses have an extremely broad host-cell range (Burns and Desrosiers, 1994). Thus, lentiviral expression of transgene in the brain is determined by selection of a neuronal specific promoter rather than specific tropism (see section "Transcriptional targeting").

Adenovirus Targeting cell entry of adenoviruses requires both the ablation of the native adenovirus tropism, especially for hepatocytes, and the incorporation of targeting ligands into the virus capsid. These aims have been achieved by complexing adenoviruses with different types of bispecific adapter molecules able to bind to specific cell-surface receptor (Barnett et al., 2002). To date, various approaches to retargeting adenoviruses (Ad) have been described. These include genetic modification strategies to incorporate peptide ligands (within fiber knob domain, fiber shaft, penton base, pIX, or hexon), pseudotyping of capsid proteins to include whole fiber substitutions or fiber knob chimeras, pseudotyping with monoclonal antibodies directed against surface-expressed target antigens and more (for review, see Coughlan et al., 2010).

Adeno-associated virus AAV serotype 2 (AAV2) was the first AAV serotype to be cloned into bacterial plasmids (Samulski et al., 1982). Since its discovery, 10 other serotypes with different capsid proteins affecting receptor-mediated endocytosis of AAV particles have been characterized (Rutledge et al., 1998). The tropism of AAV has been limited to particular cell types but can be expanded to include other cell types through modification of the capsid to target-specific cells or enhance AAV transduction. Thanks to the high degree of homology between the amino acid sequences of the different AAV serotypes and knowledge about the AAV2 crystal structure, it is possible to form a virion shell from capsid subunits of different serotypes to generate AAV that are composed of a mixture of viral capsid proteins from different serotypes. These mosaic virions exhibit a broader tissue tropism due to the combination of the tropisms from different serotypes, and also exhibit enhanced transgene expression since different serotypes may have different cellular trafficking pathways that serve to initiate transgene expression more efficiently (Cearley and Wolfe, 2006; Gao et al., 2005; Taymans et al., 2007). A recent study of neuronal infectivity for different serotypes has demonstrated that different brain regions exhibit different patterns of transduction (Taymans et al., 2007).

As for retroviruses, AAV tropism can be expanded to include other cell types by modification of receptor targeting by using an AAV-specific antibody that is chemically linked to another antibody binding specifically to a cellular receptor known to be expressed on the targeted cell surface. An example is the pioneering experiment performed with AAV2 able to recognize megakaryocyte cells (Ponnazhagan et al., 1996). Another strategy is the genetic manipulation of the capsid gene by insertion of a foreign protein sequence either from another wild-type AAV or an unrelated protein (for review, see Choi et al., 2005). A good example is the expression of the cellular receptor for retroviral envelope protein by an adeno-associated vector expressing improved viral transduction into numerous cell lines (Qing et al., 1997). Finally, it is possible to enhance AAV transduction by transcapsidation that consists in packaging an AAV genome containing an ITR from one serotype into the capsid of another serotype (Rabinowitz et al., 2002).

Herpes simplex virus HSV enter their host via mucosal epithelia, skin, or cornea. Studies have shown that HSV can infect many cell types both *in vitro* and *in vivo*. In this case, the main goal is to target a particular cell type both to increase efficiency and to potentially reduce the amount of virus that is needed. Strategies to retarget HSV vectors have been concentrated on deleting the HSP binding domain in both gB and gC in conjunction with receptor-specific ligand insertion. Another approach is based on pseudotyping of gD-deficient HSV with VSV-G to modify virus specificity (Anderson et al., 2000). These recombinant vectors should bind to the appropriate cell surface receptor but often had decreased infect rate (Grandi et al., 2002; Laquerre et al., 1998).

Transcriptional targeting

One of the major challenge in targeted gene transfer is the specificity of transgene expression only in the cell types of interest. As complex mechanisms regulate gene expression *in vivo* and most viral promoters do not have specific targeting capacities, a variety of tissue- or cell-specific promoters have been characterized. Moreover, viral promoters such as human cytomegalovirus (CMV) immediate-early gene are commonly used as regulatory element due to their strong activity in various cell types *in vitro*, but they are often not suitable for long-term expression in neurons. As discussed in previous chapters, promoters in mammals are several fold larger than their viral counterparts (up to hundred of kb), which is incompatible with the packaging capacity of all viral vectors except for the last generation HSV amplicons. To overcome this limitation, hybrid and synthetic promoters are being developed in order to improve the cell-type specificity and provide high level of transgene expression. In a recent study, it was reported on the ability of short promoter sequences to drive fluorescent protein expression in specific types of mammalian cortical inhibitory neurons using AAV and lentivirus vectors (Nathanson et al., 2009a). This group demonstrated that among fugu compact promoters PV, CR, SST, and NPY only the somatostatin and the neuropeptide promoters largely restricted expression to GABAergic neurons. Moreover, GFP expressing lentivirus vector that can accommodate larger regions of these promoters drove expression in excitatory neurons but not in inhibitory neurons consistent with the expected differences due to viral tropism (Nathanson et al., 2009b).

These results highlight the complexity of gene regulation and our lack of knowledge about regulatory elements that are required to precisely restrict expression in neurons. Nevertheless, many neuron-specific promoters have been used for transcriptional targeting by viral vectors, including those that control the expression of genes encoding neuron-specific enolase (NSE), synapsin-1 (SYN), platelet-derived growth factor (PDGF), tyrosine hydroxylase (TH), and dopamine β-hydroxylase (DBH) (Fitzsimons et al., 2002; Glover et al., 2002; Kugler et al., 2001; Paterna and Bueler, 2002).

Intersectional genetic switches

Intersectional strategies described in detail in section "Genetic modification of host genome by transgenesis approaches" are currently the most efficient approaches to achieve high cell-type specificity and have therefore been adapted for use with viral vectors. For example, a highly specific neuronal expression in rat brain was achieved using adenoviral infection (Namikawa et al., 2006). By combining an adenovirus expressing Cre recombinase under the control of a modified promoter of the superior cervical ganglion10 (SCG10) with another adenovirus vector expressing a Cre inducible EGFP flanked by loxP sites, this group was capable of mediating transgene expression at high levels both in neuronal cells of mixed cultures and in an animal model. The Cre/lox system has also been used as a random gene splicing strategy to express various combinations of fluorescent proteins in individual neurons of the brain of a transgenic mouse called brainbow mouse (Livet et al., 2007). More recently, this approach has been adapted for use with rAAV to deliver opsin gene in defined cell types (Fig. 3a). The specificity of this system is very high and relies on the introduction in the viral backbone of an ubiquitous promoter such as translational elongation factor EF1α in front of the gene encoding ChR2-YFP surrounded by two pairs of heterotypic and antiparallel loxP/lox2272 recombination sites (Atasoy et al., 2008; Kuhlman and Huang, 2008; Sohal et al., 2009). When this rAAV is stereotaxically injected in the brain of a Cre transgenic mouse (Fig. 3b), only neurons selectively expressing of Cre are able to process the cassette and therefore will express of the ChR2-YFP carried by the viral construct (Fig. 3c). Another advantage of this strategy is the low false positive background due to transcriptional readthrough observed with classic lox-STOP-lox cassette (Kuhlman and Huang, 2008).

In addition to achieving cell-type-specific expression, it would be desirable to achieve regulation of transgene expression. Here again, several regulatory presented in the section "Genetic modification of host genome by transgenesis approaches" have currently been adapted for use with viral vectors. For example, the tet system has been shown to be functional when expressed from several viral vector enabling tight regulation and inductility of transgene expression (Fotaki et al., 1997; Harding et al., 1998; Hwang et al., 1996).

Circuit mapping

The ability to visualize complex and extended neural networks is critical to understand the functional organization of the brain. Classically, brain circuit mapping has been established with chemical probes but this method lack cell-type specificity. To overcome this problem, a useful strategy uses viral tropism to target cell types based on their axonal projections. For example, different types of cortical pyramidal neurons project axons to distinct distant targets. Viruses that can efficiently infect neurons through their axon terminals can therefore be injected into a particular target structure, resulting in the selective infection of neurons that have axons in that structure. This method has been successfully

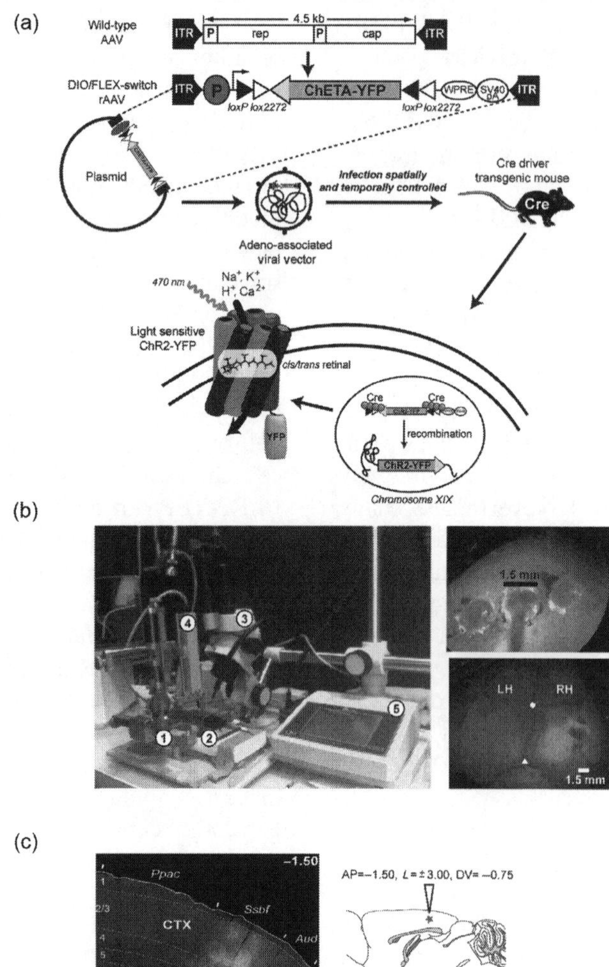

employed using HSV amplicon vectors, recombinant rabies virus, and adenovirus, as well as lentivirus pseudotyped with the rabies virus envelope protein (Mazarakis et al., 2001; Sandler et al., 2002; Tomioka and Rockland, 2006; Wickersham et al., 2007). For example, adenovirus vectors can be transported in a retrograde manner from the injection site to the projection cell bodies, following uptake at nerve terminals (Akli et al., 1993; Ridoux et al., 1994). Another interesting application of this ability of adenovirus to be retrogradely transported is that it could be used for the specific targeting of selected neuronal populations not easily accessible by direct injection, while avoiding any undesirable side effects associated with systemic administration (Finiels et al., 1995) or tissue damage due to viral toxicity at the site of injection (Cayouette and Gravel, 1996). Trans-synaptic targeting may also be achieved by expressing genetically encoded neuronal tracers such as wheat germ agglutinin (WGA) (Gradinaru et al., 2010) or tetanus toxin C (TTC) (Schwab et al., 1979). Recently, monosynaptic inputs of specific neuronal cell types have been determined by injection of glycoprotein-deleted (ΔG) rabies virus in transgenic mice that conditionally express rabies glycoprotein (Weible et al., 2010). Moreover, 12 new ΔG rabies virus variants were recently developed including ChR2-mCherry to allow selective light-controlled neuronal activation on the basis of their connectivity (Osakada et al., 2011).

Conclusion

A strategy to answer the difficult question of how the brain works is: to *know*, to *modify*, and to *control* neuronal activity. To *know* means to identify the meaningful genetic markers that are expressed in the cell types of interest within the brain. Even if this work is a long-drawn-out job, our knowledge of enhancers and promoters controlling gene expression in the CNS will be greatly improved by international projects on whole transcriptome or protein expression at single cell level. This will give the *cis*-regulating elements and/or the specific cell-surface receptors that can be targeted by a virus, depending on the strategy to express a transgene

Fig. 3. Highly specific viral mediated gene delivery. (a) Schematic representation of DIO/FLEX system to express ChR2-YFP in a specific neuronal subpopulation. This strategy is based on two components: a Cre-dependant virus containing ChR2 expression cassette under the control of a ubiquitous promoter P and a Cre driver transgenic mouse expressing Cre recombinase in a specific population of cells. After viral infection, the transgene integrates into the host genome facilitating stable expression of ChR2-YFP. (b) Experimental setup used to inject viral particles in mouse brain. (1) A stereotaxic apparatus is used to inject virions in the region of interest with high precision (±10μm) in an anesthetized mouse. (2) Body temperature is maintained around 37°C during surgery on a heat-controlled blanket. (3) For viral injection, thin holes are drilled through the skull (top right panel) under the guidance of a stereomicroscope. (4) After the micropipette is positioned in the brain parenchyma at the desired coordinates, the virus is injected with a constant speed. (5) The speed and volume of the injection is monitored by injector system (Stoelting QSI). The bottom right panel shows typical fluorescence observed after 14 days of infection by using rAAV2/1-FLEX-ChR2 in PV-Cre transgenic mouse (Hippenmeyer et al., 2005) (image is a composite of bright field and YFP epi-fluorescence images). White dot: bregma position, white triangle: lambda position, LH: left hemisphere, RH: right hemisphere. Unpublished results. (c) The right panel shows a high-resolution mosaic image in the YFP channel (consisting of around 100 individual frames) obtained through an automated Zeiss microscope, equipped with a high-precision motorized stage. The mosaic image was obtained from entire coronal brain sections cut at 40μm thickness. Level of ChR2 expression in somatosensory cortex was assessed by YFP visualization. Control Alexa568-labeled anti-parvalbumin antibody (Swant) show correct immunostaining of PV expressing interneurons (white triangle) under confoncal microscopy (bottom right panel). The top right panel indicates coordinates used for virus injection and a schematic representation of the injection site in the mouse brain (red star). Ppac: posterior parietal association cortex, Ssbf: somatosensory barrelfield, Aud: Auditory cortex, CTX: cortex, HPC: hippocampus, TH: thalamus. A. Urban et al., unpublished results. (For interpretation of the references to color in this figure legend, the reader is referred to the Web version of this chapter.)

in a chosen population of neurons. Based on this dataset, to *modify* relies on the technical ability to access the genome of these selected cells. Transgenic technologies have emerged as invaluable tools to manipulate the genes in neuroscience research. However, this method is still hampered by a relatively low efficiency and high cost in species other than mice. For this reason, the use of viral mediated gene delivery paves the way for an easier, faster, and cheaper modification of host genome and on a wider range of species. Then, *controlling* neurons will become possible thanks to optogenetics by using the most suited strategy to express opsins in the cells of interest with the greatest strength and specificity. When the first two steps will be fully mastered, optogenetic technology will extend even more our capacity to control a variety of neuronal networks with the highest possible spatiotemporal resolution, which is necessary to go further in the understanding of complex processes that occur in the brain.

References

Adesnik, H., & Scanziani, M. (2010). Lateral competition for cortical space by layer-specific horizontal circuits. *Nature*, *464*, 1155–1160.

Akli, S., Caillaud, C., Vigne, E., Stratford-Perricaudet, L. D., Poenaru, L., Perricaudet, M., et al. (1993). Transfer of a foreign gene into the brain using adenovirus vectors. *Nature Genetics*, *3*, 224–228.

Anderson, D. B., Laquerre, S., Ghosh, K., et al. (2000). Pseudotyping of glycoprotein D-deficient herpes simplex virus type 1 with vesicular stomatitis virus glycoprotein G enables mutant virus attachment and entry. *Journal of Virology*, *74*, 2481–2487.

Arenkiel, B. R., Peca, J., Davison, I. G., Feliciano, C., Deisseroth, K., Augustine, G. J., et al. (2007). In vivo light-induced activation of neural circuitry in transgenic mice expressing channelrhodopsin-2. *Neuron*, *54*, 205–218.

Ascoli, G. A., Alonso-Nanclares, L., Anderson, S. A., Barrionuevo, G., Benavides-Piccione, R., Burkhalter, A., et al. (2008). Petilla terminology: Nomenclature of features of GABAergic interneurons of the cerebral cortex. *Nature Reviews Neuroscience*, *9*, 557–568.

Atasoy, D., Aponte, Y., Su, H. H., & Sternson, S. M. (2008). A FLEX switch targets Channelrhodopsin-2 to multiple cell types for imaging and long-range circuit mapping. *The Journal of Neuroscience: The Official Journal of the Society for Neuroscience*, *28*, 7025–7030.

Awatramani, R., Soriano, P., Rodriguez, C., Mai, J. J., & Dymecki, S. M. (2003). Cryptic boundaries in roof plate and choroid plexus identified by intersectional gene activation. *Nature Genetics*, *35*, 70–75.

Bagga, R., Michalowski, S., Sabnis, R., Griffith, J. D., & Emerson, B. M. (2000). HMG I/Y regulates long-range enhancer-dependent transcription on DNA and chromatin by changes in DNA topology. *Nucleic Acids Research*, *28*, 2541–2550.

Barnett, B. G., Crews, C. J., & Douglas, J. T. (2002). Targeted adenoviral vectors. *Biochimica et Biophysica Acta*, *1575*, 1–14.

Barton, M. C., Madani, N., & Emerson, B. M. (1997). Distal enhancer regulation by promoter derepression in topologically constrained DNA in vitro. *Proceedings of the National Academy of Sciences of the United States of America*, *94*, 7257–7262.

Bellen, H. J., O'kane, C. J., Wilson, C., Grossniklaus, U., Pearson, R. K., & Gehring, W. J. (1989). P-element-mediated enhancer detection: A versatile method to study development in Drosophila. *Genes & Development*, *3*, 1288–1300.

Bergelson, J. M., Cunningham, J. A., Droguett, G., Kurt-Jones, E. A., Krithivas, A., Hong, J. S., et al. (1997). Isolation of a common receptor for Coxsackie B viruses and adenoviruses 2 and 5. *Science*, *275*, 1320–1323.

Berk, A. J. (1999). Activation of RNA polymerase II transcription. *Current Opinion in Cell Biology*, *11*, 330–335.

Berns, K. I., & Giraud, C. (1995). Adenovirus and adeno-associated virus as vectors for gene therapy. *Annals of the New York Academy of Sciences*, *772*, 95–104.

Blomer, U., Naldini, L., Kafri, T., Trono, D., Verma, I. M., & Gage, F. H. (1997). Highly efficient and sustained gene transfer in adult neurons with a lentivirus vector. *Journal of Virology*, *71*, 6641–6649.

Borrell, V., Yoshimura, Y., & Callaway, E. M. (2005). Targeted gene delivery to telencephalic inhibitory neurons by directional in utero electroporation. *Journal of Neuroscience Methods*, *143*, 151–158.

Bram, R. J., Lue, N. F., & Kornberg, R. D. (1986). A GAL family of upstream activating sequences in yeast: Roles in both induction and repression of transcription. *The EMBO Journal*, *5*, 603–608.

Broach, J. R., Guarascio, V. R., & Jayaram, M. (1982). Recombination within the yeast plasmid 2mu circle is site-specific. *Cell*, *29*, 227–234.

Bronson, S. K., Plaehn, E. G., Kluckman, K. D., Hagaman, J. R., Maeda, N., & Smithies, O. (1996). Single-copy transgenic mice with chosen-site integration. *Proceedings of the National Academy of Sciences of the United States of America*, *93*, 9067–9072.

Brough, D. E., Lizonova, A., Hsu, C., Kulesa, V. A., & Kovesdi, I. (1996). A gene transfer vector-cell line system for complete functional complementation of adenovirus early regions E1 and E4. *Journal of Virology*, *70*, 6497–6501.

Burns, D. P., & Desrosiers, R. C. (1994). Envelope sequence variation, neutralizing antibodies, and primate lentivirus persistence. *Current Topics in Microbiology and Immunology*, *188*, 185–219.

Carpenter, D. E., & Stevens, J. G. (1996). Long-term expression of a foreign gene from a unique position in the latent herpes simplex virus genome. *Human Gene Therapy*, *7*, 1447–1454.

Cauli, B., Audinat, E., Lambolez, B., Angulo, M. C., Ropert, N., Tsuzuki, K., et al. (1997). Molecular and physiological diversity of cortical nonpyramidal cells. *The Journal of Neuroscience*, *17*, 3894–3906.

Cayouette, M., & Gravel, C. (1996). Adenovirus-mediated gene transfer to retinal ganglion cells. *Investigative Ophthalmology & Visual Science*, *37*, 2022–2028.

Cearley, C. N., & Wolfe, J. H. (2006). Transduction characteristics of adeno-associated virus vectors expressing cap serotypes 7, 8, 9, and Rh10 in the mouse brain. *Molecular Therapy: The Journal of the American Society of Gene Therapy*, *13*, 528–537.

Chandler, K. J., Chandler, R. L., Broeckelmann, E. M., Hou, Y., Southard-Smith, E. M., & Mortlock, D. P. (2007). Relevance of BAC transgene copy number in mice: Transgene copy number variation across multiple transgenic lines and correlations with transgene integrity and expression. *Mammalian Genome: Official Journal of the International Mammalian Genome Society*, *18*, 693–708.

Chattopadhyaya, B., Di Cristo, G., Higashiyama, H., Knott, G. W., Kuhlman, S. J., Welker, E., et al. (2004). Experience and activity-dependent maturation of perisomatic GABAergic innervation in primary visual cortex during a postnatal critical period. *The Journal of Neuroscience: The Official Journal of the Society for Neuroscience*, *24*, 9598–9611.

Choi, V. W., Mccarty, D. M., & Samulski, R. J. (2005). AAV hybrid serotypes: Improved vectors for gene delivery. *Current Gene Therapy*, *5*, 299–310.

Conway, J. E., Zolotukhin, S., Muzyczka, N., Hayward, G. S., & Byrne, B. J. (1997). Recombinant adeno-associated virus type 2 replication and packaging is entirely supported by a herpes simplex virus type 1 amplicon expressing Rep and Cap. *Journal of Virology*, *71*, 8780–8789.

Coughlan, L., Alba, R., Parker, A. L., Bradshaw, A. C., Mcneish, I. A., Nicklin, S. A., et al. (2010). Tropism-modification strategies for targeted gene delivery using adenoviral vectors. *Viruses*, *2*, 2290–2355.

Davison, J. M., Akitake, C. M., Goll, M. G., Rhee, J. M., Gosse, N., Baier, H., et al. (2007). Transactivation from Gal4-VP16 transgenic insertions for tissue-specific cell labeling and ablation in zebrafish. *Developmental Biology*, *304*, 811–824.

Ding, W., Zhang, L., Yan, Z., & Engelhardt, J. F. (2005). Intracellular trafficking of adeno-associated viral vectors. *Gene Therapy*, *12*, 873–880.

Dymecki, S. M. (1996). Flp recombinase promotes site-specific DNA recombination in embryonic stem cells and transgenic mice. *Proceedings of the National Academy of Sciences of the United States of America*, *93*, 6191–6196.

Etienne-Julan, M., Roux, P., Carillo, S., Jeanteur, P., & Piechaczyk, M. (1992). The efficiency of cell targeting by recombinant retroviruses depends on the nature of the receptor and the composition of the artificial cell-virus linker. *Journal of General Virology*, *73*(Pt 12), 3251–3255.

Featherstone, M. (2002). Coactivators in transcription initiation: Here are your orders. *Current Opinion in Genetics and Development*, *12*, 149–155.

Feil, R., Brocard, J., Mascrez, B., Lemeur, M., Metzger, D., & Chambon, P. (1996). Ligand-activated site-specific recombination in mice. *Proceedings of the National Academy of Sciences of the United States of America*, *93*, 10887–10890.

Feng, G., Mellor, R. H., Bernstein, M., Keller-Peck, C., Nguyen, Q. T., Wallace, M., et al. (2000). Imaging neuronal subsets in transgenic mice expressing multiple spectral variants of GFP. *Neuron*, *28*, 41–51.

Fessele, S., Maier, H., Zischek, C., Nelson, P. J., & Werner, T. (2002). Regulatory context is a crucial part of gene function. *Trends in Genetics: TIG*, *18*, 60–63.

Finiels, F., Robert, J. J., Samolyk, M. L., Privat, A., Mallet, J., & Revah, F. (1995). Induction of neuronal apoptosis by excitotoxins associated with long-lasting increase of 12-O-tetradecanoylphorbol 13-acetate-responsive element-binding activity. *Journal of Neurochemistry*, *65*, 1027–1034.

Fitzsimons, H. L., Bland, R. J., & During, M. J. (2002). Promoters and regulatory elements that improve adeno-associated virus transgene expression in the brain. *Methods*, *28*, 227–236.

Fotaki, M. E., Pink, J. R., & Mous, J. (1997). Tetracycline-responsive gene expression in mouse brain after amplicon-mediated gene transfer. *Gene Therapy*, *4*, 901–908.

Friedrich, G., & Soriano, P. (1991). Promoter traps in embryonic stem cells: A genetic screen to identify and mutate developmental genes in mice. *Genes & Development*, *5*, 1513–1523.

Gao, G., Vandenberghe, L. H., & Wilson, J. M. (2005). New recombinant serotypes of AAV vectors. *Current Gene Therapy*, *5*, 285–297.

Giniger, E., Varnum, S. M., & Ptashne, M. (1985). Specific DNA binding of GAL4, a positive regulatory protein of yeast. *Cell*, *40*, 767–774.

Glover, C. P., Bienemann, A. S., Heywood, D. J., Cosgrave, A. S., & Uney, J. B. (2002). Adenoviral-mediated, high-level, cell-specific transgene expression: A SYN1-WPRE cassette mediates increased transgene expression with no loss of neuron specificity. *Molecular Therapy: The Journal of the American Society of Gene Therapy*, *5*, 509–516.

Golic, K. G., & Lindquist, S. (1989). The FLP recombinase of yeast catalyzes site-specific recombination in the Drosophila genome. *Cell*, *59*, 499–509.

Gong, S., Doughty, M., Harbaugh, C. R., Cummins, A., Hatten, M. E., Heintz, N., et al. (2007). Targeting Cre recombinase to specific neuron populations with bacterial artificial chromosome constructs. *The Journal of Neuroscience: The Official Journal of the Society for Neuroscience*, *27*, 9817–9823.

Gong, S., Yang, X. W., Li, C., & Heintz, N. (2002). Highly efficient modification of bacterial artificial chromosomes (BACs) using novel shuttle vectors containing the R6Kgamma origin of replication. *Genome Research*, *12*, 1992–1998.

Gordon, J. W., Chesa, P. G., Nishimura, H., Rettig, W. J., Maccari, J. E., Endo, T., et al. (1987). Regulation of Thy-1 gene expression in transgenic mice. *Cell*, *50*, 445–452.

Gossen, M., & Bujard, H. (1992). Tight control of gene expression in mammalian cells by tetracycline-responsive promoters. *Proceedings of the National Academy of Sciences of the United States of America*, *89*, 5547–5551.

Gradinaru, V., Thompson, K. R., Zhang, F., Mogri, M., Kay, K., Schneider, M. B., et al. (2007). Targeting and readout strategies for fast optical neural control in vitro and in vivo. *The Journal of Neuroscience: The Official Journal of the Society for Neuroscience*, *27*, 14231–14238.

Gradinaru, V., Zhang, F., Ramakrishnan, C., Mattis, J., Prakash, R., Diester, I., et al. (2010). Molecular and cellular approaches for diversifying and extending optogenetics. *Cell*, *141*, 154–165.

Graham, F. L., & Prevec, L. (1991). Manipulation of adenovirus vectors. *Methods in Molecular Biology*, *7*, 109–128.

Grandi, S., Wang, D., Schuback, V., Krasnykh, M., Spear, D., Curiel, R., et al. (2002). *Molecular Therapy*, *5*, S320.S.

Haas, K., Jensen, K., Sin, W. C., Foa, L., & Cline, H. T. (2002). Targeted electroporation in Xenopus tadpoles in vivo—From single cells to the entire brain. *Differentiation; Research in Biological Diversity*, *70*, 148–154.

Hagglund, M., Borgius, L., Dougherty, K. J., & Kiehn, O. (2010). Activation of groups of excitatory neurons in the mammalian spinal cord or hindbrain evokes locomotion. *Nature Neuroscience*, *13*, 246–252.

Han, D. D., Stein, D., & Stevens, L. M. (2000). Investigating the function of follicular subpopulations during Drosophila oogenesis through hormone-dependent enhancer-targeted cell ablation. *Development*, *127*, 573–583.

Harding, T. C., Geddes, B. J., Murphy, D., Knight, D., & Uney, J. B. (1998). Switching transgene expression in the brain using an adenoviral tetracycline-regulatable system. *Nature Biotechnology*, *16*, 553–555.

Harris, J. D., & Lemoine, N. R. (1996). Strategies for targeted gene therapy. *Trends in Genetics: TIG*, *12*, 400–405.

Hashimoto, H., Kikuchi, Y., Nogi, Y., & Fukasawa, T. (1983). Regulation of expression of the galactose gene cluster in Saccharomyces cerevisiae. Isolation and characterization of the regulatory gene GAL4. *Molecular and General Genetics: MGG*, *191*, 31–38.

Hayashi, S., Ito, K., Sado, Y., Taniguchi, M., Akimoto, A., Takeuchi, H., et al. (2002). GETDB, a database compiling expression patterns and molecular locations of a collection of Gal4 enhancer traps. *Genesis*, *34*, 58–61.

Heaney, J. D., Rettew, A. N., & Bronson, S. K. (2004). Tissue-specific expression of a BAC transgene targeted to the Hprt locus in mouse embryonic stem cells. *Genomics*, *83*, 1072–1082.

Heintz, N. (2001). BAC to the future: The use of bac transgenic mice for neuroscience research. *Nature Reviews Neuroscience*, *2*, 861–870.

Heintz, N. (2004). Gene expression nervous system atlas (GENSAT). *Nature Neuroscience*, *7*, 483.

Hildinger, M., & Auricchio, A. (2004). Advances in AAV-mediated gene transfer for the treatment of inherited disorders. *European Journal of Human Genetics: EJHG*, *12*, 263–271.

Hippenmeyer, S., Vrieseling, E., Sigrist, M., Portmann, T., Laengle, C., Ladle, D. R., et al. (2005). A developmental switch in the response of DRG neurons to ETS transcription factor signaling. *PLoS Biology*, *3*, e159.

Hippenmeyer, S., Youn, Y. H., Moon, H. M., Miyamichi, K., Zong, H., Wynshaw-Boris, A., et al. (2010). Genetic mosaic dissection of Lis1 and Ndel1 in neuronal migration. *Neuron*, *68*, 695–709.

Hirrlinger, J., Requardt, R. P., Winkler, U., Wilhelm, F., Schulze, C., & Hirrlinger, P. G. (2009). Split-CreERT2: Temporal control of DNA recombination mediated by split-Cre protein fragment complementation. *PLoS One*, *4*, e8354.

Hirrlinger, J., Scheller, A., Hirrlinger, P. G., Kellert, B., Tang, W., Wehr, M. C., et al. (2009). Split-cre complementation indicates coincident activity of different genes in vivo. *PLoS One*, *4*, e4286.

Hubel, D. H., & Wiesel, T. N. (1977). Ferrier lecture. Functional architecture of macaque monkey visual cortex. *Proceedings of the Royal Society of London. Series B, Containing Papers of a Biological Character. Royal Society*, *198*, 1–59.

Huber, D., Petreanu, L., Ghitani, N., Ranade, S., Hromadka, T., Mainen, Z., et al. (2008). Sparse optical microstimulation in barrel cortex drives learned behaviour in freely moving mice. *Nature*, *451*, 61–64.

Hwang, J. J., Scuric, Z., & Anderson, W. F. (1996). Novel retroviral vector transferring a suicide gene and a selectable marker gene with enhanced gene expression by using a tetracycline-responsive expression system. *Journal of Virology*, *70*, 8138–8141.

Indra, A. K., Warot, X., Brocard, J., Bornert, J. M., Xiao, J. H., Chambon, P., et al. (1999). Temporally-controlled site-specific mutagenesis in the basal layer of the epidermis: Comparison of the recombinase activity of the tamoxifen-inducible Cre-ER(T) and Cre-ER(T2) recombinases. *Nucleic Acids Research*, *27*, 4324–4327.

Isaka, Y., & Imai, E. (2007). Electroporation-mediated gene therapy. *Expert Opinion on Drug Delivery*, *4*, 561–571.

Itasaki, N., Bel-Vialar, S., & Krumlauf, R. (1999). 'Shocking' developments in chick embryology: Electroporation and in ovo gene expression. *Nature Cell Biology*, *1*, E203–E207.

Ito, K., Awano, W., Suzuki, K., Hiromi, Y., & Yamamoto, D. (1997). The Drosophila mushroom body is a quadruple structure of clonal units each of which contains a virtually identical set of neurones and glial cells. *Development*, *124*, 761–771.

Judkewitz, B., Rizzi, M., Kitamura, K., & Hausser, M. (2009). Targeted single-cell electroporation of mammalian neurons in vivo. *Nature Protocols*, *4*, 862–869.

Kadonaga, J. T. (2004). Regulation of RNA polymerase II transcription by sequence-specific DNA binding factors. *Cell*, *116*, 247–257.

Kakidani, H., & Ptashne, M. (1988). GAL4 activates gene expression in mammalian cells. *Cell*, *52*, 161–167.

Kaner, R. J., Baird, A., Mansukhani, A., Basilico, C., Summers, B. D., Florkiewicz, R. Z., et al. (1990). Fibroblast growth factor receptor is a portal of cellular entry for herpes simplex virus type 1. *Science*, *248*, 1410–1413.

Karagiannis, A., Gallopin, T., David, C., Battaglia, D., Geoffroy, H., Rossier, J., et al. (2009). Classification of NPY-expressing neocortical interneurons. *The Journal of Neuroscience: The Official Journal of the Society for Neuroscience*, *29*, 3642–3659.

Katzel, D., Zemelman, B. V., Buetfering, C., Wolfel, M., & Miesenbock, G. (2011). The columnar and laminar organization of inhibitory connections to neocortical excitatory cells. *Nature Neuroscience*, *14*, 100–107.

Kellendonk, C., Simpson, E. H., Polan, H. J., Malleret, G., Vronskaya, S., Winiger, V., et al. (2006). Transient and selective overexpression of dopamine D2 receptors in the striatum causes persistent abnormalities in prefrontal cortex functioning. *Neuron*, *49*, 603–615.

Kubota, Y., Hattori, R., & Yui, Y. (1994). Three distinct subpopulations of GABAergic neurons in rat frontal agranular cortex. *Brain Research*, *649*, 159–173.

Kugler, S., Meyn, L., Holzmuller, H., Gerhardt, E., Isenmann, S., Schulz, J. B., et al. (2001). Neuron-specific expression of therapeutic proteins: Evaluation of different cellular promoters in recombinant adenoviral vectors. *Molecular and Cellular Neurosciences*, *17*, 78–96.

Kuhlman, S. J., & Huang, Z. J. (2008). High-resolution labeling and functional manipulation of specific neuron types in mouse brain by Cre-activated viral gene expression. *PLoS One*, *3*, e2005.

Lai, S. L., & Lee, T. (2006). Genetic mosaic with dual binary transcriptional systems in Drosophila. *Nature Neuroscience*, *9*, 703–709.

Lambolez, B., Audinat, E., Bochet, P., Crepel, F., & Rossier, J. (1992). AMPA receptor subunits expressed by single Purkinje cells. *Neuron*, *9*, 247–258.

Laquerre, R., Argnani, D. B., Anderson, S., Zucchnil, R., Manservigi, & Glorioso, J. (1998). *Journal of Virology*, *72*, 6119–6130.

Lee, T., & Luo, L. (1999). Mosaic analysis with a repressible cell marker for studies of gene function in neuronal morphogenesis. *Neuron*, *22*, 451–461.

Lewis, T. L., Jr., Mao, T., Svoboda, K., & Arnold, D. B. (2009). Myosin-dependent targeting of transmembrane proteins to neuronal dendrites. *Nature Neuroscience*, *12*, 568–576.

Livet, J., Weissman, T. A., Kang, H., Draft, R. W., Lu, J., Bennis, R. A., et al. (2007). Transgenic strategies for combinatorial expression of fluorescent proteins in the nervous system. *Nature*, *450*, 56–62.

Lo, W. D., Qu, G., Sferra, T. J., Clark, R., Chen, R., & Johnson, P. R. (1999). Adeno-associated virus-mediated gene transfer to the brain: Duration and modulation of expression. *Human Gene Therapy*, *10*, 201–213.

Lopez-Bendito, G., Sturgess, K., Erdelyi, F., Szabo, G., Molnar, Z., & Paulsen, O. (2004). Preferential origin and layer destination of GAD65-GFP cortical interneurons. *Cerebral Cortex*, *14*, 1122–1133.

Luan, H., & White, B. H. (2007). Combinatorial methods for refined neuronal gene targeting. *Current Opinion in Neurobiology*, *17*, 572–580.

Lue, N. F., Chasman, D. I., Buchman, A. R., & Kornberg, R. D. (1987). Interaction of GAL4 and GAL80 gene regulatory proteins in vitro. *Molecular and Cellular Biology*, *7*, 3446–3451.

Luo, L., Callaway, E. M., & Svoboda, K. (2008). Genetic dissection of neural circuits. *Neuron*, *57*, 634–660.

Ma, J., Przibilla, E., Hu, J., Bogorad, L., & Ptashne, M. (1988). Yeast activators stimulate plant gene expression. *Nature*, *334*, 631–633.

Markowitz, D., Goff, S., & Bank, A. (1988). A safe packaging line for gene transfer: Separating viral genes on two different plasmids. *Journal of Virology*, *62*, 1120–1124.

Marra, M. A., Kucaba, T. A., Dietrich, N. L., Green, E. D., Brownstein, B., Wilson, R. K., et al. (1997). High throughput fingerprint analysis of large-insert clones. *Genome Research*, *7*, 1072–1084.

Mazarakis, N. D., Azzouz, M., Rohll, J. B., Ellard, F. M., Wilkes, F. J., Olsen, A. L., et al. (2001). Rabies virus glycoprotein pseudotyping of lentiviral vectors enables retrograde axonal transport and access to the nervous system after peripheral delivery. *Human Molecular Genetics*, *10*, 2109–2121.

Mcguire, S. E., Le, P. T., Osborn, A. J., Matsumoto, K., & Davis, R. L. (2003). Spatiotemporal rescue of memory dysfunction in Drosophila. *Science*, *302*, 1765–1768.

Mcmullen, N. T., Smelser, C. B., & Rice, F. L. (1994). Parvalbumin expression reveals a vibrissa-related pattern in rabbit SI cortex. *Brain Research*, *660*, 225–231.

Mertz, K. D., Weisheit, G., Schilling, K., & Luers, G. H. (2002). Electroporation of primary neural cultures: A simple method for directed gene transfer in vitro. *Histochemistry and Cell Biology*, *118*, 501–506.

Miao, C. H., Nakai, H., Thompson, A. R., Storm, T. A., Chiu, W., Snyder, R. O., et al. (2000). Nonrandom transduction of recombinant adeno-associated virus vectors in mouse hepatocytes in vivo: Cell cycling does not influence hepatocyte transduction. *Journal of Virology*, *74*, 3793–3803.

Misra, R. P., & Duncan, S. A. (2002). Gene targeting in the mouse: Advances in introduction of transgenes into the genome by homologous recombination. *Endocrine*, *19*, 229–238.

Miyazaki, S., Miyazaki, T., Tashiro, F., Yamato, E., & Miyazaki, J. (2005). Development of a single-cassette system for spatiotemporal gene regulation in mice. *Biochemical and Biophysical Research Communications*, *338*, 1083–1088.

Morsy, M. A., Gu, M., Motzel, S., Zhao, J., Lin, J., Su, Q., et al. (1998). An adenoviral vector deleted for all viral coding sequences results in enhanced safety and extended expression of a leptin transgene. *Proceedings of the National Academy of Sciences of the United States of America*, *95*, 7866–7871.

Nagayoshi, S., Hayashi, E., Abe, G., Osato, N., Asakawa, K., Urasaki, A., et al. (2008). Insertional mutagenesis by the Tol2 transposon-mediated enhancer trap approach generated mutations in two developmental genes: tcf7 and synembryn-like. *Development*, *135*, 159–169.

Nakazawa, K., Quirk, M. C., Chitwood, R. A., Watanabe, M., Yeckel, M. F., Sun, L. D., et al. (2002). Requirement for hippocampal CA3 NMDA receptors in associative memory recall. *Science*, *297*, 211–218.

Naldini, L., Blomer, U., Gallay, P., Ory, D., Mulligan, R., Gage, F. H., et al. (1996). In vivo gene delivery and stable transduction of nondividing cells by a lentiviral vector. *Science*, *272*, 263–267.

Namikawa, K., Murakami, K., Okamoto, T., Okado, H., & Kiyama, H. (2006). A newly modified SCG10 promoter and Cre/loxP-mediated gene amplification system achieve highly specific neuronal expression in animal brains. *Gene Therapy*, *13*, 1244–1250.

Nathanson, J. L., Jappelli, R., Scheeff, E. D., Manning, G., Obata, K., Brenner, S., et al. (2009a). Short promoters in viral vectors drive selective expression in mammalian inhibitory neurons, but do not restrict activity to specific inhibitory cell-types. *Frontiers in Neural Circuits*, *3*, 19.

Nathanson, J. L., Yanagawa, Y., Obata, K., & Callaway, E. M. (2009b). Preferential labeling of inhibitory and excitatory cortical neurons by endogenous tropism of adeno-associated virus and lentivirus vectors. *Neuroscience*, *161*, 441–450.

Neve, R. L., Neve, K. A., Nestler, E. J., & Carlezon, W. A. Jr. (2005). Use of herpes virus amplicon vectors to study brain disorders. *Biotechniques*, *39*, 381–391.

Nikolov, D. B., & Burley, S. K. (1997). RNA polymerase II transcription initiation: A structural view. *Proceedings of the National Academy of Sciences of the United States of America*, *94*, 15–22.

Oki, M., & Kamakaka, R. T. (2002). Blockers and barriers to transcription: Competing activities? *Current Opinion in Cell Biology*, *14*, 299–304.

Oliva, A. A., Jr., Jiang, M., Lam, T., Smith, K. L., & Swann, J. W. (2000). Novel hippocampal interneuronal subtypes identified using transgenic mice that express green fluorescent protein in GABAergic interneurons. *The Journal of Neuroscience: The Official Journal of the Society for Neuroscience*, *20*, 3354–3368.

Ornitz, D. M., Moreadith, R. W., & Leder, P. (1991). Binary system for regulating transgene expression in mice: Targeting int-2 gene expression with yeast GAL4/UAS control elements. *Proceedings of the National Academy of Sciences of the United States of America*, *88*, 698–702.

Osakada, F., Mori, T., Cetin, A. H., Marshel, J. H., Virgen, B., & Callaway, E. M. (2011). New rabies virus variants for monitoring and manipulating activity and gene expression in defined neural circuits. *Neuron*, *71*, 617–631.

Osterwalder, T., Yoon, K. S., White, B. H., & Keshishian, H. (2001). A conditional tissue-specific transgene expression system using inducible GAL4. *Proceedings of the National Academy of Sciences of the United States of America*, *98*, 12596–12601.

Parks, R. J., Chen, L., Anton, M., Sankar, U., Rudnicki, M. A., & Graham, F. L. (1996). A helper-dependent adenovirus vector system: Removal of helper virus by Cre-mediated excision of the viral packaging signal. *Proceedings of the National Academy of Sciences of the United States of America*, *93*, 13565–13570.

Paterna, J. C., & Bueler, H. (2002). Recombinant adeno-associated virus vector design and gene expression in the mammalian brain. *Methods*, *28*, 208–218.

Perier, R. C., Junier, T., Bonnard, C., & Bucher, P. (1999). The Eukaryotic Promoter Database (EPD): Recent developments. *Nucleic Acids Research*, *27*, 307–309.

Petreanu, L., Huber, D., Sobczyk, A., & Svoboda, K. (2007). Channelrhodopsin-2-assisted circuit mapping of long-range callosal projections. *Nature Neuroscience*, *10*, 663–668.

Ponnazhagan, S., Wang, X. S., Woody, M. J., Luo, F., Kang, L. Y., Nallari, M. L., et al. (1996). Differential expression in human cells from the p6 promoter of human parvovirus B19 following plasmid transfection and recombinant adeno-associated virus 2 (AAV) infection: Human megakaryocytic leukaemia cells are non-permissive for AAV infection. *Journal of General Virology*, *77*(Pt 6), 1111–1122.

Potter, C. J., Tasic, B., Russler, E. V., Liang, L., & Luo, L. (2010). The Q system: A repressible binary system for transgene expression, lineage tracing, and mosaic analysis. *Cell, 141*, 536–548.

Qing, K., Bachelot, T., Mukherjee, P., Wang, X. S., Peng, L., Yoder, M. C., et al. (1997). Adeno-associated virus type 2-mediated transfer of ecotropic retrovirus receptor cDNA allows ecotropic retroviral transduction of established and primary human cells. *Journal of Virology, 71*, 5663–5667.

Rabinowitz, J. E., Rolling, F., Li, C., Conrath, H., Xiao, W., Xiao, X., et al. (2002). Cross-packaging of a single adeno-associated virus (AAV) type 2 vector genome into multiple AAV serotypes enables transduction with broad specificity. *Journal of Virology, 76*, 791–801.

Ridoux, V., Robert, J. J., Zhang, X., Perricaudet, M., Mallet, J., & Le Gal La Salle, G. (1994). Adenoviral vectors as functional retrograde neuronal tracers. *Brain Research, 648*, 171–175.

Roth, J. A., Nguyen, D., Lawrence, D. D., Kemp, B. L., Carrasco, C. H., Ferson, D. Z., et al. (1996). Retrovirus-mediated wild-type p53 gene transfer to tumors of patients with lung cancer. *Nature Medicine, 2*, 985–991.

Rutledge, E. A., Halbert, C. L., & Russell, D. W. (1998). Infectious clones and vectors derived from adeno-associated virus (AAV) serotypes other than AAV type 2. *Journal of Virology, 72*, 309–319.

Saito, T., & Nakatsuji, N. (2001). Efficient gene transfer into the embryonic mouse brain using in vivo electroporation. *Developmental Biology, 240*, 237–246.

Samulski, R. J., Berns, K. I., Tan, M., & Muzyczka, N. (1982). Cloning of adeno-associated virus into pBR322: Rescue of intact virus from the recombinant plasmid in human cells. *Proceedings of the National Academy of Sciences of the United States of America, 79*, 2077–2081.

Samulski, R. J., Chang, L. S., & Shenk, T. (1989). Helper-free stocks of recombinant adeno-associated viruses: Normal integration does not require viral gene expression. *Journal of Virology, 63*, 3822–3828.

Sandler, V. M., Wang, S., Angelo, K., Lo, H. G., Breakefield, X. O., & Clapham, D. E. (2002). Modified herpes simplex virus delivery of enhanced GFP into the central nervous system. *Journal of Neuroscience Methods, 121*, 211–219.

Sauer, B. (1993). Manipulation of transgenes by site-specific recombination: Use of Cre recombinase. *Methods in Enzymology, 225*, 890–900.

Schiedner, G., Morral, N., Parks, R. J., Wu, Y., Koopmans, S. C., Langston, C., et al. (1998). Genomic DNA transfer with a high-capacity adenovirus vector results in improved in vivo gene expression and decreased toxicity. *Nature Genetics, 18*, 180–183.

Schwab, M. E., Suda, K., & Thoenen, H. (1979). Selective retrograde transsynaptic transfer of a protein, tetanus toxin, subsequent to its retrograde axonal transport. *The Journal of Cell Biology, 82*, 798–810.

Scott, E. K., & Baier, H. (2009). The cellular architecture of the larval zebrafish tectum, as revealed by gal4 enhancer trap lines. *Frontiers in Neural Circuits, 3*, 13.

Scott, E. K., Mason, L., Arrenberg, A. B., Ziv, L., Gosse, N. J., Xiao, T., et al. (2007). Targeting neural circuitry in zebrafish using GAL4 enhancer trapping. *Nature Methods, 4*, 323–326.

Shizuya, H., Birren, B., Kim, U. J., Mancino, V., Slepak, T., Tachiiri, Y., et al. (1992). Cloning and stable maintenance of 300-kilobase-pair fragments of human DNA in Escherichia coli using an F-factor-based vector. *Proceedings of the National Academy of Sciences of the United States of America, 89*, 8794–8797.

Sohal, V. S., Zhang, F., Yizhar, O., & Deisseroth, K. (2009). Parvalbumin neurons and gamma rhythms enhance cortical circuit performance. *Nature, 459*, 698–702.

Soriano, P. (1999). Generalized lacZ expression with the ROSA26 Cre reporter strain. *Nature Genetics, 21*, 70–71.

Stebbins, M. J., Urlinger, S., Byrne, G., Bello, B., Hillen, W., & Yin, J. C. (2001). Tetracycline-inducible systems for Drosophila. *Proceedings of the National Academy of Sciences of the United States of America, 98*, 10775–10780.

Struhl, G., & Basler, K. (1993). Organizing activity of wingless protein in Drosophila. *Cell, 72*, 527–540.

Subkhankulova, T., Yano, K., Robinson, H. P., & Livesey, F. J. (2010). Grouping and classifying electrophysiologically-defined classes of neocortical neurons by single cell, whole-genome expression profiling. *Frontiers in Molecular Neuroscience, 3*, 10.

Szuts, D., & Bienz, M. (2000). LexA chimeras reveal the function of Drosophila Fos as a context-dependent transcriptional activator. *Proceedings of the National Academy of Sciences of the United States of America, 97*, 5351–5356.

Tanahira, C., Higo, S., Watanabe, K., Tomioka, R., Ebihara, S., Kaneko, T., et al. (2009). Parvalbumin neurons in the forebrain as revealed by parvalbumin-Cre transgenic mice. *Neuroscience Research, 63*, 213–223.

Tasic, B., Hippenmeyer, S., Wang, C., Gamboa, M., Zong, H., Chen-Tsai, Y., et al. (2011). From the cover: Site-specific integrase-mediated transgenesis in mice via pronuclear injection. *Proceedings of the National Academy of Sciences of the United States of America, 108*, 7902–7907.

Taymans, J. M., Vandenberghe, L. H., Haute, C. V., Thiry, I., Deroose, C. M., Mortelmans, L., et al. (2007). Comparative analysis of adeno-associated viral vector serotypes 1, 2, 5, 7, and 8 in mouse brain. *Human Gene Therapy, 18*, 195–206.

Teruel, M. N., Blanpied, T. A., Shen, K., Augustine, G. J., & Meyer, T. (1999). A versatile microporation technique for the transfection of cultured CNS neurons. *Journal of Neuroscience Methods, 93*, 37–48.

Testa, G., Zhang, Y., Vintersten, K., Benes, V., Pijnappel, W. W., Chambers, I., et al. (2003). Engineering the mouse genome with bacterial artificial chromosomes to create multipurpose alleles. *Nature Biotechnology, 21*, 443–447.

Thomson, J. G., & Ow, D. W. (2006). Site-specific recombination systems for the genetic manipulation of eukaryotic genomes. *Genesis, 44*, 465–476.

Thyagarajan, S., Van Wyk, M., Lehmann, K., Lowel, S., Feng, G., & Wassle, H. (2010). Visual function in mice with photoreceptor degeneration and transgenic expression of channelrhodopsin 2 in ganglion cells. *The Journal of Neuroscience: The Official Journal of the Society for Neuroscience, 30*, 8745–8758.

Tjian, R., & Maniatis, T. (1994). Transcriptional activation: A complex puzzle with few easy pieces. *Cell, 77*, 5–8.

Tomioka, R., & Rockland, K. S. (2006). Improved Golgi-like visualization in retrogradely projecting neurons after EGFP-adenovirus infection in adult rat and monkey. *The Journal of Histochemistry and Cytochemistry: Official Journal of the Histochemistry Society, 54*, 539–548.

Tsien, J. Z., Huerta, P. T., & Tonegawa, S. (1996). The essential role of hippocampal CA1 NMDA receptor-dependent synaptic plasticity in spatial memory. *Cell, 87*, 1327–1338.

Vetter, D., Andrews, B. J., Roberts-Beatty, L., & Sadowski, P. D. (1983). Site-specific recombination of yeast 2-micron DNA in vitro. *Proceedings of the National Academy of Sciences of the United States of America, 80*, 7284–7288.

Walker, G. C. (1984). Mutagenesis and inducible responses to deoxyribonucleic acid damage in Escherichia coli. *Microbiological Reviews, 48*, 60–93.

Wallace, H., Ansell, R., Clark, J., & Mcwhir, J. (2000). Preselection of integration sites imparts repeatable transgene expression. *Nucleic Acids Research, 28*, 1455–1464.

Wang, H., Peca, J., Matsuzaki, M., Matsuzaki, K., Noguchi, J., Qiu, L., et al. (2007). High-speed mapping of synaptic connectivity using photostimulation in Channelrhodopsin-2 transgenic mice. *Proceedings of the National Academy of Sciences of the United States of America, 104*, 8143–8148.

Webster, N., Jin, J. R., Green, S., Hollis, M., & Chambon, P. (1988). The yeast UASG is a transcriptional enhancer in human HeLa cells in the presence of the GAL4 trans-activator. *Cell, 52*, 169–178.

Wei, F., Xia, X. M., Tang, J., Ao, H., Ko, S., Liauw, J., et al. (2003). Calmodulin regulates synaptic plasticity in the anterior cingulate cortex and behavioral responses: A microelectroporation study in adult rodents. *The Journal of Neuroscience: The Official Journal of the Society for Neuroscience, 23*, 8402–8409.

Weible, A. P., Schwarcz, L., Wickersham, I. R., Deblander, L., Wu, H., Callaway, E. M., et al. (2010). Transgenic targeting of recombinant rabies virus reveals monosynaptic connectivity of specific neurons. *The Journal of Neuroscience: The Official Journal of the Society for Neuroscience, 30*, 16509–16513.

Weiss, R. A., & Tailor, C. S. (1995). Retrovirus receptors. *Cell, 82*, 531–533.

Wickersham, I. R., Lyon, D. C., Barnard, R. J., Mori, T., Finke, S., Conzelmann, K. K., et al. (2007). Monosynaptic restriction of transsynaptic tracing from single, genetically targeted neurons. *Neuron, 53*, 639–647.

Wickham, T. J., Mathias, P., Cheresh, D. A., & Nemerow, G. R. (1993). Integrins alpha v beta 3 and alpha v beta 5 promote adenovirus internalization but not virus attachment. *Cell, 73*, 309–319.

Wu, Y., Reece, R. J., & Ptashne, M. (1996). Quantitation of putative activator-target affinities predicts transcriptional activating potentials. *The EMBO Journal, 15*, 3951–3963.

Wudunn, D., & Spear, P. G. (1989). Initial interaction of herpes simplex virus with cells is binding to heparan sulfate. *Journal of Virology, 63*, 52–58.

Zhang, F., Wang, L. P., Brauner, M., Liewald, J. F., Kay, K., Watzke, N., et al. (2007). Multimodal fast optical interrogation of neural circuitry. *Nature, 446*, 633–639.

Zhao, Y., Flandin, P., Long, J. E., Cuesta, M. D., Westphal, H., & Rubenstein, J. L. (2008). Distinct molecular pathways for development of telencephalic interneuron subtypes revealed through analysis of Lhx6 mutants. *The Journal of Comparative Neurology, 510*, 79–99.

Zincarelli, C., Soltys, S., Rengo, G., & Rabinowitz, J. E. (2008). Analysis of AAV serotypes 1–9 mediated gene expression and tropism in mice after systemic injection. *Molecular Therapy: The Journal of the American Society of Gene Therapy, 16*, 1073–1080.

T. Knöpfel and E. Boyden (Eds.)
Progress in Brain Research, Vol. 196
ISSN: 0079-6123

CHAPTER 10

Mouse transgenic approaches in optogenetics

Hongkui Zeng* and Linda Madisen

Allen Institute for Brain Science, Seattle, WA, USA

Abstract: A major challenge in neuroscience is to understand how universal behaviors, such as sensation, movement, cognition, and emotion, arise from the interactions of specific cells that are present within intricate neural networks in the brain. Dissection of such complex networks has typically relied on disturbing the activity of individual gene products, perturbing neuronal activities pharmacologically, or lesioning specific brain regions, to investigate the network's response in a behavioral output. Though informative for many kinds of studies, these approaches are not sufficiently fine-tuned for examining the functionality of specific cells or cell classes in a spatially or temporally restricted context. Recent advances in the field of optogenetics now enable researchers to monitor and manipulate the activity of genetically defined cell populations with the speed and precision uniquely afforded by light. Transgenic mice engineered to express optogenetic tools in a cell type-specific manner offer a powerful approach for examining the role of particular cells in discrete circuits in a defined and reproducible way. Not surprisingly then, recent years have seen substantial efforts directed toward generating transgenic mouse lines that express functionally relevant levels of optogenetic tools. In this chapter, we review the state of these efforts and consider aspects of the current technology that would benefit from additional improvement.

Keywords: transgenic mice; genetic manipulation; cell type; Cre; channelrhodopsin; halorhodopsin; archaerhodopsin; calcium indicator; voltage sensor.

Introduction

Transgenic mice have been widely used in neuroscience research to facilitate the deciphering of gene and cellular functions. Perhaps the greatest advantage of using a transgenic approach in such studies is that cell population-restricted transgene expression can be achieved using specific promoters, and this restricted pattern of expression can be passed on to subsequent generations fairly reproducibly. In functional studies of the mouse brain, a variety of transgenic strategies have been used to inactivate or overexpress particular genes, label specific cell populations or

*Corresponding author.
Tel.: +1-206-548-7104; Fax: +1-206-548-7083
E-mail: hongkuiz@alleninstitute.org

DOI: 10.1016/B978-0-444-59426-6.00010-0

their subcellular compartments, and manipulate the activity or function of specific cell populations (Luo et al., 2008). For example, the strong, neuronally restricted expression of fluorescent reporter in *Thy1*-EYFP mice has made possible studies of morphology, connectivity, electrophysiology, and mRNA content of a single neuron and has permitted long-term *in vivo* imaging of neurons (Feng et al., 2000; Micheva et al., 2010; Sugino et al., 2006). Further, by incorporating a strategy for combinatorial expression of fluorescent proteins, BrainBow mice have enabled the simultaneous mapping of projections and connectivity among multiple neurons (Livet et al., 2007). Given the wealth of information transgenic mice have yielded in past studies of neural circuits, it is not surprising that considerable efforts have been expended to establish lines in which the activity of populations of neurons can be both easily observed and reliably and reversibly manipulated.

One of the most exciting recent advances in experimental neuroscience has been the development of genetically encoded light-sensitive proteins, giving rise to the burgeoning field of optogenetics. In its broadest sense, optogenetic tools include both optical indicators of neuronal activity, such as genetically encoded calcium or voltage sensors, and optical actuators of neuronal activity, such as light-activated membrane channels and pumps. Although both types of tools are of intense interest to the neuroscience community, the latter group of molecules has been especially pursued given the opportunity they offer for being able to activate and inactivate particular neurons in live, behaving animals. Recent work incorporating three of these molecules, the neural-activating cation channel, channelrhodopsin-2 (ChR2) (Boyden et al., 2005; Nagel et al., 2003), the neural-silencing chloride transporter, halorhodopsin (NpHR) (Han and Boyden, 2007; Zhang et al., 2007), and the neural-silencing proton pump, archaerhodopsin (Arch) (Chow et al., 2010), has demonstrated the power of these tools to activate or silence neurons with unparalleled specificity and temporal precision on a millisecond scale. In addition, ChR2 has already been widely used in rodents to map circuits between defined neuronal populations.

Optogenetic actuators, such as ChR2, function by regulating the membrane potential of excitable cells. To generate sufficient membrane depolarization for light-induced action potentials, functional ChR2 protein must be expressed on the cell membrane at very high level or density due to the low single-channel conductance. In the past, such high-level expression has routinely been achieved using strategies that rely on delivering high copy numbers of transgene to cells, such as by *in utero* electroporation or viral infection. With a few notable exceptions, it has proven more difficult to obtain transgenic mice that express these genetic tools both robustly and widely enough to allow for probing the functionality of a wide range of cell types. For example, although the *Thy1* promoter directed sufficient ChR2 expression to investigate the cortical and olfactory circuits (Arenkiel et al., 2007; Wang et al., 2007), this promoter is sensitive to inhibitory positional effects when randomly integrated into the genome, and it lacks ubiquitous neuronal expression. Clearly, to exploit the full potential of current and future optogenetic tools for elucidating neural circuitry, transgenic lines need to be developed that will allow for high-level transgene expression in any specific cell type of interest. Recently developed Cre-dependent reporter mouse lines with the ability to robustly express a variety of opsins proffer great promise to fulfill this need.

Optogenetic indicators, such as genetically encoded calcium indicators (GECIs) (also called fluorescent calcium indicator proteins) and voltage sensitive fluorescent proteins (VSFPs), have had a longer history of development than the optogenetic actuators. In particular, a variety of GECIs have been engineered, based on combinations of different types of calcium-binding proteins and fluorescent proteins (Mank and Griesbeck, 2008). Major advantages of using genetically encoded optical sensors over synthetic indicators to monitor cell activity include their suitability for long-term tracking of particular cells over time, as well as their ability to target specific cell types or populations. Although recent work has improved upon the sensitivity and stability of early generation optical indicators, current versions of both GECIs and VSFPs still require very high-level expression to present changes in relative fluorescence at a sufficiently high signal-to-noise ratio (SNR). Thus, there is a continued need to improve both the SNR and the sensitivity to subthreshold and single spike-induced changes in calcium or voltage. Similar to optogenetic actuators, genetically encoded optical indicators are most commonly delivered through viral or DNA plasmid transduction in functional studies. The most recent versions of these indicators, such as GCaMP3 (Tian et al., 2009), possess greatly improved properties over earlier iterations, making the transgenic approach feasible for their application. Indeed, promising mouse lines that express these molecules have been developed and are currently being characterized.

In the following sections, we will review common strategies for generating transgenic mice, discuss the significant progress that has been made over the past few years in developing transgenic lines that express optogenetic molecules to functional levels, and consider what improvements to current technologies are needed to allow transgenic lines to capitalize on the exceedingly powerful tools offered by optogenetics.

General transgenic approaches

There are two general strategies for expressing a transgene in a cell population-specific manner. The first is to express the transgene directly under a promoter that is active in only particular cell types. The second is to use a binary system, in which expression of the transgene is regulated by another "driver" gene, whose own expression is controlled by a specific promoter.

Approaches based on a single transgenic line

The single transgenic approach (Fig. 1) relies on one of several different methods to achieve promoter-specific transgene expression. In the simplest approach, a defined promoter that is active in a specific population of cells is directly assembled with the transgene of interest. Upon pronuclear injection into zygotic eggs, the transgenic construct randomly integrates into the mouse genome. Pronuclear injection of DNA plasmids often results in concatemerization of multiple copies of the transgenic construct and their cointegration into the same genomic locus, which can result in high-level expression of the transgene. In other instances, however, tandem arrays of cointegrated transgenes have been subject to silencing that is mediated by a heterochromatin-like complex (Henikoff, 1998). Examples of transgenic lines commonly used in neurobiology that were generated by this approach include *Thy1*-YFP mice (Feng et al., 2000), *Gad67*-GFP mice (Ma et al., 2006), etc. A major limitation to the simple pronuclear injection of a promoter-transgene assembly is that for many genes, regulation of cell type-specific gene expression is poorly understood. *cis*-Acting enhancer elements that contribute to the specificity of expression can be located far away from the gene's transcriptional start site and may not yet have been identified. In these instances, faithful recapitulation of a particular pattern of gene expression by a relatively short promoter surrounding the transcriptional start site may not be possible. Indeed, to date, there have been only a limited number of promoters successfully used to drive cell type-specific expression in neurons of transgenic mice. In addition, because these transgenes are randomly integrated, their expression can be greatly influenced by activating or inactivating positional effects, which may result in ectopic expression that is unrelated to the promoter in use or suppression of expression in relevant cells (sometimes in a mosaic manner). Although generally thought to be undesirable, positional effects and ectopic expression can

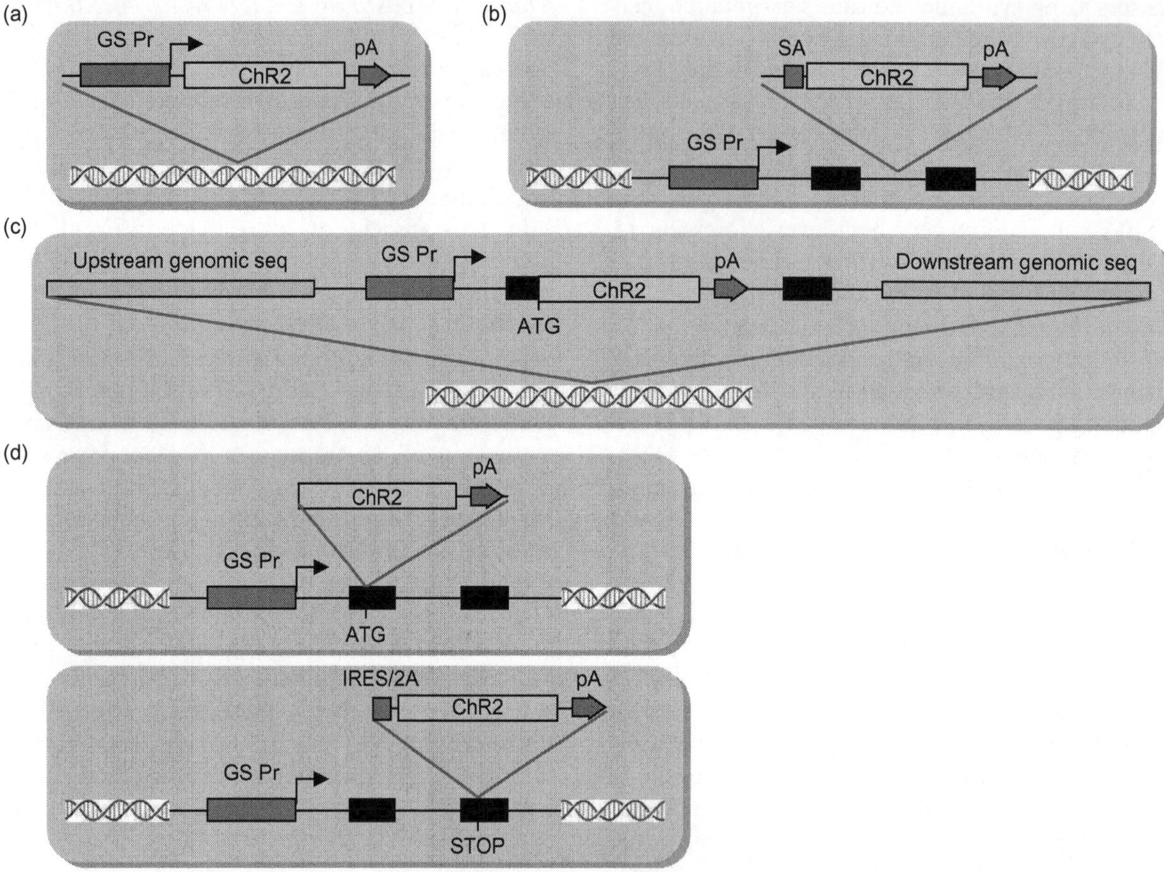

Fig. 1. Single transgenic approaches to expressing optogenetic tools (using ChR2 as an example). (a) Conventional transgenic approach, in which an expression cassette contains a promoter and the transgene and is randomly integrated into the genome. GS Pr, gene-specific promoter. pA, polyA signal. (b) Gene trap approach, in which a promoterless cassette containing the transgene is randomly integrated into the genome, and the transgene expression is determined by a "trapped" nearby endogenous promoter. SA, splice acceptor. Black boxes indicate endogenous gene exons. (c) BAC transgenic approach, in which the transgene is inserted into the locus of the gene-of-interest contained within a BAC clone, and this BAC clone is randomly integrated into the genome. (d) Knock-in approach, in which the transgene is targeted to the endogenous locus of the gene-of-interest by homologous recombination. The targeting site can be either at the ATG start codon (upper panel) or at the STOP codon (lower panel). (For color version of this figure, the reader is referred to the Web version of this chapter.)

occasionally direct novel and more restricted patterns of expression compared to that directed by the specific promoter.

A second method for attaining population-specific gene expression in transgenic mice is gene trapping. In this approach, a DNA cassette that contains an RNA splicing acceptor site, a promoterless transgene, and a polyadenylation sequence is introduced into mouse embryonic stem (ES) cells by transfection, viral infection, or transposition. Integration of the promoterless transgenic cassette into the genome is generally random, but when integrated into an intron of an expressed gene, the promoterless cassette is transcribed from the "trapped" endogenous promoter. When mice are generated from a sufficiently large gene trap library (Nord et al., 2006), it is possible to obtain a variety of highly specific transgene expression patterns. On the other hand, many trapped events result in random, nonspecific expression of the transgene, as the landing site may not be perfectly situated (e.g., too upstream, too downstream, in between genes, etc.) to capture the relevant, specific regulatory elements of a nearby gene. Therefore, tremendous effort is needed to sort out the meaningful and useful specific patterns of expression from the random, nonspecific ones.

In recent years, transgenic vectors based on bacterial artificial chromosomes (BACs) have become increasingly popular for attempting to generate patterns of restricted transgene expression. In this approach, an endogenous gene is selected as exhibiting the desired pattern of specific expression. A BAC clone that contains the selected endogenous gene near its center is identified, and a transgene of interest is inserted into the BAC, typically at the translation start site of the endogenous gene. Consequently, BAC constructs typically contain very large regions (~30–100 kb) of both 5′- and 3′genomic sequences, which flank the specific promoter-linked transgene. As a result, they likely contain many of the *cis*-acting regulatory elements required to direct cell type-specific gene expression, increasing the probability that the transgene will be expressed in the same cells in which the "targeted" promoter is normally active. The extremely long stretches of genomic sequence present in BAC vectors may also help buffer the transgene from activating or repressing positional effects encountered after random integration into the genome. In the large-scale BAC transgenic project, GENSAT (http://www.gensat.org/index.html), thousands of BAC-GFP mouse lines have been created in an effort to recapitulate the restricted expression patterns directed by hundreds of high-interest gene promoters (Gong et al., 2003). Characterization of GFP expression in these lines has revealed a variety of specific and nonspecific patterns, some of which either faithfully or partially recapitulate the endogenous gene's expression pattern as reported in the Allen Mouse Brain Atlas gene expression database (http://www.brain-map.org/) (Lein et al., 2007), or in a retinal cell type screening effort (Siegert et al., 2009). GFP labeling in some lines appears very restricted and is limited to a small population of either known or novel cell types. However, GFP expression in other GENSAT lines appears to be nonspecific, demonstrating that significant variations in gene expression still occur when using the BAC transgenic strategy.

An alternative to the randomly integrating transgenic or BAC transgenic approaches is the so-called site-specific transgenesis approach (Monetti et al., 2011; Tasic et al., 2011). In this approach, a docking site containing recombinase recognition site(s) (e.g., loxP, FRT, AttP, etc.) is created in a permissible genomic locus through homologous recombination, and then a promoter-transgene vector or BAC-transgene vector of interest is integrated into the docking site through recombinase-mediated cassette exchange (RMCE). The RMCE process can be carried out either in the ES cells or in the embryos carrying the docking site through cotransfection or comicroinjection of the transgene vector and a recombinase vector. Commonly used genomic loci include the well-characterized *Rosa26* and *Hprt* loci, as well as newly identified ones (Tasic et al., 2011). Site-specific transgenesis has the advantage of creating single-copy transgenics in a defined, consistent genomic locus, overcoming positional effect and multi-copy-induced gene silencing. On the other hand, its expression specificity is still dependent on the particular promoter fragment or BAC clone used.

Perhaps the most faithful method for generating single transgenic mice with promoter-specific gene expression is the knock-in approach. In this strategy, the transgene is inserted into a target gene's genomic locus through homologous recombination. As a result, the transgene becomes located within the exact genomic context of the target gene, and transgene expression is controlled entirely by the target gene's promoter and *cis*-acting regulatory elements. Depending on the precise experimental goals, transgenes can be knocked-in at either the target gene's ATG start codon, for direct expression from a monocistronic transcript, or anywhere downstream of the start codon, where transgene expression is usually mediated by an IRES or 2A sequence from a bicistronic transcript. A drawback of the knock-in approach is that for transgenes inserted at the ATG, the endogenous gene locus is perturbed, leading to the disruption of the target gene's transcription; this problem is theoretically avoided by use of an IRES or 2A sequence to place the transgene at the 3′end of the endogenous gene. In either case, it is important to evaluate heterozygous mice for deleterious gene dosage effects, which could arise from having one wild-type and one altered allele. The utility of the knock-in approach is further limited by being inherently laborious. Each genetic marker or tool one would like to have expressed in a specific population requires isolation of a correct homologous recombination event in mouse ES cells, a task that has been difficult to realize when targeting some neuronal genes.

Regardless of the above method used to generate transgenic mice, the level and pattern of transgene expression depend tremendously on both the promoter linked to the transgene, either through design or through trapping, and the genomic landscape the transgene integrates into, which may confer unpredictable transcriptional or epigenetic regulation. For this reason, transgene expression and functionality cannot be assumed; rather, they need to be carefully examined in each transgenic line.

The binary systems

The binary approach (Fig. 2) for expressing a transgene in a cell population-specific manner is based on the requirement of having two components, each essential for expression, in the same cell. Generally, these two components are brought into cells by the breeding of two separate mouse lines, which are often called driver and reporter lines. Driver lines express a master control "driver" gene from a chosen specific promoter and are generated via any of the above-mentioned single transgenic approaches. Reporter lines carry the transgene of interest in a cassette, whose expression is regulated by driver gene activity. A major strength of the binary system is the flexibility conferred by having these two autonomous components combined to direct transgene expression. Assortments of unique driver and reporter lines can be generated and optimized independently of one another. Once cell type specificity is achieved in a collection of driver lines, and robust and functional expression of different genetic probes or tools is achieved in reporter lines, the lines can be merged in numerous combinations to direct a variety of cell type-specific genetic labeling methods and manipulations.

Binary expression systems established both *in vitro* and *in vivo* have generally incorporated one of two main kinds of driver genes: site-specific recombinases (SSRs) and transcriptional activators. The SSR Cre protein mediates recombination between loxP sites within DNA, and, due to being highly effective in mammalian cells, it is the most commonly used SSR in transgenic mice. When Cre functions as the driver in a binary system, transgene expression in the reporter line is generally initiated from a robust and ubiquitous promoter but is then blocked by a loxP-flanked (floxed) transcriptional stop cassette located between the promoter and the transgene (i.e., promoter-loxP-STOP-loxP-transgene). Cell type-specific expression of the reporter occurs when Cre, produced in a limited cell population defined by its own linked promoter, mediates recombination between the two loxP sites within the reporter locus, resulting in deletion of the intervening stop cassette and activation of transgene expression. The Cre/lox recombinase system has proven extremely useful for manipulating the mouse genome in many ways, and it is used by researchers from a broad range of scientific

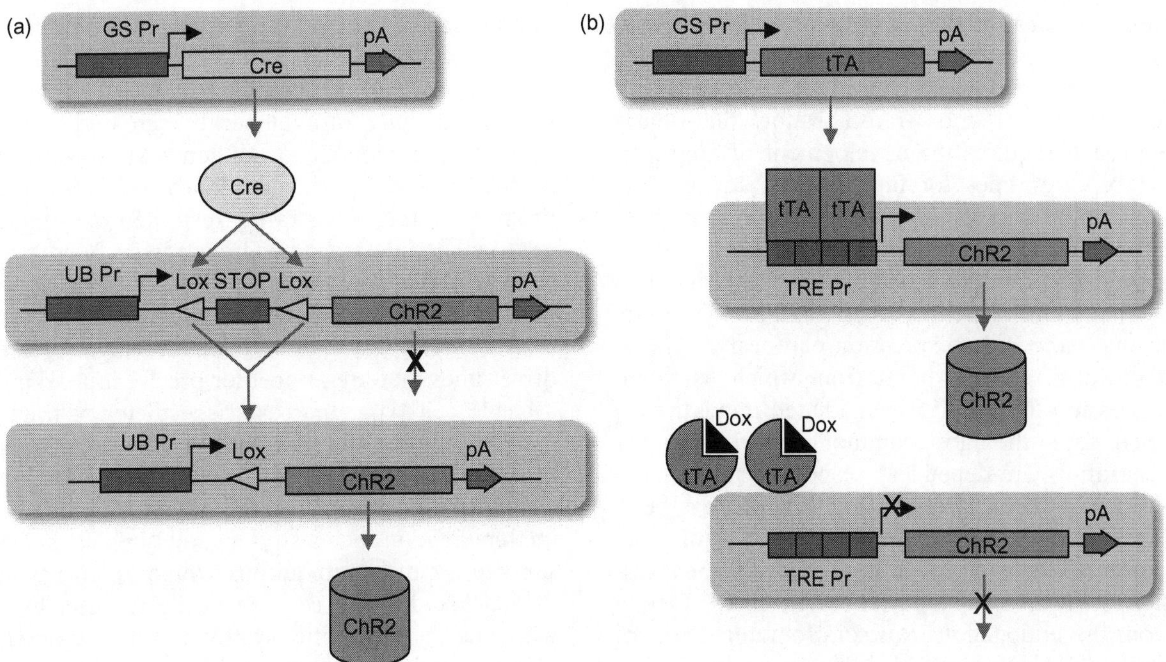

Fig. 2. The binary transgenic systems for expressing optogenetic tools (using ChR2 as an example). (a) The Cre/lox system, in which the driver line expresses Cre under the control of a gene-specific promoter, and the reporter line directs Cre-dependent expression of the transgene. Cre-mediated recombination between the two loxP sites deletes the STOP cassette and hence induces the transgene expression. UB Pr, ubiquitous promoter. (b) The Tet-inducible system, in which the driver line expresses tTA under the control of a gene-specific promoter, and the reporter line expresses the transgene under the TRE promoter. tTA binds to the TRE promoter (TRE Pr) to activate transcription of the transgene. Upon binding to tetracycline or doxycycline (Dox), tTA is released from the TRE promoter and transcription stops (bottom panel). (For color version of this figure, the reader is referred to the Web version of this chapter.)

disciplines. As a result, numerous Cre-driver mouse lines, generated in individual labs and through large-scale efforts (Gong et al., 2007; Madisen et al., 2010), have been established to drive specific gene expression in a variety of cell types or populations throughout the nervous system.

In binary approaches reliant on a transcriptional activator as the driver, the tetracycline/doxycycline-regulated tTA protein is the most commonly used activator for studies of neuronal function in the mouse brain. In this variation of the system, transgene expression in the reporter line is controlled by a TRE (Tet-regulated element)-containing promoter, which is activated by the binding of tTA dimers. In the presence of tetracycline or doxycycline, tTA dimers undergo a conformational change that prevents them from binding to the TRE. As a result, transgene expression stops. Although tTA-regulated strategies have been used successfully to direct cell type-restricted gene expression in mice, there are far fewer neuronally specific tTA driver lines available than driver lines based on Cre recombinase. Further, in the lines that do exist, tTA has often been expressed from exceedingly strong neuronal promoters, such as αCaMKII (Mayford et al., 1996), NSE (Chen et al., 1998), and OMP (Yu et al., 2004). It remains unclear whether lower levels of tTA expression initiated from weaker promoters will be sufficient to stimulate TRE-dependent reporter expression efficiently.

Regardless of which type of driver gene is chosen, recombinase or transactivator, the binary approach requires the development of appropriate reporter lines for functionality. Ideally, the regulatable transgenic cassette within a reporter line would exhibit certain characteristics, such as having the capability being highly expressed in as many cell types as possible. An inherently strong, pancellular expression platform would be a powerful starting point from which to refine expression through driver-mediated regulation.

To date, the most commonly used locus for generating Cre-dependent responder mice is the *Gt(ROSA)26Sor* (*Rosa26*) locus, which has been shown to be a fairly permissive and ubiquitously expressed locus (Soriano, 1999). However, expression of fluorescent reporters (e.g., GFP) directly from the endogenous *Rosa26* promoter (Srinivas et al., 2001) is poor in the adult mouse brain. Other Cre-dependent reporter lines have been made that include strong exogenous promoters, but since they have been integrated into random genomic loci (e.g., Z/EG (Novak et al., 2000) and BrainBow (Livet et al., 2007)), the reporters have generally not been universally expressed. The creation of the MADM (Zong et al., 2005) and mT/mG (Muzumdar et al., 2007) mice demonstrated an approach for achieving higher-level, universal expression by introducing a robust exogenous promoter into the *Rosa26* locus. A further modification of this strategy, the incorporation of a woodchuck hepatitis virus posttranscriptional regulatory element (WPRE) into the reporter cassette, resulted in consistently higher expression of fluorescent proteins that efficiently label fine neuronal structures (Madisen et al., 2010).

The tTA-dependent reporter lines have been most commonly generated by random integration, which could often introduce positional effects. A so-called TIGRE locus, identified from a screen of hundreds of ES clones, showed reliable tTA-dependent expression with low basal activity and high inducibility (Zeng et al., 2008).

As described above, the pattern of driver gene expression largely dictates the specificity of transgene expression in the binary approach. Although Cre-driver lines are generally created using promoters or genomic loci of genes whose activity is strongest in particular kinds of cells, the promoters utilized are rarely restricted to a single neuronal population in one brain region. More frequently, the promoters are active in several brain regions in either similar or different types of neurons. As a result, it has been difficult to create driver lines that are altogether precise in marking a single cell type. Increased specificity of transgene expression may be achieved by employing an intersectional strategy of regulation. There are a variety of approaches (Fig. 3) that could be implemented in such a strategy, the simplest being substitution of the ubiquitous promoter, upstream of the floxed-STOP cassette in the reporter line, with a cell type-specific promoter. Another possible approach incorporates the requirement for two different drivers, which are expressed in overlapping cell populations, to trigger reporter expression. For example, the Cre and Flp SSRs can be combined to control a "double reporter" line that carries dual stop cassettes (promoter-loxP-STOP-loxP-Frt-STOP-Frt) (Farago et al., 2006). A similar approach can be taken using different kinds of driver genes, such as Cre and tTA. In this case, transgene expression in the "double reporter" line would be regulated by a TRE-containing promoter in combination with a loxP-STOP-loxP cassette. Expression of the two driver genes in these strategies could be regulated by two different cell type-specific promoters; alternatively, one of the two drivers could be delivered virally, in a region-specific manner.

Transgenic expression of optogenetic tools

Successful application of optogenetic tools to *in vivo* studies requires very high-level expression of these genes in the cells whose activity is to be manipulated. For this reason, investigators have

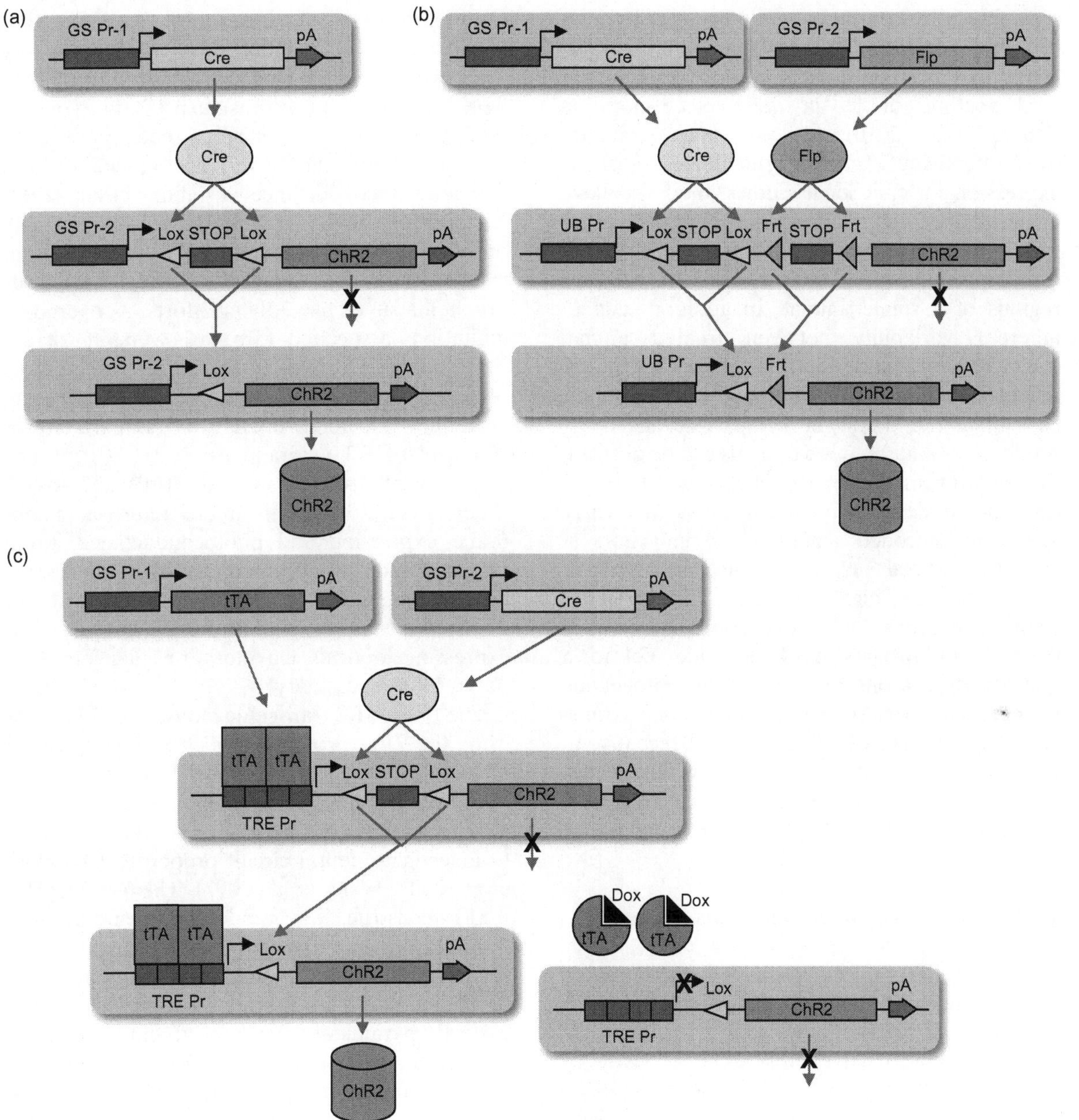

Fig. 3. The intersectional approaches to expressing optogenetic tools (using ChR2 as an example) to higher specificity. (a) A simple intersectional approach, in which the driver line uses the gene-specific promoter 1 (GS Pr-1), and the reporter line uses the gene-specific promoter 2 (GS Pr-2). (b) A Cre/Flp dual recombinase intersectional approach, in which the Cre-driver line uses gene-specific promoter 1, and the Flp driver line uses gene-specific promoter 2. The double reporter line is both Cre and Flp dependent. (c) A Cre/tTA intersectional approach, in which the tTA driver line uses gene-specific promoter 1, and the Cre-driver line uses gene-specific promoter 2. The double reporter uses the TRE promoter and is also Cre dependent. (For color version of this figure, the reader is referred to the Web version of this chapter.)

typically expressed optogenetic proteins by methods that deliver high copy numbers of transgenes to target cells, such as with recombinant viral vectors, or by *in utero* electroporation (Zhang et al., 2010). Although these strategies can effectively achieve sufficient transgene expression for functional studies, they also have encumbering limitations. Both methods suffer from directing incomplete coverage and variable transgene expression across cells in targeted regions of a single animal. In addition, due to inherent variability between treated animal subjects, use of these methods often necessitates laborious validation of transgene expression for each animal, and still the comparison and interpretation of data between animals can be difficult. To circumvent these complications, it seems essential to develop transgenic lines that carry genetically encoded actuators and indicators in cassettes whose expression is both tightly regulatable and highly inducible. The past 10 years have seen considerable efforts put toward this task, and progress has been made. Yet, for a number of reasons, expression of optogenetic molecules at levels sufficient for function in transgenic mice remains challenging until recently. In the following sections, we will review the current state of transgenic mouse lines that express these genetically encoded light-responsive proteins.

Transgenic expression of optical actuators

Individual opsin molecules induce relatively small changes in membrane polarization following photostimulation. Because of this, initial attempts to develop mice that express effective levels of the optical activating molecule, ChR2, in specific neuronal cell populations were based on directly linking the ChR2 gene to very strong cell type-specific promoters, such as those from *Thy1* (Arenkiel et al., 2007; Wang et al., 2007), *Vglut2* (Hagglund et al., 2010), *Omp* (Dhawale et al., 2010), and *Chat* (Ren et al., 2011), or to the Tet-inducible promoter (Chuhma et al., 2011). Characterization of these mice revealed that membrane depolarization and spiking activity were evoked in predicted cell types following blue light stimulation. In contrast, expression of optical inhibitory molecules in mice has been more problematic, mainly due to protein aggregation and low-current conductance (Chuhma et al., 2011; Zhao et al., 2008). To date, there has been only one report of functional NpHR expression in transgenic mice, using the orexin promoter (Tsunematsu et al., 2011). Efforts to overcome limitations associated with early versions of the silencing opsins have led to the development of second- and third-generation optical silencing molecules, including eNpHR (Gradinaru et al., 2008), eNpHR3.0 (Gradinaru et al., 2010), various forms of Arch (Chow et al., 2010), and ArchT (Han et al., 2011). With greatly improved membrane expression and photoconductance, these reengineered opsins embody a major step toward reliable genetic silencing. Below, we provide an overview of the published transgenic mouse lines expressing optical activators or silencers (see Table 1 for a summary).

The first ChR2 transgenic mouse line was made using the *Thy1* promoter, which had been shown to enable extremely high-level brain expression (Caroni, 1997; Feng et al., 2000), and it demonstrated for the first time the *in vivo* potential of ChR2 to investigate neural circuit properties (Arenkiel et al., 2007; Wang et al., 2007). The *Thy1*-ChR2-EYFP construct was randomly integrated into the mouse genome via pronuclear injection, and several transgenic lines with differential levels and patterns of expression were described. In the cortical pyramidal neurons of line 18 mice (Wang et al., 2007), which showed the highest CNS expression of ChR2, maximal peak photocurrents of 500–600 pA were reached with large-area-applied blue light pulses (~10 mW/mm^2 of 5–10 ms duration). Action potentials could be induced by light as low as 0.2 mW/mm^2, with ~6 ms average latency from light onset. Action potentials fired reliably following light pulses up to 30 Hz. These photoexcitation properties are

Table 1. Transgenic mouse lines expressing optical activators or silencers

Name	Method of generation	Promoter used	Cell type examined	Light power required to evoke spiking	Peak photocurrent	Reference
Thy1-ChR2-EYFP (line 18)	Conventional transgenic	Thy1.2	Cortical pyramidal neurons	0.2mW/mm^2	500–600pA under ~10mW/mm^2 light	Wang et al. (2007)
			MOB mitral cells	Unknown	~200pA	Arenkiel et al. (2007)
Omp-ChR2-EYFP (line ORC-M)	Conventional transgenic	Omp	MOB mitral cells (postsynaptic to the ChR2-expressing OSNs)	<2mW/mm^2 to activate postsynaptic cells	Unknown	Dhawale et al. (2010)
Vglut2-ChR2-YFP	BAC transgenic	Vglut2	Spinal cord Vglut2-positive neurons	Unable to evoke spiking (35mW/mm^2 light only induces depolarization)	Unknown	Hagglund et al. (2010)
Chat-ChR2-EYFP	BAC transgenic	Chat	MHb CHAT+ neurons	20mW/mm^2	~500pA under 20mW/mm^2 light	Ren et al. (2011)
Mrgprd-ChR2-Venus	Knock-in	Mrgprd	DRG Mrgprd+ neurons	Unknown	Unknown	Wang and Zylka (2009)
BTR (bidirectional tetO promoter driven ChR2-mCherry on one side and HaloR-EGFP on the other side) (line BTR6)	Conventional transgenic, line BTR6 crossed to aCaMKII-tTA	tetO	Dorsal striatal medium spiny neurons	Unknown	~350pA for ChR2 (HaloR aggregated, not functional)	Chuhma et al. (2011)
Thy1-NpHR-YFP	Conventional transgenic	Thy1.2	Hippocampal or cortical pyramidal neurons (NpHR-YFP formed bright intracellular blebs)	186mW/mm^2 to suppress spiking completely	~20pA under 23mW/mm^2 light ~−5mV under 23mW/mm^2 light	Zhao et al. (2008)
Orexin-Halo-GFP	Conventional transgenic	Human prepro-orexin promoter	Hypothalamic orexin neurons	4mW light through objective lens to suppress spiking	~6pA under 4mW light ~−11mV under 4mW light	Tsunematsu et al. (2011)
R26::ChR2 (H134R)-EGFP	Knock-in to the *Rosa26* locus	CAG promoter (with floxed-Neo-stop cassette)	Cortical interneurons	2mW laser light (20 ms duration)	~190pA under 2mW, 20ms laser light	Katzel et al. (2011)
Ai27 (ChR2H134R-tdTomato)	Knock-in to the *Rosa26* locus	CAG promoter (with floxed-STOP cassette and WPRE)	Cortical pyramidal neurons			Madisen et al. (in revision)
Ai32 (ChR2H134R-EYFP)	Knock-in to the *Rosa26* locus	CAG promoter (with floxed-STOP cassette and WPRE)	Cortical pyramidal neurons			Madisen et al. (in revision)
Ai35 (Arch-EGFP-ER2)	Knock-in to the *Rosa26* locus	CAG promoter (with floxed-STOP cassette and WPRE)	Cortical pyramidal neurons			Madisen et al. (in revision)
Ai39 (eNpHR3.0-EYFP)	Knock-in to the *Rosa26* locus	CAG promoter (with floxed-STOP cassette and WPRE)	Cortical pyramidal neurons			Madisen et al. (in revision)

similar to those of ChR2-positive neurons generated by viral transduction or *in utero* electroporation. In line 9 mice, a two-dimensional array light scanning method over the surface of a cell was used to map local synaptic inputs to ChR2-negative cells. Mapping data revealed that excitatory synaptic input maps to pyramidal neurons were qualitatively different from those to interneurons, with the former being much larger and more irregularly shaped. Further, it was demonstrated for the first time that ChR2-positive cortical layer 5 cells can be activated *in vivo* by light on a millisecond scale (Arenkiel et al., 2007). Brief light pulses (3.5 ms), delivered through a 200-μm diameter optical fiber positioned directly above the surface of the brain, reliably evoked responses from individual neurons with an average of ~10 ms spike latency. Neurons followed spike trains at high fidelity at frequencies as high as 40 Hz.

In the olfactory bulb of line 18 *Thy1*-ChR2-EYFP mice (Arenkiel et al., 2007), only mitral cells have ChR2 expression, and in these cells, photocurrents (mean of ~200 pA) and action potentials could be reliably induced by blue light *in vitro*. Illuminating the dorsal surface of the olfactory bulb could induce spiking in mitral cells. A circuit mapping study was conducted by stimulating mitral cells in the bulb and recording postsynaptic responses in piriform cortex. Large area (600 μm) light stimuli on the bulb drove the firing of piriform neurons much more effectively than smaller area (300 and 100 μm) light stimuli, supporting the model that multiple distant mitral cells or glomerular inputs are required for the activation of piriform cortical cells.

The 12-kb *Omp* promoter, another strong neuronal promoter, drives expression in all olfactory sensory neurons (OSNs), and several *Omp*-ChR2-EYFP transgenic mouse lines have been generated by pronuclear injection (Dhawale et al., 2010). Two transgenic lines had ChR2-EYFP expression restricted to the vomeronasal organ (VNO) and the accessory olfactory bulb. A third line (termed ORC-M) that expressed ChR2-EYFP in the olfactory epithelium and the main olfactory bulb, as well as in the VNO and accessory olfactory bulb, was used for study. Light stimulation in the glomerular layer activated ChR2 in the OSN axon terminals, which led to glutamate release and generated postsynaptic currents in the mitral cells. Light stimulation over the olfactory bulb surface induced a rapid and reliable increase of firing of the mitral/tufted cells through presynaptic activation (as opposed to the direct activation observed using cells from the *Thy1*-ChR2-EYFP mice above). Titration of light intensity (to lower than 2 mW/mm^2) enabled mapping of a single glomerulus to each recorded single-unit mitral or tufted cell. Tetrad recordings coupled with light stimulation identified neighboring mitral/tufted cell pairs that are either sister cells (i.e., cells receiving inputs from the same glomerulus) or nonsister cells (i.e., cells receiving inputs from different glomeruli). It was found that sister mitral cells showed correlated changes in their firing rate, whereas nonsister cells did not. Intriguingly, however, odor presentation desynchronized sister cells, suggesting that the odor response properties of sister cells are not redundant and could be differentially affected by local circuitry.

The *Vglut2*-ChR2-YFP transgenic mouse line was generated using a BAC transgenic strategy in which human codon-optimized hChR2-YFP was inserted in frame at the start codon of *Vglut2* (also known as *Slc17a6*) in the BAC (Hagglund et al., 2010). In these mice, ChR2 is specifically expressed in *Vglut2* expressing glutamatergic neurons, including those in the hindbrain and spinal cord, with variable intensities. Spinal cord ChR2-positive neurons responded to continuous blue light (~35 mW/mm^2) with rapid depolarization of variable degrees (1–15 mV). The relatively small and variable depolarization might be related to the low and variable expression of ChR2 in these transgenic mice. Nonetheless, light stimulation of the glutamatergic neurons in the spinal cord or the hindbrain from the ventral side was sufficient to generate rhythmic locomotor-like activities, demonstrating the direct involvement of these glutamatergic neurons in intrinsic rhythm generation.

The *Chat*-ChR2-EYFP mouse line was also generated using the BAC transgenic strategy (Ren et al., 2011). It was used to study the cholinergic transmission in the habenulo-interpeduncular pathway. ChR2-EYFP fluorescence was found in the ventral two-thirds of the medial habenula (MHb) where CHAT+cells are located, as well as the entire MHb–fr–IPN axon projection tract. Brief (5ms) or continuous blue light pulses ($20 mW/mm^2$) could evoke rapid action potential firing and large photocurrents (~500pA, peak amplitude) in MHb neurons. Light stimulation of the axonal terminals in the projection target area, interpeduncular nucleus (IPN), elicited postsynaptic responses in IPN neurons tested. Brief (5ms) light pulses evoked fast EPSCs that were blocked only by glutamate antagonists. Prolonged light stimulation evoked an additional component, the slow inward currents that were not affected by glutamate antagonists but were reduced by nAchR blockers to about one-third of their original levels, thus revealing the dual transmission nature of the MHb cholinergic neurons.

The *Mrgprd*-ChR2-Venus knock-in mice were generated by inserting the mammalian codon-optimized ChR2(H134R)-Venus gene into the *Mrgprd* gene in frame at the start codon through homologous recombination (Wang and Zylka, 2009). *Mrgprd* molecularly marks ~75% of all IB4+nonpeptidergic nociceptive neurons in the dorsal root ganglia (DRGs) and trigeminal ganglia, and these Mrgprd+neurons exclusively innervate skin and terminate in lamina II (the substantia gelatinosa or SG) of dorsal spinal cord. Light stimulation evoked action potentials in 68% heterogyzous and 94% homozygous *Mrgprd*-ChR2-Venus+DRG cells in dissociated culture. The latency between light onset and action potential peak was ~20ms, and the spike jitter (average standard deviation of the light-evoked spike latency from each neuron) was ~1.3ms. In spinal cord slices, light-evoked excitatory postsynaptic currents ($EPSC_L$) could be generated in ~50% of SG neurons. Using spike jitter and pre–post synaptic proximity as criteria, the connections between the *Mrgprd*-ChR2-Venus+DRG cells and these light-responsive SG neurons were classified into monosynaptic or polysynaptic. And it was found that monosynaptic connections made up of ~50% of the light responsive SG cells and included almost all known SG cell types except for the islet cells.

Conditional expression of ChR2 using the Tet system was achieved in the BTR mice (Chuhma et al., 2011), in which a bidirectional tetO (also known as TRE) promoter was used to drive ChR2-mCherry in one direction and HaloR-EGFP (HaloR being the same as NpHR) in the other direction. After pronuclear injection, three founder lines were obtained and crossed to a line of αCaMKII-tTA with striatal tTA expression restricted to medium spiny neurons (MSNs). Only one of the three crosses (αCaMKIIa-tTA::BTR6) had ChR2-mCherry expression and was subsequently used in the study. Within the dorsal striatum (dStr), less than 10% MSNs were fluorescently labeled and they were randomly distributed in a mosaic pattern. While mCherry fluorescence was seen in the membrane, outlining the cell soma, and extended into the processes, EGFP accumulated in aggresomes in MSN cell bodies and was not seen in the plasma membrane, suggesting that ChR2, but not HaloR, would be functional. While yellow light illumination elicited marginal HaloR responses, blue light illumination evoked large ChR2 photocurrents (~350 pA) and depolarization (~30mV). ChR2 activation was therefore used to map functional connections among different cell types in both specificity and strength (as measured in the size of inhibitory postsynaptic currents (IPSCs)). Photostimulation of MSN presynaptic terminals expressing ChR2 elicits $GABA_A$ synaptic responses from local collaterals in the dStr, as well as in their projections to globus pallidus (GP) and substantia nigra (SN). It was found that MSN synaptic connections are quite specific. In the dStr, MSNs connect robustly to other MSNs and less robustly to tonically active neurons, but not to fast-spiking interneurons. In the GP, MSNs connect strongly to type B/C neurons and practically not to type A neurons. In the SN, MSNs make their strongest connections with SNr GABAergic neurons, but no connections with SNc dopamine neurons.

Cre-dependent conditional expression of ChR2 was utilized in the creation of a floxed-STOP ChR2 mouse line (R26::ChR2-EGFP) (Katzel et al., 2011), in which a mammalian codon-optimized ChR2(H134R)-EGFP driven by the CMV early enhancer/chicken β-actin (CAG) promoter and a floxed-STOP cassette was targeted to the *Rosa26* locus by homologous recombination. After Cre-mediated excision of the floxed-STOP cassette, recombination brings the ChR2-EGFP coding sequence in frame with an initiating ATG in the loxP site, resulting in translation of ChR2-EGFP with an 11-amino-acid N-terminal "loxP tag" (MYAIRSYELAT). After crossing to a Gad2-ires-CreERT2 mouse line, ChR2-EGFP was expressed in all main subclasses of cortical GABAergic interneurons, but expression was undetectable in αCaMKII-positive pyramidal neurons following tamoxifen induction. The relatively low-level expression of ChR2 required the use of homozygous mice that were fed with retinol, as well as longer stimulating laser light pulses (20ms, 2mW) to evoke spiking than those used to activate virally transduced or *in utero* transfected neurons. In these ChR2-positive cortical interneurons, under individual light pulses, average peak photocurrents were ~190pA and average spike latency was ~16ms. The light pulses were only able to elicit spiking in perisomatic areas, not in dendritic or axonal arborizations. Optical raster stimulation was used to map synaptic inputs from ChR2-positive interneurons to ChR2-negative pyramidal neurons in three cortical regions (M1, S1, and V1). It was found that the most common circuit motif is the lateral, intralaminar inhibition, supporting the view that inhibition is largely local, intralaminar, and uniform across areas. However, rarer translaminar inhibitory inputs to subsets of layer 2/3 or layer 5 pyramidal neurons were also identified, predominantly in area V1.

The first functional Halo (NpHR) transgenic mice (*orexin*/Halo) were generated using the 3.2 kb human prepro-orexin promoter by pronuclear injection (Tsunematsu et al., 2011). Three founder lines showed sufficiently strong expression of Halo, and among these, line 5 showed the highest expression rate (94% of orexin neurons) and was used for the study. Interestingly, orexin neurons expressing Halo did not show blebbing or other features indicative of inappropriate trafficking. Maximum orange light illumination (586nm, 4mW) generated hyperpolarization of ~11mV and photocurrents of ~6pA in orexin neurons and was able to inhibit either spontaneous or depolarizing current injection-induced action potential firing. Acute inhibition of orexin neurons *in vivo* using orange LED light through optical fibers results in time-of-day-dependent induction of slow-wave sleep and in reduced firing rate of dorsal raphe neurons in an efferent projection site.

More recently, we have employed a knock-in approach to create four new mouse lines with high-level and Cre-dependent expression of ChR2 (H134R) (Nagel et al., 2005) fused to either tdTomato (Ai27; ChR2(H134R)-tdTomato) or EYFP (Ai32; ChR2(H134R)-EYFP), as well as a modified version of Arch (Ai35; Arch-EGFP-ER2) (Chow et al., 2010) or eNpHR3.0 (Ai39; eNpHR3.0-EYFP) (Gradinaru et al., 2010). These lines were generated using a transgenic expression strategy previously employed to generate a set of fluorescent reporter lines (Madisen et al., 2010), in which the expression cassette, which includes (from left to right) the CAG promoter, the floxed-STOP cassette, the transgene, and a WPRE sequence, was targeted to the *Rosa26* locus by homologous recombination. After crossing to various Cre lines, including *Emx1*-Cre, *Camk2a*-CreERT2, *Chat*-Cre, and *Pvalb*-ires-Cre, all optogenetic reporter genes were found to be strongly expressed, by *in situ* hybridization and native fluorescence, in Cre-defined areas and cell types. Peak photocurrents reached nanoampere range in both Ai27 and Ai32 ChR2 mice and 100–300pA range in Ai35 Arch-ER2 and Ai39 eNpHR3.0 mice (Madisen et al., manuscript in revision). The yet-to-published results demonstrate that robust, selective optical activation and silencing can be achieved in different neuronal cell types from different brain regions in all these mice, using a variety of photostimulation paradigms both *in vitro* on brain slices and *in vivo* in awake, behaving animals.

Transgenic expression of optical indicators

GECIs offer unique opportunities to study calcium dynamics and cell signaling under a variety of physiologically relevant conditions. Since their inception, GECIs have undergone several rounds of optimization and are being increasingly used to monitor neuronal activities. In most applications, the GECIs have been produced in cells using viral approaches. Although several groups were successful in creating transgenic lines that expressed early generation GECIs in nonneuronal tissues (Hara et al., 2004; Ji et al., 2004), the first mouse lines to express a functional GECI in the brain were two in which the indicators [inverse pericam (IP) and camgaroo-2 (Cg-2)] were transcribed from TRE-containing promoters following mating to the αCaMKII-tTA mouse line (Hasan et al., 2004). Two different transgenic lines designed to express G-CaMP2 in the brain were also generated and have been used to monitor the activities of different cell populations, including cerebellar granule cells (Diez-Garcia et al., 2005, 2007) and cells in the olfactory system (Chaigneau et al., 2007; He et al., 2008). A troponin C (TnC)-based sensor, CerTN-L15, was also expressed under the *Thy1* promoter in transgenic mice and exhibited Ca^{2+} responses in cortical pyramidal neurons both *in vitro* and *in vivo* (Heim et al., 2007). Elucidation of the crystal structure of G-CaMP2 led to rationally designed changes in the molecule, resulting in GCaMP3, which demonstrates increased brightness, stability, and dynamic range (Tian et al., 2009) as compared to G-CaMP2. No transgenic lines have yet been published that describe expression of GCaMP3 in any cell type. However, we have recently generated a Cre-dependent GCaMP3 reporter mouse that utilizes the same *Rosa26*-based expression strategy, as that employed in the Ai27, Ai32, Ai35 and Ai39 mice above. Preliminary data indicate that large calcium transients were observed with neuronal firing (Hatim Zariwala et al., manuscript in revision). Below, we provide an overview of the published transgenic mouse lines expressing GECIs (see Table 2 for a summary).

Among the first transgenics to express a GECI in the brain were two series of mice that carried randomly integrated cassettes for Tet-inducible expression of Cg-2 or IP (Hasan et al., 2004). Following mating to αCaMKII-tTA mice, many low-expressing lines exhibited punctate fluorescence inside cell bodies, while a few high-expressing lines showed more uniform cellular fluorescence. Functional Ca^{2+} responses in some of the highly expressing lines were demonstrated in cells from several tissues. In hippocampal and cortical slices, short trains of stimuli evoked fluorescence changes ($\Delta F/F$) in both Cg-2 and IP mice in the range of 2–8% by wide-field (WF) microscopy (with a CCD camera) and 10–100% by 2-photon (2P) microscopy. In retina, light-evoked fluorescence changes were seen in whole-mount preps of IP-expressing mice (10% 2P), whereas no changes were detected in preps of Cg-2 mice. Finally, *in vivo* imaging of the olfactory bulb showed odor-evoked fluorescence changes in both Cg-2- and IP-expressing mice (1–8% WF).

An improved version of the indicator yellow cameleon, YC3.60, created by using a circularly permuted YFP (cpYFP), exhibited a larger dynamic range and SNR in biochemical assays than previous versions of the molecule (Nagai et al., 2004). A transgenic mouse line was made in which a modified CAG promoter (CAGGS) was used to drive expression of membrane-localized YC3.60. Upon tetanic stimulation of the Schaffer collateral/commissural pathway in these mice, a significant increase in FRET signal ($[Ca^{2+}]$) was evoked in area CA1 ($\Delta F/F$ of 2–3% by WF microscopy), and oscillatory $[Ca^{2+}]$ in area DG ($\Delta F/F$ of 1–2% by WF microscopy). However, it was noted that the dynamic range of the indicator in cells of the CNS of these mice was greatly reduced from what had been observed in *in vitro* studies.

Table 2. Transgenic mouse lines expressing genetically encoded calcium indicators

Name	Method of generation	Promoter used	Cell type examined	$\Delta F/F$ (or $\Delta R/R$)	Reference
Ptetbi-Cg2/luc (lines MTH-Cg2-14 and MTH-Cg2-19)	Conventional transgenic, crossed to αCaMKII-tTA	tetO	Hippocampal or cortical pyramidal neurons	~4% (wide-field) or ~10% (2-photon) in response to short trains of action potentials 2–8% (wide-field) or 20–100% (2-photon) in response to synaptic stimulation	Hasan et al. (2004)
Ptetbi-IP/luc (line MTH-IP-12)	Conventional transgenic, crossed to αCaMKII-tTA	tetO	Hippocampal or cortical pyramidal neurons Retina whole mount Olfactory bulb *in vivo*	−2% (wide-field) or −30% (2-photon) in response to synaptic stimulation −10% (2-photon) in response to light −8% (wide-field) in response to odor	Hasan et al. (2004)
pCAGGS-YC3.60pm	Conventional transgenic	CAGGS	Hippocampal CA1 region	2–3% (wide-field) in response to synaptic stimulation	Nagai et al. (2004)
Kv3.1-G-CaMP2 (line 846)	Conventional transgenic	Kv3.1	Cerebellar granule cells	5–25% (wide-field at 60×) or ~50% (2-photon) in response to 10 stimuli at 100Hz	Diez-Garcia et al. (2005, 2007)
tetO-G-CaMP2	Conventional transgenic, crossed to OMP-ires-tTA	tetO	VNO olfactory sensory neurons	20% to >100% (2-photon) in response to pheromone stimuli	He et al. (2008)
Thy1-CerTN-L15 (line C)	Conventional transgenic	Thy1.2	Cortical layer 2/3 pyramidal neurons	~4% (2-photon) per action potential, 30–60% (2-photon) in response to iontophoretic glutamate stimulations	Heim et al. (2007)
Ai38 (GCaMP3)	Knock-in to the *Rosa26* locus	CAG promoter (with floxed-STOP cassette and WPRE)	Cortical layer 2/3 neurons		Zariwala et al. (in revision)

The Kv3.1 potassium channel promoter was used to direct expression of G-CaMP2 in a defined subpopulation of neurons of the transgenic mouse line 846 (Diez-Garcia et al., 2005). In the cerebellar cortex, G-CaMP2 was expressed exclusively in granule cells, where it reported presynaptic Ca^{2+} signals. In cerebellar slices, electrical stimulation in the molecular layer induced an increase in fluorescence in a beam-like area along the parallel fibers ($\Delta F/F$ of ~0.14% for a single stimulus, ~3% for 8 stimuli, and ~4% for 30 stimuli, WF microscopy). Stimulation at the granular layer induced both a local response and a beam-like response in the molecular layer. At high magnification (60×) in which brightest fluorescence was better localized, stimulations (10 pulses at 100 Hz) at either the molecular layer (antidromic activation) or the granular layer (orthodromic activation) both resulted in 5–25% $\Delta F/F$ in the target areas.

In subsequent studies on a subline of 846 (846HB) (Diez-Garcia et al., 2007), direct stimulation (10 pulses at 100 Hz) of parallel fibers in the molecular layer of cerebellar slices evoked a ~50% $\Delta F/F$ through 2P laser-scanning microscopy, which was larger than what was obtained with 1-photon laser-scanning microscopy (~30% $\Delta F/F$). Ca^{2+} signals were also detected in the cerebellar molecular layer *in vivo* by both whole-field and 2P fluorescence imaging. Stimulations of parallel fibers in the molecular layer (10 pulses at 100 Hz) induced a ~3% whole-field fluorescence change that was clearly distinguishable from wild-type mice, as well as up to 50% $\Delta F/F$ by 2P imaging across responsive areas. Further, using these transgenic mice, Ca^{2+} transients in the parallel fibers demonstrated presynatically expressed long-term plasticity (both preLTP and preLTD) at the PF—Purkinje neuron synapses (Qiu and Knopfel, 2007, 2009).

The Kv3.1-G-CaMP2 mouse line (846) has also been used in studies of the olfactory system, where G-CaMP2 was shown to be expressed in the mitral cells, tufted cells, and some juxtaglomerular cells in the olfactory bulb (Chaigneau et al., 2007; Fletcher et al., 2009). 2P imaging detected odor-induced Ca^{2+} responses in the glomeruli that were odor specific, concentration dependent, and that could be blocked by superfusion of glutamate receptor antagonists. Glomeruli Ca^{2+} signals reflected activation of multiple mitral cells synchronized during population bursts, and stimulation of individual external tufted (ET) cells could drive population bursts of mitral cells within the same glomerulus (De Saint Jan et al., 2009). These mice were employed to establish an odor-evoked sensory map with single glomerulus resolution, which reflected exclusively the activity of olfactory bulb neurons postsynaptic to sensory afferents (Fletcher et al., 2009). In these G-CaMP2-based postsynaptic odor maps, different odorants activated distinct but overlapping sets of glomeruli. Increasing odor concentration increased both the response amplitude of individual glomeruli and the total number of activated glomeruli. Further, the G-CaMP2 response displayed a fast time course that enabled analysis of the temporal dynamics of odor maps over consecutive sniff cycles.

In another transgenic mouse line that employed G-CaMP2, the indicator was expressed from the Tet-inducible promoter (He et al., 2008). Following mating of this line with the OMP-ires-tTA line, G-CaMP2 expression was restricted to the OSNs in both the main olfactory epithelium and the VNO. In VNO slices, diluted female or male urine samples evoked large Ca^{2+} transients ($\Delta F/F$ of 20% to >100%) in individual VNO neurons by 2P imaging. Diverse combinatorial activation patterns of VNO neurons were observed in response to gender-, strain-, or individual-specific pheromone stimuli.

TnC, the Ca^{2+} sensor protein in skeletal and cardiac muscle, was used as the basis to engineer a modified calcium sensor named CerTN-L15. Transgenic mice were generated to express CerTN-L15 under the *Thy1* promoter (Heim et al., 2007). In transgenic line C, which had the highest level expression of the transgene, the indicator was widely expressed in many types of neurons, most prominently in the pyramidal neurons of the hippocampus and the neocortex. In cortical slices through 2P imaging, although fluorescence changes caused by single action potentials were not reliably detected, brief trains of multiple action potentials evoked clear changes in the ratio of citrine/cerulean fluorescence ($\Delta R/R$), with a linear relationship between the number of action potentials and the $\Delta R/R$ (extrapolated $\Delta R/R = \sim 4\%$ per action potential). Iontophoretic glutamate applications in layer 2/3 neurons of the visual cortex *in vivo* evoked cellular Ca^{2+} signals that were similar to those evoked in slices, as well as dendritic Ca^{2+} transients with $\sim 49\%$ $\Delta R/R$. The TnC-based sensors may be advantageous compared to those based on calmodulin (CaM), in that CaM interacts extensively with other intracellular neuronal proteins, which could make the *in vivo* functionality of CaM-based Ca^{2+} sensors unpredictable.

Future directions

Tremendous progress has been made toward the goal of creating functionally relevant transgenic mice for optogenetic studies. The characterization of early generation transgenic lines carrying light-activatable molecules has provided insight as to the prerequisites for an effective transgenic approach to applying these tools. Key among these is the requirement for targeted cells to be more light sensitive and to exhibit more rapid on/off kinetics. Increased light sensitivity of targeted cells can be achieved through a combination of reengineering current opsin molecules to have an increased response per light unit and improving current transgene expression strategies. Targeted cells that could be triggered by lower light conditions would be less vulnerable to both short- and long-term photo damages. At the same time, since cells located in deeper tissue layers would be more reactive to low-light stimulation, any particular amount of applied light would have the potential to recruit a larger group of cells for the study. Increased light sensitivity is also often associated with a shorter latency of response (i.e., more rapid onset). For optical actuators, shorter latency could enable distinguishing direct light-induced activation or silencing from secondary effects mediated by synaptic transmission. For optical indicators, more rapid onset could facilitate faster tracking of spiking events.

Recently, a number of new optogenetic variants with novel or superior light-response properties have emerged, including the activating opsin molecules CatCh (Kleinlogel et al., 2011), ChR2-ET/TC (Berndt et al., 2011), and C1V1 (Yizhar et al., 2011); the inhibitory opsin molecules ArchT (Han et al., 2011) and Halo57 (E. Boyden, personal communications); the GECIs GCaMP5 and GCaMP6 (L. Looger, personal communications); and the VSFP-Butterfly (T. Knopfel, personal communications). Continued efforts will be needed to assess the functionality of these new variants in a transgenic context. Also important is the development of methods aimed at increasing the overall level of transgene expression. Next generation transgenics may be based on stronger transcriptional promoters, whether ubiquitous or cell type-specific, or on novel genomic loci that enable highly stable and highly inducible expression of inserted exogenous genes. The latter would be especially beneficial for application of the tTA system in mice, where precise and nonleaky control of TRE-promoter activity has been difficult to achieve. TRE-promoter cassettes have often not been well expressed in mice and are prone to undergo chromatin silencing modifications over time. Targeting TRE-promoter-driven reporter genes to genomic loci less susceptible to epigenetic modification might increase the utility of the tTA approach for transgenesis (e.g., Zeng et al., 2008).

Elucidating the role individual components play in a complex neural circuit will require the ability to monitor and manipulate cell activity with extreme specificity. Expression strategies that rely on combinatorial or intersectional regulation should make possible the refined cell type- and region-specific transgene expression patterns required. Double reporter lines that demand multiple events to induce optogenetic molecule expression need to be generated, as will driver lines based on "drivers" other than Cre recombinase. In some instances, the ability to limit optogenetic protein activity to particular cellular compartments (such as axons, dendrites, or soma) will be essential for understanding neuronal function. Such specificity can be attained by fusing known subcellular targeting domains with optogenetic molecules (Lewis et al., 2009, 2011). Further, the use of sculpted light for photostimulation can also improve the specificity of cell activation, independent of the strategy for transgene expression (Andrasfalvy et al., 2010; Papagiakoumou et al., 2010).

Finally, a researcher's dream is to be able to use a combined optogenetic approach to activate and silence different cell types at the same time while monitoring all cell activities. We can work on steps toward this goal. By generating a plethora of transgenic tools, people can pick and choose the combination to manipulate different cell types. With continued efforts in genetic engineering, protein engineering, and optical/electronic engineering, we may not need to wait long to realize this dream.

Acknowledgments

This work was funded by the Allen Institute for Brain Science and NIH grants (MH085500, DA028298) to H. Z. The authors wish to thank the Allen Institute founders, Paul G. Allen and Jody Allen, for their vision, encouragement, and support.

References

Andrasfalvy, B. K., Zemelman, B. V., Tang, J., & Vaziri, A. (2010). Two-photon single-cell optogenetic control of neuronal activity by sculpted light. *Proceedings of the National Academy of Sciences of the United States of America, 107*, 11981–11986.

Arenkiel, B. R., Peca, J., Davison, I. G., Feliciano, C., Deisseroth, K., Augustine, G. J., et al. (2007). *In vivo* light-induced activation of neural circuitry in transgenic mice expressing channelrhodopsin-2. *Neuron, 54*, 205–218.

Berndt, A., Schoenenberger, P., Mattis, J., Tye, K. M., Deisseroth, K., Hegemann, P., et al. (2011). High-efficiency channelrhodopsins for fast neuronal stimulation at low light levels. *Proceedings of the National Academy of Sciences of the United States of America, 108*, 7595–7600.

Boyden, E. S., Zhang, F., Bamberg, E., Nagel, G., & Deisseroth, K. (2005). Millisecond-timescale, genetically targeted optical control of neural activity. *Nature Neuroscience, 8*, 1263–1268.

Caroni, P. (1997). Overexpression of growth-associated proteins in the neurons of adult transgenic mice. *Journal of Neuroscience Methods, 71*, 3–9.

Chaigneau, E., Tiret, P., Lecoq, J., Ducros, M., Knopfel, T., & Charpak, S. (2007). The relationship between blood flow and neuronal activity in the rodent olfactory bulb. *Journal of Neuroscience, 27*, 6452–6460.

Chen, J., Kelz, M. B., Zeng, G., Sakai, N., Steffen, C., Shockett, P. E., et al. (1998). Transgenic animals with inducible, targeted gene expression in brain. *Molecular Pharmacology, 54*, 495–503.

Chow, B. Y., Han, X., Dobry, A. S., Qian, X., Chuong, A. S., Li, M., et al. (2010). High-performance genetically targetable optical neural silencing by light-driven proton pumps. *Nature, 463*, 98–102.

Chuhma, N., Tanaka, K. F., Hen, R., & Rayport, S. (2011). Functional connectome of the striatal medium spiny neuron. *Journal of Neuroscience, 31*, 1183–1192.

De Saint Jan, D., Hirnet, D., Westbrook, G. L., & Charpak, S. (2009). External tufted cells drive the output of olfactory bulb glomeruli. *Journal of Neuroscience, 29*, 2043–2052.

Dhawale, A. K., Hagiwara, A., Bhalla, U. S., Murthy, V. N., & Albeanu, D. F. (2010). Non-redundant odor coding by sister mitral cells revealed by light addressable glomeruli in the mouse. *Nature Neuroscience, 13*, 1404–1412.

Diez-Garcia, J., Akemann, W., & Knopfel, T. (2007). *In vivo* calcium imaging from genetically specified target cells in mouse cerebellum. *NeuroImage, 34*, 859–869.

Diez-Garcia, J., Matsushita, S., Mutoh, H., Nakai, J., Ohkura, M., Yokoyama, J., et al. (2005). Activation of cerebellar parallel fibers monitored in transgenic mice expressing a fluorescent Ca2+ indicator protein. *European Journal of Neuroscience, 22*, 627–635.

Farago, A. F., Awatramani, R. B., & Dymecki, S. M. (2006). Assembly of the brainstem cochlear nuclear complex is revealed by intersectional and subtractive genetic fate maps. *Neuron, 50*, 205–218.

Feng, G., Mellor, R. H., Bernstein, M., Keller-Peck, C., Nguyen, Q. T., Wallace, M., et al. (2000). Imaging neuronal subsets in transgenic mice expressing multiple spectral variants of GFP. *Neuron, 28*, 41–51.

Fletcher, M. L., Masurkar, A. V., Xing, J., Imamura, F., Xiong, W., Nagayama, S., et al. (2009). Optical imaging of postsynaptic odor representation in the glomerular layer of the mouse olfactory bulb. *Journal of Neurophysiology*, *102*, 817–830.

Gong, S., Doughty, M., Harbaugh, C. R., Cummins, A., Hatten, M. E., Heintz, N., et al. (2007). Targeting Cre recombinase to specific neuron populations with bacterial artificial chromosome constructs. *Journal of Neuroscience*, *27*, 9817–9823.

Gong, S., Zheng, C., Doughty, M. L., Losos, K., Didkovsky, N., Schambra, U. B., et al. (2003). A gene expression atlas of the central nervous system based on bacterial artificial chromosomes. *Nature*, *425*, 917–925.

Gradinaru, V., Thompson, K. R., & Deisseroth, K. (2008). eNpHR: A Natronomonas halorhodopsin enhanced for optogenetic applications. *Brain Cell Biology*, *36*, 129–139.

Gradinaru, V., Zhang, F., Ramakrishnan, C., Mattis, J., Prakash, R., Diester, I., et al. (2010). Molecular and cellular approaches for diversifying and extending optogenetics. *Cell*, *141*, 154–165.

Hagglund, M., Borgius, L., Dougherty, K. J., & Kiehn, O. (2010). Activation of groups of excitatory neurons in the mammalian spinal cord or hindbrain evokes locomotion. *Nature Neuroscience*, *13*, 246–252.

Han, X., & Boyden, E. S. (2007). Multiple-color optical activation, silencing, and desynchronization of neural activity, with single-spike temporal resolution. *PLoS One*, *2*, e299.

Han, X., Chow, B. Y., Zhou, H. H., Klapoetke, N. C., Chuong, A., Rajimehr, R., et al. (2011). A high-light sensitivity optical neural silencer: Development and application to optogenetic control of non-human primate cortex. *Frontiers in Systems Neuroscience*, *5*, 1.

Hara, M., Bindokas, V., Lopez, J. P., Kaihara, K., Landa, L. R., Jr., Harbeck, M., et al. (2004). Imaging endoplasmic reticulum calcium with a fluorescent biosensor in transgenic mice. *American Journal of Physiology*, *287*, C932–C938.

Hasan, M. T., Friedrich, R. W., Euler, T., Larkum, M. E., Giese, G., Both, M., et al. (2004). Functional fluorescent Ca2+ indicator proteins in transgenic mice under TET control. *PLoS Biology*, *2*, e163.

He, J., Ma, L., Kim, S., Nakai, J., & Yu, C. R. (2008). Encoding gender and individual information in the mouse vomeronasal organ. *Science (New York, NY)*, *320*, 535–538.

Heim, N., Garaschuk, O., Friedrich, M. W., Mank, M., Milos, R. I., Kovalchuk, Y., et al. (2007). Improved calcium imaging in transgenic mice expressing a troponin C-based biosensor. *Nature Methods*, *4*, 127–129.

Henikoff, S. (1998). Conspiracy of silence among repeated transgenes. *Bioessays*, *20*, 532–535.

Ji, G., Feldman, M. E., Deng, K. Y., Greene, K. S., Wilson, J., Lee, J. C., et al. (2004). Ca2+-sensing transgenic mice: Postsynaptic signaling in smooth muscle. *Journal of Biological Chemistry*, *279*, 21461–21468.

Katzel, D., Zemelman, B. V., Buetfering, C., Wolfel, M., & Miesenbock, G. (2011). The columnar and laminar organization of inhibitory connections to neocortical excitatory cells. *Nature Neuroscience*, *14*, 100–107.

Kleinlogel, S., Feldbauer, K., Dempski, R. E., Fotis, H., Wood, P. G., Bamann, C., et al. (2011). Ultra light-sensitive and fast neuronal activation with the Ca(2)+-permeable channelrhodopsin CatCh. *Nature Neuroscience*, *14*, 513–518.

Lein, E. S., Hawrylycz, M. J., Ao, N., Ayres, M., Bensinger, A., Bernard, A., et al. (2007). Genome-wide atlas of gene expression in the adult mouse brain. *Nature*, *445*, 168–176.

Lewis, T. L., Jr., Mao, T., & Arnold, D. B. (2011). A role for myosin VI in the localization of axonal proteins. *PLoS Biology*, *9*, e1001021.

Lewis, T. L., Jr., Mao, T., Svoboda, K., & Arnold, D. B. (2009). Myosin-dependent targeting of transmembrane proteins to neuronal dendrites. *Nature Neuroscience*, *12*, 568–576.

Livet, J., Weissman, T. A., Kang, H., Draft, R. W., Lu, J., Bennis, R. A., et al. (2007). Transgenic strategies for combinatorial expression of fluorescent proteins in the nervous system. *Nature*, *450*, 56–62.

Luo, L., Callaway, E. M., & Svoboda, K. (2008). Genetic dissection of neural circuits. *Neuron*, *57*, 634–660.

Ma, Y., Hu, H., Berrebi, A. S., Mathers, P. H., & Agmon, A. (2006). Distinct subtypes of somatostatin-containing neocortical interneurons revealed in transgenic mice. *Journal of Neuroscience*, *26*, 5069–5082.

Madisen, L., Zwingman, T. A., Sunkin, S. M., Oh, S. W., Zariwala, H. A., Gu, H., et al. (2010). A robust and high-throughput Cre reporting and characterization system for the whole mouse brain. *Nature Neuroscience*, *13*, 133–140.

Mank, M., & Griesbeck, O. (2008). Genetically encoded calcium indicators. *Chemical Reviews*, *108*, 1550–1564.

Mayford, M., Bach, M. E., Huang, Y. Y., Wang, L., Hawkins, R. D., & Kandel, E. R. (1996). Control of memory formation through regulated expression of a CaMKII transgene. *Science (New York, NY)*, *274*, 1678–1683.

Micheva, K. D., Busse, B., Weiler, N. C., O'Rourke, N., & Smith, S. J. (2010). Single-synapse analysis of a diverse synapse population: Proteomic imaging methods and markers. *Neuron*, *68*, 639–653.

Monetti, C., Nishino, K., Biechele, S., Zhang, P., Baba, T., Woltjen, K., et al. (2011). PhiC31 integrase facilitates genetic approaches combining multiple recombinases. *Methods (San Diego, Calif)*, *53*, 380–385.

Muzumdar, M. D., Tasic, B., Miyamichi, K., Li, L., & Luo, L. (2007). A global double-fluorescent Cre reporter mouse. *Genesis*, *45*, 593–605.

Nagai, T., Yamada, S., Tominaga, T., Ichikawa, M., & Miyawaki, A. (2004). Expanded dynamic range of fluorescent indicators for Ca(2+) by circularly permuted yellow fluorescent proteins. *Proceedings of the National Academy of Sciences of the United States of America*, *101*, 10554–10559.

Nagel, G., Brauner, M., Liewald, J. F., Adeishvili, N., Bamberg, E., & Gottschalk, A. (2005). Light activation of channelrhodopsin-2 in excitable cells of Caenorhabditis elegans triggers rapid behavioral responses. *Current Biology*, *15*, 2279–2284.

Nagel, G., Szellas, T., Huhn, W., Kateriya, S., Adeishvili, N., Berthold, P., et al. (2003). Channelrhodopsin-2, a directly light-gated cation-selective membrane channel. *Proceedings of the National Academy of Sciences of the United States of America*, *100*, 13940–13945.

Nord, A. S., Chang, P. J., Conklin, B. R., Cox, A. V., Harper, C. A., Hicks, G. G., et al. (2006). The International Gene Trap Consortium Website: A portal to all publicly available gene trap cell lines in mouse. *Nucleic Acids Research*, *34*, D642–D648.

Novak, A., Guo, C., Yang, W., Nagy, A., & Lobe, C. G. (2000). Z/EG, a double reporter mouse line that expresses enhanced green fluorescent protein upon Cre-mediated excision. *Genesis*, *28*, 147–155.

Papagiakoumou, E., Anselmi, F., Begue, A., de Sars, V., Gluckstad, J., Isacoff, E. Y., et al. (2010). Scanless two-photon excitation of channelrhodopsin-2. *Nature Methods*, *7*, 848–854.

Qiu, D. L., & Knopfel, T. (2007). An NMDA receptor/nitric oxide cascade in presynaptic parallel fiber-Purkinje neuron long-term potentiation. *Journal of Neuroscience*, *27*, 3408–3415.

Qiu, D. L., & Knopfel, T. (2009). Presynaptically expressed long-term depression at cerebellar parallel fiber synapses. *Pflügers Archiv*, *457*, 865–875.

Ren, J., Qin, C., Hu, F., Tan, J., Qiu, L., Zhao, S., et al. (2011). Habenula "cholinergic" neurons co-release glutamate and acetylcholine and activate postsynaptic neurons via distinct transmission modes. *Neuron*, *69*, 445–452.

Siegert, S., Scherf, B. G., Del Punta, K., Didkovsky, N., Heintz, N., & Roska, B. (2009). Genetic address book for retinal cell types. *Nature Neuroscience*, *12*, 1197–1204.

Soriano, P. (1999). Generalized lacZ expression with the ROSA26 Cre reporter strain. *Nature Genetics*, *21*, 70–71.

Srinivas, S., Watanabe, T., Lin, C. S., William, C. M., Tanabe, Y., Jessell, T. M., et al. (2001). Cre reporter strains produced by targeted insertion of EYFP and ECFP into the ROSA26 locus. *BMC Developmental Biology*, *1*, 4.

Sugino, K., Hempel, C. M., Miller, M. N., Hattox, A. M., Shapiro, P., Wu, C., et al. (2006). Molecular taxonomy of major neuronal classes in the adult mouse forebrain. *Nature Neuroscience*, *9*, 99–107.

Tasic, B., Hippenmeyer, S., Wang, C., Gamboa, M., Zong, H., Chen-Tsai, Y., et al. (2011). Site-specific integrase-mediated transgenesis in mice via pronuclear injection. *Proceedings of the National Academy of Sciences of the United States of America*, *108*, 7902–7907.

Tian, L., Hires, S. A., Mao, T., Huber, D., Chiappe, M. E., Chalasani, S. H., et al. (2009). Imaging neural activity in worms, flies and mice with improved GCaMP calcium indicators. *Nature Methods*, *6*, 875–881.

Tsunematsu, T., Kilduff, T. S., Boyden, E. S., Takahashi, S., Tominaga, M., & Yamanaka, A. (2011). Acute optogenetic silencing of orexin/hypocretin neurons induces slow-wave sleep in mice. *Journal of Neuroscience*, *31*, 10529–10539.

Wang, H., Peca, J., Matsuzaki, M., Matsuzaki, K., Noguchi, J., Qiu, L., et al. (2007). High-speed mapping of synaptic connectivity using photostimulation in Channelrhodopsin-2 transgenic mice. *Proceedings of the National Academy of Sciences of the United States of America*, *104*, 8143–8148.

Wang, H., & Zylka, M. J. (2009). Mrgprd-expressing polymodal nociceptive neurons innervate most known classes of substantia gelatinosa neurons. *Journal of Neuroscience*, *29*, 13202–13209.

Yizhar, O., Fenno, L. E., Prigge, M., Schneider, F., Davidson, T. J., O'Shea, D. J., et al. (2011). Neocortical excitation/inhibition balance in information processing and social dysfunction. *Nature*, *477*, 171–178.

Yu, C. R., Power, J., Barnea, G., O'Donnell, S., Brown, H. E., Osborne, J., et al. (2004). Spontaneous neural activity is required for the establishment and maintenance of the olfactory sensory map. *Neuron*, *42*, 553–566.

Zeng, H., Horie, K., Madisen, L., Pavlova, M. N., Gragerova, G., Rohde, A. D., et al. (2008). An inducible and reversible mouse genetic rescue system. *PLoS Genetics*, *4*, e1000069.

Zhang, F., Gradinaru, V., Adamantidis, A. R., Durand, R., Airan, R. D., de Lecea, L., et al. (2010). Optogenetic interrogation of neural circuits: Technology for probing mammalian brain structures. *Nature Protocols*, *5*, 439–456.

Zhang, F., Wang, L. P., Brauner, M., Liewald, J. F., Kay, K., Watzke, N., et al. (2007). Multimodal fast optical interrogation of neural circuitry. *Nature*, *446*, 633–639.

Zhao, S., Cunha, C., Zhang, F., Liu, Q., Gloss, B., Deisseroth, K., et al. (2008). Improved expression of halorhodopsin for light-induced silencing of neuronal activity. *Brain Cell Biology*, *36*, 141–154.

Zong, H., Espinosa, J. S., Su, H. H., Muzumdar, M. D., & Luo, L. (2005). Mosaic analysis with double markers in mice. *Cell*, *121*, 479–492.

T.Knöpfel and E.Boyden (Eds.)
Progress in Brain Research, Vol. 196
ISSN: 0079-6123

CHAPTER 11

Optogenetics in the nonhuman primate

Xue Han*

Department of Biomedical Engineering, Boston University, Boston, MA, USA

Abstract: The nonhuman primate brain, the model system closest to the human brain, plays a critical role in our understanding of neural computation, cognition, and behavior. The continued quest to crack the neural codes in the monkey brain would be greatly enhanced with new tools and technologies that can rapidly and reversibly control the activities of desired cells at precise times during specific behavioral states. Recent advances in adapting optogenetic technologies to monkeys have enabled precise control of specific cells or brain regions at the millisecond timescale, allowing for the investigation of the causal role of these neural circuits in this model system. Validation of optogenetic technologies in monkeys also represents a critical preclinical step on the translational path of new generation cell-type-specific neural modulation therapies. Here, I discuss the current state of the application of optogenetics in the nonhuman primate model system, highlighting the available genetic, optical and electrical technologies, and their limitations and potentials.

Keywords: monkey; genetic manipulation; optical; channelrhodopsin; archaerhodopsin; halorhodopsin; rat.

Introduction

Optogenetic technologies utilize light to control the activity patterns of neurons that are genetically modified to express light-activated opsin proteins. Recent advances in improving the functions of opsin proteins have made it possible to effectively activate or silence many types of brain cells with light at the millisecond timescale. The efficiency of optogenetic control depends upon a number of factors, including the intrinsic physiological properties of the cell, the architecture of the neural network, the number of opsin proteins present on the cell membrane, the optical response kinetics of opsins, and the amount of light that reaches the cell. The major challenge in using this technology in genetically intractable animals, like primates, is the ability to target specific cells or anatomical connections. Here, I discuss details on genetic transduction methods, optical illumination strategies, and electrical monitoring techniques, with a main focus on the use of

*Corresponding author.
Tel.: 617-358-6189, Fax: 617-353-6766
E-mail: xuehan@bu.edu

DOI: 10.1016/B978-0-444-59426-6.00011-2

optogenetics in the genetically intractable nonhuman primate system.

Genetic transduction of brain cells

A major advantage of optogenetic control technology is the ability to control specific genetically modified cells that express light-activated opsin proteins. In genetically intractable animal models, such as the nonhuman primate, the technique to transduce specific cells has been largely limited to viral methods. The most commonly used viruses are lentivirus and adeno-associated virus (AAV). Other virus types, such as adenovirus and herpes simplex virus (HSV) that are effective in transducing brain cells in many model systems, have not yet been well adapted to monkeys, perhaps due to the concerns about the potential adverse immune responses and the limited duration of transgene expression. Although lentivirus and AAVs have been successfully used to transduce brain cells, there has been limited success in targeting specific cell types, and it remains to be a major challenge in realizing the full potential of optogenetics in this model system.

Lentivirus

Lentivirus is an enveloped retrovirus with a single-stranded RNA genome. Current recombinant lentivectors are derived from human immunodeficiency virus (HIV) and other nonhuman lentivirus, such as feline immunodeficiency virus (FIV) and equine infectious anemia virus (EIAV). The potential use of lentivirus in human gene therapy has led to major advances in improving the safety of these vectors. With over 95% of the parental viral genome removed, these recombinant lentiviral vectors induce minimal inflammatory responses. Lentivectors have a modest packaging capacity of ~8kb, large enough to deliver many genes of interest for gene therapy and basic research (Federico, 2003; Thomas et al., 2003).

Effective transduction of the target cells by lentivirus is influenced at each step from virus entry into the cytosol to gene expression within the nucleus. The tropism of a lentivirus is determined by the interaction of the glycoproteins on its viral envelop and the cell surface receptors on the target cell. To facilitate the entry into target cells, lentivirus can be pseudotyped with different viral envelop glycoproteins that recognize membrane surface receptors on a broad range of cell types. However, pseudotyped lentivirus is often unstable and cannot be easily manufactured to produce high titer virus. Over the past two decades, a few lentiviruses have been successfully pseudotyped including HIV, FIV, and EIAV. For example, lentivirus pseudotyped with the glycoprotein (G) from *Vesicular stomatitis virus* (VSV-G) is stable and can be easily concentrated to a titer of 10^9 in a laboratory and can transduce a wide range of cell types. Upon entering a target cell, lentiviral genomic RNA is reverse transcribed into DNA in the cytoplasm, which is then actively transported into the nucleus and integrated into the host genome. Lentivirus tends to insert themselves to the genomic regions undergoing active transcription, which has raised the concerns for mutagenesis, especially in targeting proliferating cells, such as hematopoietic cells (Schroder et al., 2002). However, in the brain, where most cells are terminally differentiated, the chance of inducing brain cancer is extremely low (Thomas et al., 2003).

It has been well established that lentivirus can mediate widespread and long-term gene expression in the brain, capable of transducing neurons with high efficiency (Blomer et al., 1997). Recently, we have successfully used lentivirus pseudotyped with VSV-G to deliver channelrhodopsin (ChR2), Archaerhodopsin-3 (Arch), and Archaerhodopsin-TP009 (ArchT) into the monkey cortex (Han et al., 2009, 2011). For example, lentivirus is able to mediate ChR2-GFP expression in widespread and healthy neurons months after viral injections, and the expression of ChR2-GFP is well targeted to the plasma membrane (Fig. 1a and b). Detailed examination of the cell-type specificity revealed that lentivirus with a CaMKII promoter preferentially labels

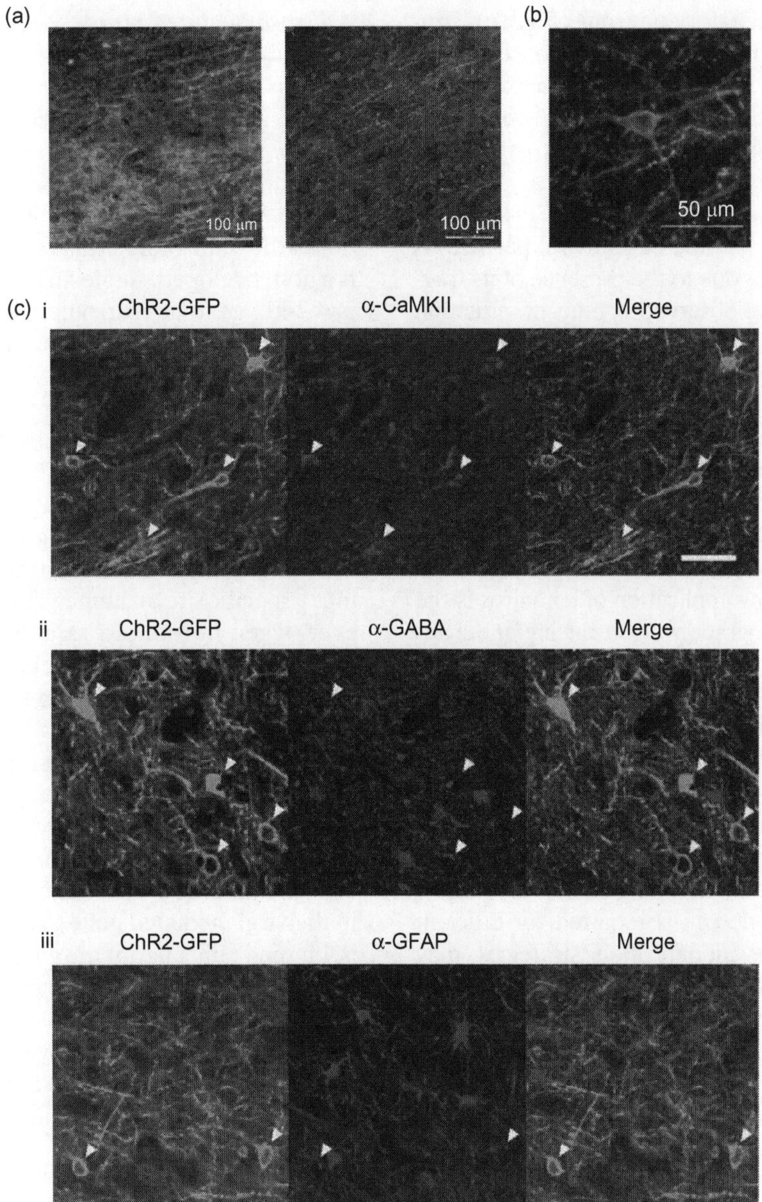

Fig. 1. Cell-type-specific transduction of monkey cortical neurons with light-activated opsin proteins. (a) Widespread and healthy neurons expressing ChR2-GFP (green) months after viral transduction (a-i, red, To-Pro3 nuclear DNA staining; a-ii, red, anti NeuN neuronal staining). (b) ChR2 expression is well targeted to the neuronal plasma membrane and processes. (c) Expression of ChR2-GFP by lentivirus with a CaMKII promoter is restricted to the excitatory neurons. Images of anti-GFP fluorescence (left) and immunofluorescence of cell-type makers, CaMKII (i), GABA (ii), and GFAP (iii; middle), and their overlay with anti-GFP fluorescence (right). Arrowheads indicate ChR2-GFP positive cell bodies. (Adapted from Han et al., 2009.) (For interpretation of the references to color in this figure legend, the reader is referred to the Web version of this chapter.)

monkey cortical excitatory neurons (Fig. 1c), consistent with that observed in the mouse cortex (Nathanson et al., 2009b). Various tests examining immune responses, tissue pathology revealed no detectable damages on neural tissues expressing high levels of opsin molecules (Han et al., 2009).

High titer lentivirus is easy to produce in a laboratory. However, the lifetime of lentiviral particles is rather short, perhaps due to the presence of its fragile envelop. We have observed significant reduction in viral titer after storing them at −80 °C for only half a year. In addition, the infectivity is easily destroyed by repeated freeze–thaw processes or simply sitting at room temperature for an extended period of time. Thus, precautions have to be made when using lentivirus. We found it necessary to use fresh stocks of lentivirus, to avoid repeated freeze–thaw process, and to limit the exposure to room temperature.

Another promising application of lentivirus is the capability of retrogradely transducing neurons projecting to the area of viral injections. Most promisingly, EIAV pseudotyped with rabies glycoprotein can be transported through axons, labeling large number of cells retrogradely (Mazarakis et al., 2001). However, it remains to be determined whether this retrograde strategy is capable of introducing sufficient amount of opsin molecules for optical modulation. Since significant number of opsins within a single cell is required for efficient optical modulation, amplification strategies may be helpful in amplifying opsin expression in retrogradely labeled projecting cells. For example, one could retrogradely express Cre enzymes by injecting rabies glycoprotein-pseudotyped EIAV and then transduce the Cre-expressing projecting neurons in the upstream brain regions with VSV-G-pseudotyped lentivirus through a second local viral injection to express opsins under the control of Cre enzymes. This way, the retrograde expression of a small number of Cre enzymes can be used to initiate the production of a large number of opsins in these projecting cells. However, as it is already difficult to perform genetic modification with a single virus in monkeys, the feasibility of performing such two-step viral transduction method remains to be tested.

Adeno-associated virus

AAV is becoming a common vector of choice for human gene therapy or basic research because of its low pathogenicity, low immunogenicity, high efficiency, and long-lasting transgene expression (Muzyczka, 1992; Peel and Klein, 2000). AAV, a nonpathogenic human parvovirus, ~20 nm in size, is a nonenveloped, single-stranded DNA virus with a 4.7-kb genome surrounded by coat proteins. AAVs are present in ~80% of human adults but cause no known pathology. AAV has been developed as a human gene therapy vector since 1984 (Hermonat and Muzyczka, 1984). Current recombinant AAV virus has ~96% of its viral genome removed, resulting in a greater reduction of possible immune responses for gene therapy. In non-active, nonamplifying conditions, AAV integrates into a specific site on human chromosome 19. However, when introduced as a gene therapy vector, only ~10% of AAV particles integrate into the genome and in a random nonspecific fashion (Thomas et al., 2003).

It has been of great interest to engineer different serotypes for effective and cell-type-specific transduction with AAVs (Muzyczka and Warrington, 2005). Infection efficiency of AAVs to target cells can be influenced by a variety of factors involved in the viral mediated gene expression process, that is, binding efficiency of the viral particles to specific cell surface receptors, rate of virus endocytosis, intracellular trafficking to cell nucleus, removing of the viral coat proteins, synthesis of the second strand of the viral genome, and transcription and translation of the gene of interest in the nucleus. Engineering novel AAV serotypes has been mostly through modifying AAV capsids, the protein shell, because such approaches are expected to have minimal influence on vector assembly, packaging, and particle stability. So far, over 100 unique AAV capsid sequences have been identified, among which AAV1–9 and Rh10 have been characterized in greater detail.

To produce AAVs, the capsid sequence for each serotype is often engineered into a separate vector, in addition to the genomic vector that contains the gene of interest. Thus, the same genomic vector can be packaged with different capsids, enabling direct comparison of the efficiency of each capsid serotype in transducing specific cells. Note, since in most protocols, the genomic vectors used contain the inverted terminal repeats (ITRs) from AAV2, the pseudotyped AAV virus is often called AAV2/*, such as AAV2/5, often simplified as AAV5 in the literature, as well as in the following text, meaning the virus is made from the genomic vector containing AAV2 ITRs but pseudotyped with AAV5 capsid.

Tropism of different AAVs varies between developmental stages, cell types, and species. So far, AAV1, 2, 5, 7, 8, 9, and Rh10 seem to transduce adult brain cells with various efficiencies. In the adult rodent brain, direct comparison of AAV2, 4, 5 revealed that AAV5 has the highest efficiency in transducing striatum, both neurons and glia (Davidson et al., 2000). (Note, in neonatal mice brain, AAV1 seems to be better than AAV2 in transducing brain cells, whereas AAV5 failed to achieve a significant amount of transduction (Passini et al., 2003).) Direct comparison of AAV1, 2, 5, 7, and 8 suggested that AAV5, 7, 8 are able to transduce comparable brain volume at high titers, but at lower titers AAV5 and 7 transduced larger brain volume than AAV8 (Taymans et al., 2007). Direct comparison of AAV7, 8, 9, and Rh10 revealed that they are all quite efficient in transducing neurons, but not glia, in the cortex, hippocampus, striatum, and thalamus. Their transduction efficiencies slightly vary in different brain structures, with AAV9 being most effective in the hippocampus and cortex, and Rh10 being most effective in the thalamus (Cearley and Wolfe, 2006). Cearley further screened and identified new serotypes hu.32, hu.37, pi.2, hu.11, rh.8, and hu.48R3 that are all more efficient than AAV9 (Cearley et al., 2008). In addition, Lawlor et al. demonstrated that cy5, rh20, and rh39 are more efficient than AAV8 (Lawlor et al., 2009). Retrograde transport of AAVs has also been observed for various serotypes in different brain structures. Most prominently, Cearley et al. observed that AAV9 and Rh10 are effectively transported retrogradely in many brain structures. AAV9 seems to result in more efficient retrograde transduction in the hippocampus and septal nuclei, whereas Rh10 is more effective in the thalamus (Cearley and Wolfe, 2006). With the identification of over 100 new serotypes, detailed characterization of these different serotypes should identify more efficient and more specific serotypes for different types of cells in different brain structures (Gao et al., 2002).

The recent discovery that AAVs can pass the blood–brain barrier has created tremendous enthusiasm in the use of these vectors. Duque et al. found that AAV9 with double-stranded genome can pass the blood–brain barrier when delivered intravenously in neonatal mice and resulted in the transduction of motor neurons in the spinal cord and all brain structures being tested including cortex, striatum, thalamus, hippocampus, cerebellum, and brain stem (Duque et al., 2009; Zhang et al., 2011). However, intravenous delivery of AAV9 in adult mice almost exclusively transduces astrocytes throughout the entire CNS, with little neuronal transduction (Foust et al., 2009). Another study showed that Rh10 is comparable if not more potent than AAV9 in transducing brain cells when delivered through systematic intravenous delivery (Zhang et al., 2011). Although it remains to be determined whether such delivery route will be effective in monkeys, it is highly plausible that some or many of the new AAV serotypes will be able to transduce brain cells through systemic injections.

In primates, direct comparison of AAV1–6 showed that AAV5 is the most efficient vector in substantia nigra and striatum, which transduces both neurons and glia (Markakis et al., 2010). AAV5 seems to be more efficient than AAV8 (Dodiya et al., 2010). AAV1 can effectively transduce both neurons and glia in monkey brains but unfortunately seemed to induce strong humoral and cell-mediated immune responses (Hadaczek et al., 2009). AAV1 has been successfully used to transduce striate cortex in monkeys for two photon imaging (Heider et al., 2010; Stettler et al., 2006). Recently, Diester et al. used AAV5 with a synapsin promoter or a Thy-1 promoter to express ChR2, ChR2-C128S mutant, and eNpHR2.0 in the monkey cortex (Diester et al., 2011). These viruses induced strong expression in cortical neurons, and the transduced neurons responded well to light modulation. Unexpectedly, Diester et al. found no superficial layers of the cortex expressed any of these opsins regardless of promoters. It remains unclear whether this lack of labeling is due the specific serotype used, the promoters used, or the cortical regions transduced.

In summary, AAV1, 5, 8, 9, and Rh10 are all effective in transducing neurons. At lower titers, AAV5 seems to be more efficient than AAV1 and 8, but AAV5 seems to transduce both neurons and glia. In contrast, AAV8, 9, and Rh10 seem to be more specific to neurons. Perhaps due to the lack of lipid envelop, AAV can be easily concentrated to a high titer of 10^{12} or higher. AAV is also much more stable during storage, and the titer of AAV does not seem to decline noticeable even after a couple years of storage at −80°C. It is difficult to manufacture AAVs in a research laboratory due to the complicated procedures in purifying and concentrating AAVs, which may even influence the tropism of the virus (Klein et al., 2008). However, many viral core facilities are providing standard service for packaging AAVs with different serotypes with typical titers of 10^{12} and above, such as University of North Carolina gene therapy center (http://genetherapy.unc.edu/), University of Florida Powell gene therapy center (http://www.gtc.ufl.edu/), and University of Pennsylvania gene therapy program (http://www.med.upenn.edu/gtp/).

Cell-type specificity achieved with viral methods

Genetic targeting of specific cell types has been successful in transgenic mice, facilitated by the bacteria Cre–Lox system (Yizhar et al., 2011). Most recently, several new lines of transgenic mice have been made to facilitate the targeting of specific cell types using the Cre–Lox system, in which opsin expression is regulated by Cre enzymes (Katzel et al., 2011). These advances and the continued effort in improving the expression levels of opsins in particular cell types in transgenic mice will certainly revolutionize the analysis of the causal role of specific cells in neural circuit functions in rodent models.

Transgenic monkeys, on the other hand, are costly and time consuming to generate and maintain. So far, there are two lines of transgenic rhesus monkeys reported, one expressing GFP alone (Chan et al., 2001), and the other expressing human huntingtin gene (Yang et al., 2008). In addition, transgenic marmosets have also been successfully generated (Sasaki et al, 2009). In monkeys, and other genetically intractable models, viral methods will remain the main methods of expressing opsins. However, limited success has been made in targeting specific cell types with viral technologies. This could be due to the limited packaging capacity of lentivirus or AAV, the lack of understanding of the interaction between target cells and viral particles, and the difficulty of predicting cellular regulation mechanisms of viral mediated gene expression. The glycoproteins on the envelops of the lentivirus or the capsid proteins of the AAV will determine the entry of viral particles to specific cell types through interactions with the surface receptors presented on target cells. So far, the commonly used lentivirus pseudotyped with VSV-G or AAV pseudotyped with the capsids from AAV1, 5, 8, 9, and Rh10 are able to effectively transduce neurons in rodents and monkeys. AAV1 and 5 are able to transduce both neurons and glia, whereas AAV 8, 9, and Rh10 seem to be more specific toward neurons. Lentivirus can infect both neurons and glia but prefers neurons at lower titers.

Detailed analysis of AAV1 tropism revealed that the transduction efficiency of excitatory neurons, inhibitory neurons, and nonneuronal glia cells depends upon promoters, viral titers, and cortical layers. For example, AAV1 with a CAG promoter can transduce both neurons and glia cells effectively, but AAV1 with hSyn promoter preferentially transduce layers 2/3 inhibitory neurons at lower titer of $\sim 3\times10^{11}$ and preferentially transduce excitatory neurons at a higher titer of $\sim 8\times10^{12}$ (Nathanson et al., 2009b). In monkeys, AAV1 selectively labeled neurons (Heider et al., 2010), whereas AAV5 showed no specificity toward different cell types, transducing both excitatory, inhibitory neurons and glia (Diester et al., 2011).

Lentivirus, with its ability to package larger transgenes, has the potential to carry slightly larger promoter sequences for cell-specific targeting. Lentivirus with a CaMKII promoter has been demonstrated to selectively label excitatory neurons in mouse (Dittgen et al., 2004) and monkey cortex (Fig. 1c; Han et al., 2009). Such specificity may be partly attributed to the tropism of lentivirus particles themselves, as it has been suggested that lentivirus with hSyn promoter can selectively label excitatory neurons in the rodent cortex (Nathanson et al., 2009b). However, we have observed that high titer lentivirus with a CAG promoter readily labeled GFAP positive glia cells in the primate frontal cortex. In addition, the purification procedure during lentivirus production may influence the tropism of lentivirus.

There are a few successes in identifying promoters specific for particular cell types. A synthetic promoter or enhanced promoter based on the promoter sequence of human dopamine β-hydroxylase promoter, when used in adenovirus, was able to mediate efficient and specific transduction of noradrenergic neurons in rat locus coeruleus (Hwang et al., 2001, 2005). Lentivirus with a 3-kb promoter region of neuroactive peptide cholecystokinin (CCK) was found to be specific in transducing CCK positive cells (Chhatwal et al., 2007). Recently, a detailed systematic analysis was performed by the Callaway group in examining over 20 short promoters derived from fugu, mouse, human, and synthetic ones, packaged in AAV1 or lentivirus (Nathanson et al., 2009a). They identified a number of short promoters that showed selectivity toward inhibitory neurons versus excitatory neurons in rodents. However, none of these promoters targeted a specific set of inhibitory neurons that can be coordinated with our current classification of inhibitory neurons base on protein markers, that is, parvalbumin, somatostatin, CCK, calretinin, etc. It remains to be determined whether these short promoters will be able to achieve higher specificity in monkeys.

In vivo optical control of transduced cells

Success in the adaption of optogenetic techniques to control the activities of specific cells in monkeys has raised significant excitement in the field (Diester et al., 2011; Han et al., 2009, 2011; Fig. 2). The effectiveness of optically controlling neural activities is determined by the absolute amount of light reaching the neuron, the number of opsin molecules presented on the neuronal plasma membrane, and the sensitivity of the opsins to light. The final output of the control precision, the time course and the magnitude of light modulation, is also influenced by the intrinsic physiological properties of the neuron and its surrounding neural network environment, which cannot be controlled by experimenters. In monkeys, a few unique challenges remain to be overcome to achieve effective optical control of larger brain volume and to perform long-term recordings that is often required for monkey experiments. The tissue volume that can be controlled optically is determined by the optical properties of the brain, the biophysical properties of the opsins, and the efficiency of genetic transduction of cells. Long-term repeated optogenetic experiments within the same brain region require minimal tissue damage during each recording and optical control session.

Optical properties of the brain

The absolute power of light reaching a neuron depends upon the input of light power at the optical fiber tip and the pattern of light propagation in the tissue. Light propagation is determined by

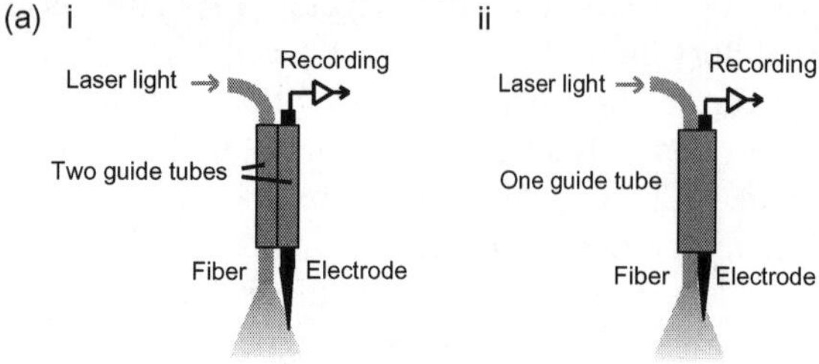

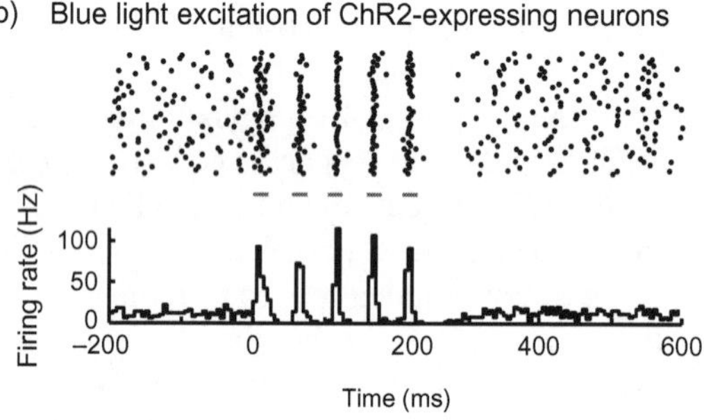

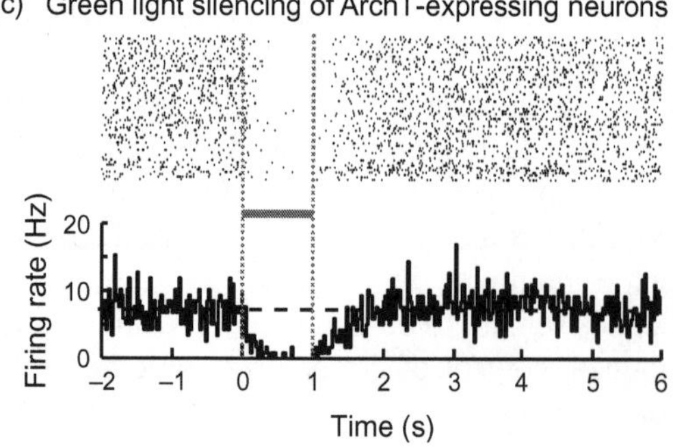

Fig. 2. Temporal precise optical activation or silencing of cortical neurons. (a) Two configurations of coupling an optical fiber to a recording electrode. (a-i) The optical fiber and the electrode are guided separately in two individual guide tubes. The relative distance between the optical fiber and the electrode can be easily adjusted during experiments. (a-ii) The optical fiber is directly glued to the electrode and is guided together within the same guide tube. (b) Temporal precise optical activation of ChR2-expressing cortical excitatory neurons. Top, spike raster plot displaying each spike as a black dot and each trial as a horizontal row; bottom, the histogram of instantaneous firing rate, averaged across all trails. Periods of light illumination are indicated by horizontal blue dashes. (c) Temporal precise optical silencing of ArchT-expressing cortical excitatory neurons. Top, spike raster plot; bottom, histogram of instantaneous firing rate. Periods of light illumination are indicated by horizontal green dashes. (Adapted from Han et al., 2009, 2011.) (For interpretation of the references to color in this figure legend, the reader is referred to the Web version of this chapter.)

the optical properties of brain tissue, with major considerations being tissue absorption and tissue scattering (Mobley and Vo-Dinh, 2003). Brain tissues are heterogeneous, with spatial variations in their optical properties. This spatial variation and the density of this variation make brain tissues strong scatterers. Tissue scattering and absorption will reduce the intensity of light as it propagates within the tissue, away from the light source. At locations within the close proximity of the light source, where light intensity is at the saturation level for opsin functions, the efficiency of optically controlling neurons will not vary with locations. But at locations further away from the light source, where the light intensity falls below the saturation level, the power to control neural activities will reduce with distance, and eventually at locations where the light power falls below the threshold of opsin activation, light will be unable to modulate neural activities. In addition, the presence of electrodes and optical fibers will influence the pattern of light propagation.

Brain tissues have been successfully modeled as a two-component system, a homogeneous continuum with randomly positioned scattering and absorbing particles (Bevilacqua et al., 1999; Yaroslavsky et al., 2002). Tissue absorption is determined by its molecular compositions. In the brain, hemoglobins, both oxygenated (HbO_2) and deoxygenated (Hb), are the major light absorbers at visible wavelength relevant for optical control (Fig. 3, summarized by Scott Prahl, http://omlc.ogi.edu/spectra). Current available classes of opsins, channelrhodopsins, halorhodopsins, and archaerhodopsins, are mainly excited by visible light of 450–600nm. Hb/HbO_2 absorbs highly in this wavelength range. To increase light propagation, it is useful to develop novel opsin molecules that can be sensitized with red light, that is, >650nm, where both Hb and HbO_2 absorption coefficients are drastically reduced. Development of these novel optogenetic molecules would be particular useful in experiments in monkeys where controlling larger brain volumes may be necessary to perturb enough neurons to influence information processing.

Monte Carlo simulation of light propagation

To model light propagation in brain tissue, a radiation transport model has been established to simulate photon energy transport that explicitly ignores the complex multiple scattering effects. To simulate light transport and to visualize the distribution of light in tissue, the widely accepted and most commonly used method is Monte Carlo simulation (Mobley and Vo-Dinh, 2003). Monte Carlo methods include a broad class of computational algorithms that employ random numbers in simulating complex systems. In predicting light propagation in tissues, the Monte Carlo method tracks the trajectory of each photon and calculates the light intensity at each position within the tissue based on the distributions of photons. Photon trajectory is estimated based on the random walk that each photon performs in a specific tissue, and the specific parameters for each step, that is, length and direction, are calculated using random numbers. The accuracy of the simulation increases with the number of photons launched. Typically more than 100,000 photons are required for three-dimensional simulations of light propagation in the brain.

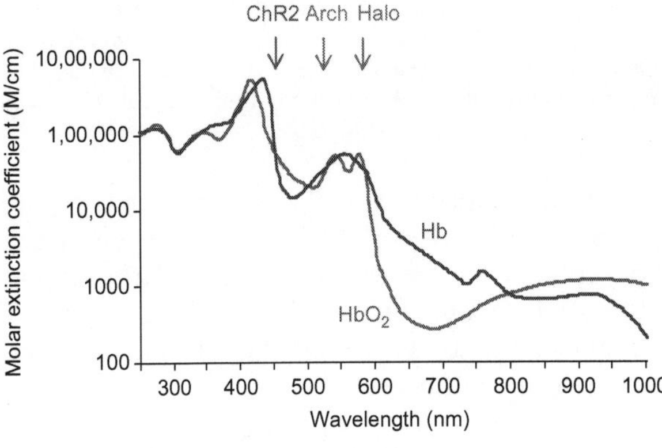

Fig. 3. Hemoglobin, oxygenated (HbO_2) and deoxygenated (Hb), absorption spectrum. Values are based on that summarized by Scott Prahl, http://omlc.ogi.edu/spectra. The peak excitation wavelengths for ChR2, Arch, and Halo are indicated. (For color version of this figure, the reader is referred to the Web version of this chapter.)

Monte Carlo simulation for both blue and yellow light propagation in the brain was performed by Ed Boyden's lab, for both light emitted out of LEDs and optical fibers (Bernstein et al., 2008; Chow et al., 2010). For example, they simulated the trajectory of yellow light emitted from the end of an optical fiber in a cube of gray matter of $200 \times 200 \times 200$ grid of voxels, corresponding to $10\,mm \times 10\,mm \times 10\,mm$ in dimension, using previously published parameters and algorithms (Binzoni et al., 2006; Wang et al., 1995; Yaroslavsky et al., 2002). In this simulation, a scattering coefficient of $13\,mm^{-1}$ and an absorption coefficient of $0.028\,mm^{-1}$ were used from the interpolated data in Yaroslavsky et al. (2002). 5×10^6 photons were launched in a pattern through a model fiber with a numerical aperture of 0.48 (Optran 0.48 HPCS, Thorlabs; Wang et al., 1995). The anisotropic scattering model based on the Henyey–Greenstein phase function with an anisotropy parameter of 0.89 was first used before randomizing the photon trajectories (Binzoni et al., 2006; Yaroslavsky et al., 2002). In the simulation, the photon was absorbed according to the distance it traveled for each step. When a photon enters a voxel, the stimulation probabilistically calculates the forecasted traveling distance for the next step, and the direction of the photon packet propagation is randomly chosen according to the Henyey–Greenstein function. Using this model, the estimated light propagation in brain tissue can be plotted (Fig. 4a). The simulated light distribution generally agrees with that measured in brain slices (Wang et al., 1995).

Experimental examination of light propagation on optical control efficiency

To estimate the influence of light propagation on the efficiency of optical modulation *in vivo*, we have performed detailed analysis of light modulation of neural activities recorded with the same electrode at the same location in the brain of awake monkeys (Han et al., 2009). First, we systematically reduced the light power out of the fiber tip while

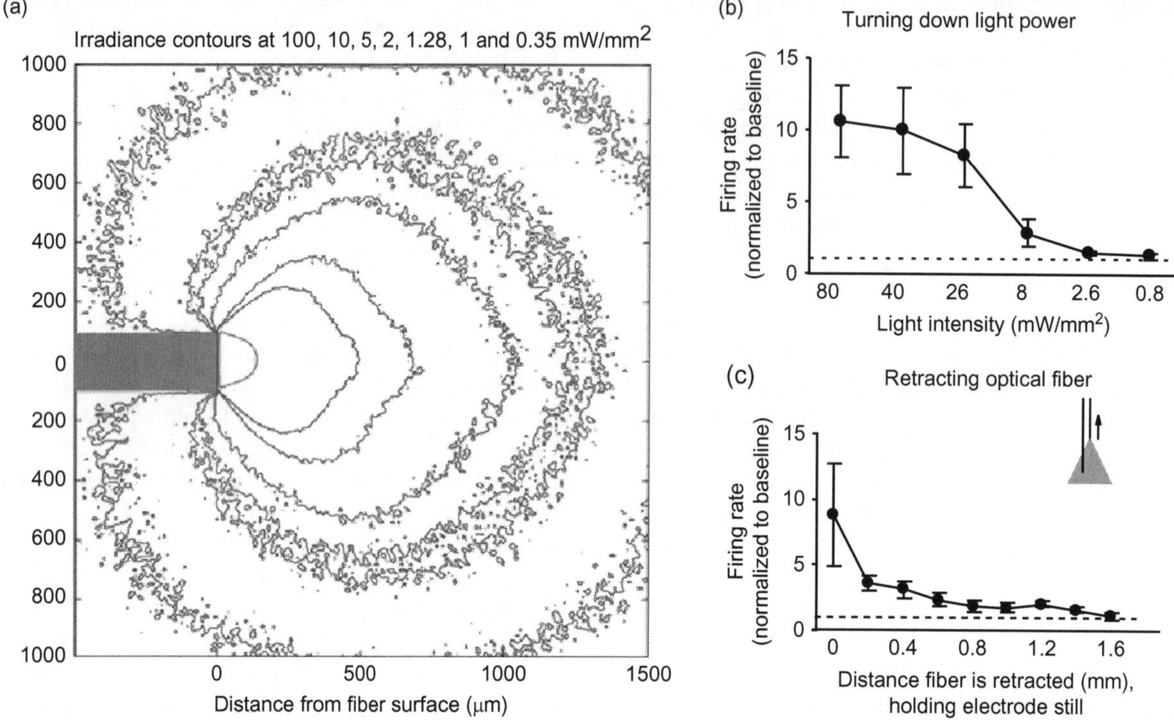

Fig. 4. Influence of light propagation on optogenetic control. (a) Monte Carlo simulation of light propagation in the brain. 593nm yellow light irradiance in the brain gray matter is plotted as a function of location with respect to the tip of an optical fiber (200μm in diameter, represented by the gray rectangle shown at left; the irradiance at the surface of the fiber tip was set at 200mW/mm^2). The contours display 100, 10, 5, 2, 1.28, 1, and 0.35mW/mm^2 irradiances. (b) Light intensity dependence of optical control. Optical modulation of neural activity driven by ChR2-mediated activation of excitatory neuron is determined upon illumination with different light power intensities. (c) Distance dependence of optical control. Modulation of neural activities is determined when optical fibers were retracted in 200μm steps, while holding the electrode still. (Adapted from Han et al., 2009.) (For interpretation of the references to color in this figure legend, the reader is referred to the Web version of this chapter.)

keeping the distance between the optical fiber and the recording electrode constant (Fig. 4b). Reduction of light intensity reduced the efficiency of light modulation. However, the effect of light power on modulation efficiency is not linear. Efficiency of light modulation drops sharply from 26 to 8mW/mm^2, and the effect of optical modulation is nearly abolished at ~2.6mW/mm^2. In a second experiment, we kept the light power constant but systematically retracted the optical fiber away from the recording electrode in 200μm steps (Fig. 4c). The change in optical modulation drops sharply when the optical fiber is retracted from the recording electrode, with most of the reduction happening in the first 200μm step. Together, our experimental observations agree with the Monte Carlo simulation of light propagation, in which light falls off nonlinearly and falls to ~1% at a location ~1mm away from the fiber tip.

Tissue damages from device insertion and heat

Since many experiments conducted in awake behaving monkeys are chronic, often extending to several years, a major consideration for optogenetic experiments is tissue damage introduced by device insertion, that is, viral injection cannula and optical fibers, and the tissue damage from heat produced by light. Viral injection-induced tissue damage is a somewhat minor concern because viral mediated gene expression, either with lentivirus or with AAV, is long lasting. One successful injection of a few microliter of virus is often sufficient to label a spherical volume of a few millimeters in diameter. A critical consideration is the success of injecting virus into targeted areas in the monkey brain. MRI or electrophysiology mapping would be helpful in targeting the desired locations. To perform an accurate injection, it is important to eliminate as much as possible the dead space between the syringe that holds the virus and the needle tip in the brain. It would be optimal to position the syringe directly over the injection site and connect the needle directly to the syringe without using tubing in between. Many commercially available syringe pumps are compact and can be easily mounted onto a manipulator, for example, UltraMicroPump from World Precision Instruments. However, if tubing is necessary, it is important to use thin and nonelastic tubing to efficiently transduce the force from syringe pump to the tip of the needle. Before and after each injection, it is useful to check possible leaks.

To reduce mechanical damage from optical fibers, ideally one would want to use optical fibers as thin as possible. Typical electrodes, even with a shank size of 200 μm, have fine tips within 10 μm. A 200-μm optical fiber with a blunt tip is thus orders of magnitude larger than an electrode tip. We have tried to taper the optical fiber to address this issue. But, it is difficult to polish a tapered fiber tip, and the distribution of light out of the fiber tip is different from that modeled out of a blunt end fiber tip. With these potential concerns, it may still be advantageous to use tapered fibers, since the variability in viral injections and the uncertainty of targeting the injection sites in monkeys are often more variable than the variability in light emission out of a tapered fiber tip. A potential strategy to avoid repeated penetration of brain tissue with optical fiber is to leave the optical fiber in the brain for as long as possible. Another strategy of reducing the mechanical tissue damage is to use arrays of small optical fibers. For example, a single optical fiber of 200 μm in diameter is equivalent in volume to four optical fibers of 100 μm in diameter. But four optical fibers of 100 μm in diameter are capable of illuminating a much larger volume than a single 200 μm fiber. Adaption of high-density fiber arrays, such as those developed in Ed Boyden's lab, will be helpful to reduce mechanical tissue damage (http://syntheticneurobiology.org/).

Heat produced by shining light into the brain is another consideration for tissue damage. Measuring heat dissipation within the brain is rather difficult, since the introduction of the measuring device itself will influence heat dissipation. It has been observed that heating of the metal recording electrode with strong laser light will in turn activate wild-type neurons expressing no opsins, though this heat-induced neural activation has a much slower time constant (personal communication with Michael Fee). Thus, whenever possible, the power of laser light used should be limited to what is sufficient to drive neurons. Typically, a few hundred mW/mm^2 of irradiance or a few mW of total light power does not seem to produce detrimental damaging effects.

Neuronal and behavioral modulation with optogenetic control

Electrical recording of neural activities in monkeys has been a major driving force in our understanding of the neural basis of many brain functions, for example, sensation, action, decision making, attention, emotion, etc. Electrophysiological methods can establish a precise correlative relationship between neural activity patterns and brain state or behavioral phenomena with a high temporal resolution. However, optogenetic control provides a unique approach in examining the causal role of specific neural activity patterns in neural information processing and behavior. Often during optogenetic experiments, simultaneous optical control and electrophysiology or optical imaging methods are performed to directly link the changes in neural activity patterns to neural network dynamics or behavior. Here, I focus on challenges in simultaneous electrophysiological recording during light illumination and the possibilities for current failures in modulating behavior with optogenetic control in monkeys.

Optical artifact on metal electrodes

Coupling an optical fiber to a metal electrode is a simple and reliable way to measure light modulation of brain activities. However, we and others have observed strong voltage deflection artifact when laser light was directed onto metal electrode tips, in the brain or in saline (Fig. 5; Ayling et al., 2009; Han et al., 2009, 2011). This effect was clearly observed when the electrode tip was positioned in the blue laser beam in saline. This artifact was also evident in the brain with a radiant flux of 80mW/mm^2, a light intensity that is often needed for *in vivo* optogenetic experiments, when the tip of the optical fiber is 0.5–1mm away from the electrode tip. It is possible that part of the voltage defection recorded in the brain reflects physiological changes in local field potential (LFP) upon optical stimulation of transduced neurons. However, it is not yet possible to isolate light evoked physiological responses from the optical artifact.

The light-induced artifact is slow evolving. Upon illumination with a long light pulse of 200 ms, the voltage defection slowly reaches its peak after tens of milliseconds, which can be easily eliminated with a high-pass filter that electrophysiologists typically use for isolating spikes during extracellular recordings. For example, this artifact is completely removed by the band pass filter of 170–8000Hz used in Plexon data acquisition system. However, precautions are needed when laser light is pulsed at higher frequencies, since high-frequency artifact produced by brief light pulses cannot be removed by simply filtering the signal with high-pass filters. But if the artifact produced by brief high-frequency light pulse trains is significantly different from the spikes recorded, it is possible to isolate the light artifact waveforms through spike sorting. In contrast, LFP that measures slow voltage fluctuations at lower frequencies in the range of Hz to tens of Hz cannot be isolated from slow evolving light-induced artifact. Thus, while this artifact typically does not influence the ability to record spikes at the site of illumination, it does prevent accurate measure of LFP at the site of illumination.

The magnitude of the artifact depends upon the precise power of light illumination, the relative position of the light source and the electrode, the properties of the electrode tip, and the optical properties of the brain tissue between the light source and the electrode. We have observed that the magnitude of the artifact is proportional to the power of light illumination but varies with the wavelength of the light. For example, we observed stronger voltage deflection artifact with 472nm blue light than with 532nm green light or 589nm yellow light.

The observed light-induced artifact is consistent with the Becquerel effect. The Becquerel effect describes a classical photoelectrochemical phenomenon first demonstrated by French scientist Becquerel in 1839 (Gratzel, 2001; Honda, 2004). Becquerel demonstrated that exposing metal electrodes, such as platinum, gold, and silver to sunlight produced very small electric current when these metals were positioned in electrolyte. This phenomenon has inspired major research interests in improving this photoelectrochemical effect in converting sun light to electrical powers. However, for the neuroscientists applying optogenetic techniques, it remains a critical challenge to minimize such photoelectrochemical effects.

Consistent with the generality of the Becquerel effect, we observed such artifact with metal electrode wires made of stainless steel, platinum–iridium, silver/silver chloride, gold, nichrome, copper, or silicon. However, we have never observed such

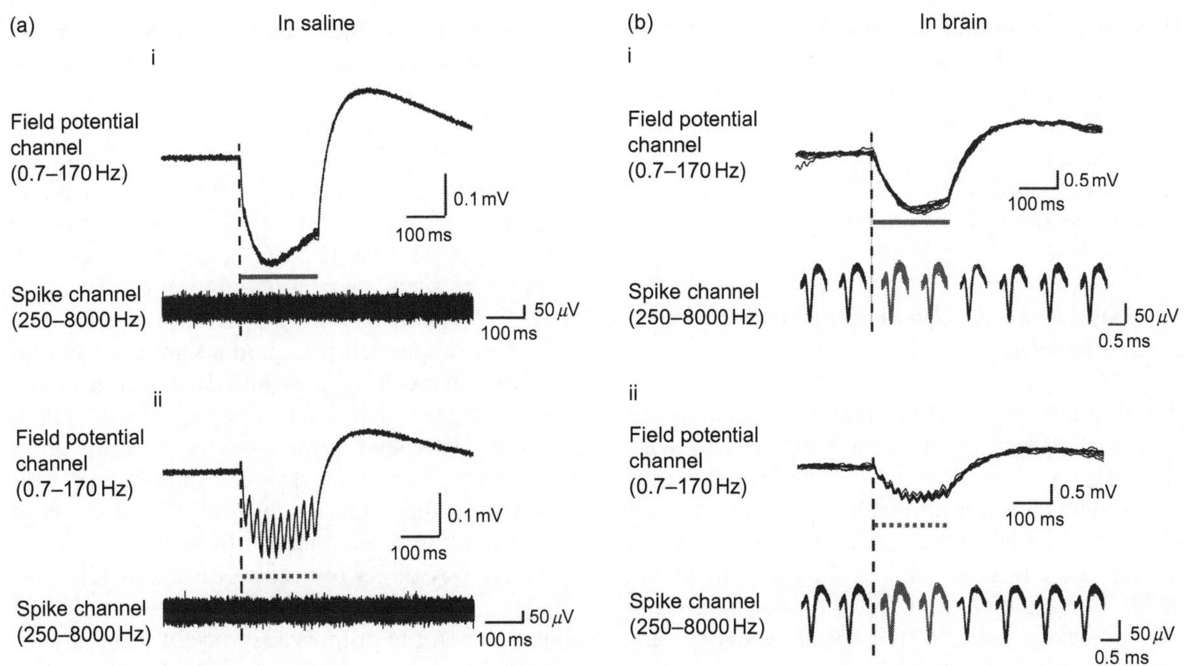

Fig. 5. Optical artifact observed on tungsten electrodes immersed in saline (a) or brain (b) upon tip exposure to 200ms blue light pulses (i) or trains of 10ms blue light pulses delivered at 50Hz (ii). Light pulses are indicated by blue dashes. Electrode data were hardware filtered using two data acquisition channels operating in parallel, yielding a low-frequency component (field potential channel) and a high-frequency component (spike channel). For the "spike channel" traces taken in brain (b), spikes were grouped into 100ms bins, and then the binned spikes were displayed beneath corresponding parts of the simultaneously acquired "field potential channel" signal. (Shown are the spikes in eight such bins—the two bins before light onset, the two bins during the light delivery period, and the four bins after light cessation.) (Adapted from Han et al., 2009.) (For interpretation of the references to color in this figure legend, the reader is referred to the Web version of this chapter.)

artifact with hollow glass microelectrodes (Boyden et al., 2005; Han and Boyden, 2007). A few cautions have to be made with glass electrodes in optogenetic experiments. For example, if laser light reaches the Ag/AgCl wire that is in direct contact with the solution inside the glass electrode, light will induce artifact. Since the Ag/AgCl wire is typically tens of millimeters away from the tip of the glass electrode under illumination, this can be easily controlled. Similarly, if light reaches the metal ground electrode, this optical artifact will also be picked up by glass electrodes. Even though glass electrodes offer a good way to circumvent the artifact problem with recording LFP at the site of illumination, it is - difficult to record from multiple glass electrodes, and the use of glass electrodes in monkeys are limited, in particular in chronic awake experiments where breaking of the electrode tip in the brain would lead to significant tissue damage at the recording sites.

It might be possible to develop computational methods to remove this artifact, since the amplitude and the time course of this artifact are stable with repeated light illumination when the electrode and the optical fiber remain at the same location and the light intensity remains the same. However, to isolate or average out the real physiological effects, some computational/experimental methods have to be used, which is yet to be developed. More promisingly, optimization of the electrode tip surface or electrode material may be proven useful in eliminating optical artifact.

Recently, Zorzos and Boyden eliminated this artifact by coating the surface of the electrodes with conducting material indium tin oxide (ITO; Zorzos et al., 2009). Continued advance in improving electrode coating strategy or developing novel electrodes is critical in enabling measurement of LFP at the site of illumination.

Homeostatic neural dynamics upon perturbing specific neurons

Altering the activities of a small set of neurons or even a single neuron can induce complex network changes, as demonstrated elegantly by recent experiments through intracellular current injection via whole cell patch clamp electrodes in anesthetized rats (Brecht et al., 2004; Li et al., 2009). For example, stimulating a single neuron in the superficial layers of the visual or somatosensory cortex can switch global cortical states from slow wave-like to rapid-eye-movement-sleep-like states (Li et al., 2009), whereas stimulating a single pyramidal cell in layer six of the motor cortex can evoke whisker movement (Brecht et al., 2004). Since light cannot be easily directed to only one cell as with a patch electrode, the major advantage of optogenetics is to control a set of genetically identified cell types. It might be possible to stimulate just a few cells when light is directed into areas with sparsely transduced cells.

With the readily available genetic techniques in transgenic mice, rapid progress is being made in assessing the functions of specific cell types in transgenic mouse models, such as, parvalbumin-positive cells (Cardin et al., 2010; Sohal et al., 2009), hypothalamus POMC-positive neurons (Aponte et al., 2011), cholinergic neurons (Witten et al., 2010), specific dopamine receptor-expressing neurons (Kravitz et al., 2010; Lobo et al., 2010), and retina ganglion cells (Thyagarajan et al., 2010). However, the examination of the functional significance of specific cell types in genetically intractable animals is limited by the available genetic tools as described above. So far, we are only able to target cortical excitatory neurons as recently demonstrated with a lentivirus with a CaMKII promoter in monkeys (Han et al., 2009).

When a population of pyramidal cells expressing ChR2 was excited with blue laser light in the monkey brain, we observed a major population of excited cells, and a significant minority of suppressed cells (Fig. 6a and b). In contrast, temporary silencing of a population of pyramidal cells expressing ArchT resulted in a major population of suppressed cells, and a significant minority of excited cells (Fig. 6c and d). Together, these results suggest that when a population of cells is directly controlled, either excited or silenced, a secondary component from a minority of cells reacted in the opposite fashion, which could in part homeostatically balance the network activity produced by direct optical perturbation. The secondary component responded with a longer latency than the primary response. Further, when a perturbation is induced repeatedly, the set of cells that undergo the secondary response does not change from trial to trial but instead retains its identity.

Together, these results suggest a novel homeostatic principle for the cortex governing changes upon the control of the activity of specific cell populations. It is possible that specific neural circuit elements, such as inhibitory neurons, are important in balancing the overall network responses upon perturbation of pyramidal neurons. For example, the silencing of pyramidal neurons decreases their excitatory drive to inhibitory neurons, which in turn disinhibit downstream targeted cortical neurons; conversely, exciting pyramidal neurons increases the drive to inhibitory neurons, decreasing activity in targeted cortical neurons. It is important to point out that interactions between cortical areas may also contribute to the observed opponent responses. The complex neural network architecture also predicts that the secondary responses may also modulate the primary responses. For example, a neuron may increase its activity as a direct response to optical modulation. However, it also receives synaptic inputs from neurons responding in the opposite

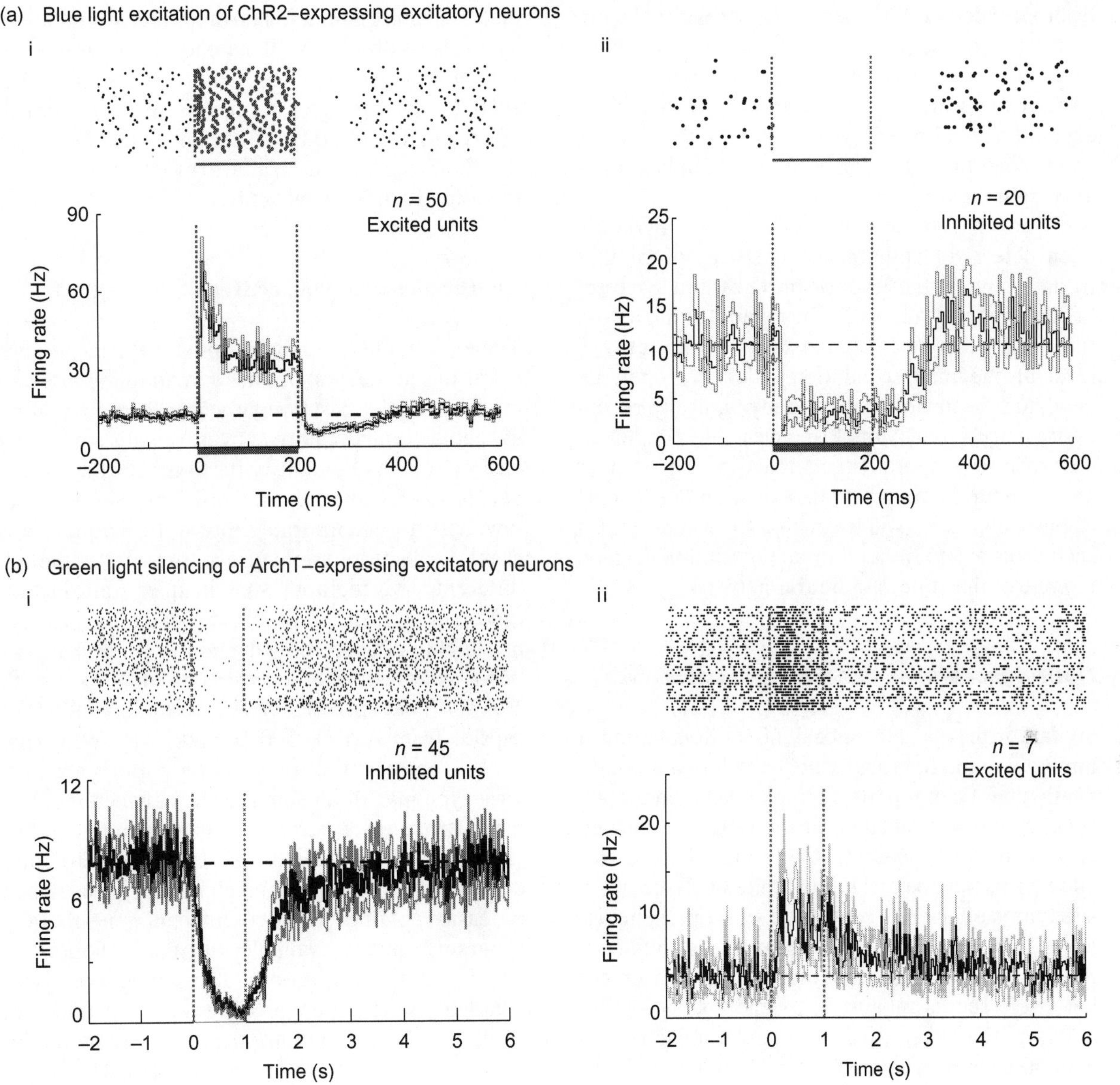

Fig. 6. Homeostatic neural network responses upon optical activation or silencing of cortical excitatory neurons. (a-i) Top, increase in spiking activity in one neuron during blue light illumination of ChR2-expressing neurons. Spike raster plot displays each spike as a black dot and each trial as a horizontal row; periods of blue light illumination are indicated by horizontal blue dashes. Bottom, instantaneous firing rate, averaged across all excited units recorded upon 200ms blue light exposure (black line, mean; gray lines, mean$\pm$standard error) (n=50 units). (a-ii) Top, decreases in spiking activity in one neuron during blue light illumination of ChR2-expressing neurons. Bottom, instantaneous firing rate averaged across all suppressed units upon 200ms blue light exposure (n=20 units). (b-i) Top, decreases in spiking activity in one neuron during green light illumination of ArchT-expressing neurons. Bottom, instantaneous firing rate averaged across all suppressed neurons upon 1s light exposure (n=45 units). (b-ii) Top, increases in spiking activity in one neuron during green light illumination of ArchT-expressing neurons. Bottom, instantaneous firing rate averaged across all excited neurons upon 1s light exposure (n=7 units). (Adapted from Han et al., 2009, 2011.) (For interpretation of the references to color in this figure legend, the reader is referred to the Web version of this chapter.)

fashion, which would lessen, slow, or reverse the primary responses. Indeed, we have observed complex time course upon optical stimulation, in particular in ChR2-expressing cortical neurons. Often a brief excitation is followed by a long-lasting inhibition, which sometimes last for hundreds of milliseconds after light illumination.

Given that brain homeostasis has previously been detected chiefly on the timescale of minutes to days, and at the level of proteins and synapses (Thiagarajan et al., 2005; Turrigiano et al., 1998), the ability to detect homeostasis at the network level at the millisecond timescale may open up new studies on how networks dynamically reconfigure during behavior. Principles that predict or govern how a neural circuit reacts to a particular kind of neural manipulation, activation or silencing of particular cells, will be increasingly important if such tools are to be used in an algorithmic fashion to control the state of a neural network.

Optical modulation of behaviors in monkeys

So far, with a single optical fiber illuminating a limited volume of brain tissue, no behavioral modification has been reported in monkeys, even though small electrical simulation at the same site elicited clear behaviors (Diester et al., 2011; Han et al., 2009). This may be due to the limited volume of tissue that is effectively controlled with optogenetics, as compared to microelectrical stimulation. Although optogenetic modulation of behavior has been very successful in rodents, the volume of illumination that is required for modulating behaviors in monkeys may be much larger than that in rodents, or it could be due to the inherent difference in optical and electrical simulation methods. For example, ChR2 fails to drive high-frequency firing that is often needed for evoking a movement behavior by microstimulating cortex, and the temporal precision with optogenetic stimulation is not as high as with electrical stimulation. In addition, optogenetic modulation relies mainly on the intrinsic physiological properties of the cell, in contrast to the artificial electrical pulse activation through microstimulation. The lack of observable behavioral effect in the two studies employing optogenetic stimulation published to date remains unknown. With continued effort and tremendous enthusiasm in the field, I predict that scientists will be able to modulate behaviors in monkeys in the near future.

Conclusions and perspectives

The excitement in applying optogenetic techniques in various model systems has spread to the monkey model over the past couple years. With increasing available commercial resources, scientists can now easily obtain high-quality virus to genetically modify neurons and can purchase hardware at reasonably low cost to incorporate optical techniques into classical electrophysiology experiments. Although different viral methods vary in their transduction efficiency for different brain structures, reliable and highly efficient transduction of brain cells has been achieved with lentivirus pseudotyped with VSV-G coat protein, AAV pseudotyped with capsids from AAV1, 5, 8, 9, and Rh10. While the exact efficiency and transduction pattern for each virus type may differ significantly between monkey brain structures, it remains advisable to test different types of virus for the specific brain structure of interest. Fortunately, high-quality viruses can be easily obtained from commercial gene therapy programs, many available in small aliquots at low cost, for test injections. The simple strategy of coupling optical fibers to recording electrodes can be used to reliably monitor the effect of light modulation on neural activities in monkeys. The major advantages of optogenetic control in monkeys, at the current state, are the ability to directly stimulate neurons, instead of involving nonspecific antidromic stimulation of axon terminals as with electrical stimulations, and to silence neurons with unprecedented millisecond time resolution. It is also advantageous to activate or silence specific pathways in the monkey brain by expressing opsins retrogradely.

A few challenges remain in conducting long-term optogenetic control experiments. Specifically, development or implementation of semichronic or chronic optical fiber implants could eliminate mechanical tissue damage from repeated optical fiber insertions. Adaption of artifact free electrodes for *in vivo* recordings in monkeys will enable accurate measure of LFPs at the site of light illumination. Adaption of optical fiber arrays in monkeys or development of novel opsins with higher light sensitivity, larger photocurrents, and red shift action spectrum will increase the tissue volume effectively controlled by light, which may be necessary for perturbing enough brain tissues. Finally, and most challengingly, the development of new viral technologies is needed to target specific cell types in the monkey brain.

The success of using optogenetic molecules in monkeys has pointed to the serious translational potentials. Indeed, opsins have been shown to be functional when expressed in human *ex vivo* retinas last year (Busskamp et al., 2010). Several groups are hopeful in conducting clinical trials in the near future on the treatment of blindness (i.e., the groups in Switzerland (Busskamp et al., 2010) and Eos Neuroscience of California, USA (Doroudchi et al., 2011)) and spinal cord injury (Case Western Reserve University in Ohio, USA). It is exciting to see the rapid progress from identifying the first opsin in 2005 (Boyden et al., 2005), to the realization of its translational potential today. There is realistic potential that the continued rapid progress in the field will eventually lead to novel cell type specific neuromodulation therapies.

Acknowledgments

X. H. acknowledges funding from NIH (R00MH085944), Alfred P. Sloan Foundation, and Boston University Photonic Center. X. H thanks Dr. Thomas Knopfel for helpful comments on the manuscript.

References

Aponte, Y., Atasoy, D., et al. (2011). AGRP neurons are sufficient to orchestrate feeding behavior rapidly and without training. *Nature Neuroscience*, *14*(3), 351–355.

Ayling, O. G., Harrison, T. C., et al. (2009). Automated light-based mapping of motor cortex by photoactivation of channelrhodopsin-2 transgenic mice. *Nature Methods*, *6*(3), 219–224.

Bernstein, J. G., Han, X., et al. (2008). Prosthetic systems for therapeutic optical activation and silencing of genetically-targeted neurons. *Proceedings—Society of Photo-Optical Instrumentation Engineers*, *6854*, 68540H.

Bevilacqua, F., Piguet, D., et al. (1999). In vivo local determination of tissue optical properties: Applications to human brain. *Applied Optics*, *38*(22), 4939–4950.

Binzoni, T., Leung, T. S., et al. (2006). The use of the Henyey-Greenstein phase function in Monte Carlo simulations in biomedical optics. *Physics in Medicine and Biology*, *51*(17), N313–N322.

Blomer, U., Naldini, L., et al. (1997). Highly efficient and sustained gene transfer in adult neurons with a lentivirus vector. *Journal of Virology*, *71*(9), 6641–6649.

Boyden, E. S., Zhang, F., et al. (2005). Millisecond-timescale, genetically targeted optical control of neural activity. *Nature Neuroscience*, *8*(9), 1263–1268.

Brecht, M., Schneider, M., et al. (2004). Whisker movements evoked by stimulation of single pyramidal cells in rat motor cortex. *Nature*, *427*(6976), 704–710.

Busskamp, V., Duebel, J., et al. (2010). Genetic reactivation of cone photoreceptors restores visual responses in retinitis pigmentosa. *Science*, *329*(5990), 413–417.

Cardin, J. A., Carlen, M., et al. (2010). Targeted optogenetic stimulation and recording of neurons in vivo using cell-type-specific expression of Channelrhodopsin-2. *Nature Protocols*, *5*(2), 247–254.

Cearley, C. N., Vandenberghe, L. H., et al. (2008). Expanded repertoire of AAV vector serotypes mediate unique patterns of transduction in mouse brain. *Molecular Therapy*, *16*(10), 1710–1718.

Cearley, C. N., & Wolfe, J. H. (2006). Transduction characteristics of adeno-associated virus vectors expressing cap serotypes 7, 8, 9, and Rh10 in the mouse brain. *Molecular Therapy*, *13*(3), 528–537.

Chan, A. W., Chong, K. Y., et al. (2001). Transgenic monkeys produced by retroviral gene transfer into mature oocytes. *Science*, *291*(5502), 309–312.

Chhatwal, J. P., Hammack, S. E., et al. (2007). Identification of cell-type-specific promoters within the brain using lentiviral vectors. *Gene Therapy*, *14*(7), 575–583.

Chow, B. Y., Han, X., et al. (2010). High-performance genetically targetable optical neural silencing by light-driven proton pumps. *Nature*, *463*(7277), 98–102.

Davidson, B. L., Stein, C. S., et al. (2000). Recombinant adeno-associated virus type 2, 4, and 5 vectors: Transduction of variant cell types and regions in the mammalian central nervous system. *Proceedings of the National Academy of Sciences of the United States of America, 97*(7), 3428–3432.

Diester, I., Kaufman, M. T., et al. (2011). An optogenetic toolbox designed for primates. *Nature Neuroscience, 14*(3), 387–397.

Dittgen, T., Nimmerjahn, A., et al. (2004). Lentivirus-based genetic manipulations of cortical neurons and their optical and electrophysiological monitoring in vivo. *Proceedings of the National Academy of Sciences of the United States of America, 101*(52), 18206–18211.

Dodiya, H. B., Bjorklund, T., et al. (2010). Differential transduction following basal ganglia administration of distinct pseudotyped AAV capsid serotypes in nonhuman primates. *Molecular Therapy, 18*(3), 579–587.

Doroudchi, M. M., Greenberg, K. P., et al. (2011). Virally delivered channelrhodopsin-2 safely and effectively restores visual function in multiple mouse models of blindness. *Molecular Therapy, 19*(7), 1220–1229.

Duque, S., Joussemet, B., et al. (2009). Intravenous administration of self-complementary AAV9 enables transgene delivery to adult motor neurons. *Molecular Therapy, 17*(7), 1187–1196.

Federico, M. (2003). From lentiviruses to lentivirus vectors. In: M. Federico (Ed.), *Lentivirus gene engineering protocols. Methods in molecular biology* (vol. 229), (pp. 3–15). Totowa, NJ: Humana Press Inc.

Foust, K. D., Nurre, E., et al. (2009). Intravascular AAV9 preferentially targets neonatal neurons and adult astrocytes. *Nature Biotechnology, 27*(1), 59–65.

Gao, G. P., Alvira, M. R., et al. (2002). Novel adeno-associated viruses from rhesus monkeys as vectors for human gene therapy. *Proceedings of the National Academy of Sciences of the United States of America, 99*(18), 11854–11859.

Gratzel, M. (2001). Photoelectrochemical cells. *Nature, 414* (6861), 338–344.

Hadaczek, P., Forsayeth, J., et al. (2009). Transduction of nonhuman primate brain with adeno-associated virus serotype 1: Vector trafficking and immune response. *Human Gene Therapy, 20*(3), 225–237.

Han, X., & Boyden, E. S. (2007). Multiple-color optical activation, silencing, and desynchronization of neural activity, with single-spike temporal resolution. *PloS One, 2*, e299.

Han, X., Chow, B., et al. (2011). A high-light sensitivity optical neural silencer: Development and application to optogenetic control of non-human primate cortex. *Frontiers in Systems Neuroscience, 5*, 18. doi:10.3389/fnsys.2011.00018.

Han, X., Qian, X., et al. (2009). Millisecond-timescale optical control of neural dynamics in the nonhuman primate brain. *Neuron, 62*(2), 191–198.

Heider, B., Nathanson, J. L., et al. (2010). Two-photon imaging of calcium in virally transfected striate cortical neurons of behaving monkey. *PloS One, 5*(11), e13829.

Hermonat, P. L., & Muzyczka, N. (1984). Use of adeno-associated virus as a mammalian DNA cloning vector: Transduction of neomycin resistance into mammalian tissue culture cells. *Proceedings of the National Academy of Sciences of the United States of America, 81*(20), 6466–6470.

Honda, K. (2004). Dawn of the evolution of photo-electrochemistry. *Journal of Photochemistry and Photobiology A: Chemistry., 166*, 63–68.

Hwang, D. Y., Carlezon, W. A. Jr., et al. (2001). A high-efficiency synthetic promoter that drives transgene expression selectively in noradrenergic neurons. *Human Gene Therapy, 12*(14), 1731–1740.

Hwang, D. Y., Hwang, M. M., et al. (2005). Genetically engineered dopamine beta-hydroxylase gene promoters with better PHOX2-binding sites drive significantly enhanced transgene expression in a noradrenergic cell-specific manner. *Molecular Therapy, 11*(1), 132–141.

Katzel, D., Zemelman, B. V., et al. (2011). The columnar and laminar organization of inhibitory connections to neocortical excitatory cells. *Nature Neuroscience, 14*(1), 100–107.

Klein, R. L., Dayton, R. D., et al. (2008). AAV8, 9, Rh10, Rh43 vector gene transfer in the rat brain: Effects of serotype, promoter and purification method. *Molecular Therapy, 16*(1), 89–96.

Kravitz, A. V., Freeze, B. S., et al. (2010). Regulation of parkinsonian motor behaviours by optogenetic control of basal ganglia circuitry. *Nature, 466*(7306), 622–626.

Lawlor, P. A., Bland, R. J., et al. (2009). Efficient gene delivery and selective transduction of glial cells in the mammalian brain by AAV serotypes isolated from nonhuman primates. *Molecular Therapy, 17*(10), 1692–1702.

Li, C. Y., Poo, M. M., et al. (2009). Burst spiking of a single cortical neuron modifies global brain state. *Science, 324* (5927), 643–646.

Lobo, M. K., Covington, H. E. 3rd, et al. (2010). Cell type-specific loss of BDNF signaling mimics optogenetic control of cocaine reward. *Science, 330*(6002), 385–390.

Markakis, E. A., Vives, K. P., et al. (2010). Comparative transduction efficiency of AAV vector serotypes 1–6 in the substantia nigra and striatum of the primate brain. *Molecular Therapy, 18*(3), 588–593.

Mazarakis, N. D., Azzouz, M., et al. (2001). Rabies virus glycoprotein pseudotyping of lentiviral vectors enables retrograde axonal transport and access to the nervous system after peripheral delivery. *Human Molecular Genetics, 10* (19), 2109–2121.

Mobley, J., & Vo-Dinh, T. (2003). Optical properties of tissue. In T. Vo-Dinh (Ed.), *Biomedical photonics handbook* (pp. 1–72). Boca Raton: CRC Press.

Muzyczka, N. (1992). Use of adeno-associated virus as a general transduction vector for mammalian cells. *Current Topics in Microbiology and Immunology, 158*, 97–129.

Muzyczka, N., & Warrington, K. H. Jr. (2005). Custom adeno-associated virus capsids: The next generation of recombinant vectors with novel tropism. *Human Gene Therapy*, *16* (4), 408–416.

Nathanson, J. L., Jappelli, R., et al. (2009a). Short promoters in viral vectors drive selective expression in mammalian inhibitory neurons, but do not restrict activity to specific inhibitory cell-types. *Frontiers in Neural Circuits*, *3*, 19.

Nathanson, J. L., Yanagawa, Y., et al. (2009b). Preferential labeling of inhibitory and excitatory cortical neurons by endogenous tropism of adeno-associated virus and lentivirus vectors. *Neuroscience*, *161*(2), 441–450.

Passini, M. A., Watson, D. J., et al. (2003). Intraventricular brain injection of adeno-associated virus type 1 (AAV1) in neonatal mice results in complementary patterns of neuronal transduction to AAV2 and total long-term correction of storage lesions in the brains of beta-glucuronidase-deficient mice. *Journal of Virology*, *77*(12), 7034–7040.

Peel, A. L., & Klein, R. L. (2000). Adeno-associated virus vectors: Activity and applications in the CNS. *Journal of Neuroscience Methods*, *98*(2), 95–104.

Sasaki, E., Suemizu, H., et al. (2009). Generation of transgenic non-human primates with germline transmission. *Nature*, *459*(7246), 523–527.

Schroder, A. R., Shinn, P., et al. (2002). HIV-1 integration in the human genome favors active genes and local hotspots. *Cell*, *110*(4), 521–529.

Sohal, V. S., Zhang, F., et al. (2009). Parvalbumin neurons and gamma rhythms enhance cortical circuit performance. *Nature*, *459*(7247), 698–702.

Stettler, D. D., Yamahachi, H., et al. (2006). Axons and synaptic boutons are highly dynamic in adult visual cortex. *Neuron*, *49*(6), 877–887.

Taymans, J. M., Vandenberghe, L. H., et al. (2007). Comparative analysis of adeno-associated viral vector serotypes 1, 2, 5, 7, and 8 in mouse brain. *Human Gene Therapy*, *18*(3), 195–206.

Thiagarajan, T. C., Lindskog, M., et al. (2005). Adaptation to synaptic inactivity in hippocampal neurons. *Neuron*, *47*(5), 725–737.

Thomas, C. E., Ehrhardt, A., et al. (2003). Progress and problems with the use of viral vectors for gene therapy. *Nature Reviews Genetics*, *4*(5), 346–358.

Thyagarajan, S., van Wyk, M., et al. (2010). Visual function in mice with photoreceptor degeneration and transgenic expression of channelrhodopsin 2 in ganglion cells. *Journal of Neuroscience*, *30*(26), 8745–8758.

Turrigiano, G. G., Leslie, K. R., et al. (1998). Activity-dependent scaling of quantal amplitude in neocortical neurons. *Nature*, *391*(6670), 892–896.

Wang, L., Jacques, S. L., et al. (1995). MCML—Monte Carlo modeling of light transport in multi-layered tissues. *Computer Methods and Programs in Biomedicine*, *47*(2), 131–146.

Witten, I. B., Lin, S. C., et al. (2010). Cholinergic interneurons control local circuit activity and cocaine conditioning. *Science*, *330*(6011), 1677–1681.

Yang, S. H., Cheng, P. H., et al. (2008). Towards a transgenic model of Huntington's disease in a non-human primate. *Nature*, *453*(7197), 921–924.

Yaroslavsky, A. N., Schulze, P. C., et al. (2002). Optical properties of selected native and coagulated human brain tissues in vitro in the visible and near infrared spectral range. *Physics in Medicine and Biology*, *47*(12), 2059–2073.

Yizhar, O., Fenno, L. E., et al. (2011). Optogenetics in neural systems. *Neuron*, *71*(1), 9–34.

Zhang, H., Yang, B., et al. (2011). Several rAAV vectors efficiently cross the blood–brain barrier and transduce neurons and astrocytes in the neonatal mouse central nervous system. *Molecular Therapy*, *19*(8), 1440–1448.

Zorzos, A. N., Dietrich, A., et al. (2009). *Light-proof neural recording electrodes Society for Neuroscience*, 388.12/GG107.

T. Knöpfel and E. Boyden (Eds.)
Progress in Brain Research, Vol. 196
ISSN: 0079-6123

CHAPTER 12

Optogenetic reporters: Fluorescent protein-based genetically encoded indicators of signaling and metabolism in the brain

Mathew Tantama, Yin Pun Hung and Gary Yellen*

Department of Neurobiology, Harvard Medical School, Boston, MA, USA

Abstract: Fluorescent protein technology has evolved to include genetically encoded biosensors that can monitor levels of ions, metabolites, and enzyme activities as well as protein conformation and even membrane voltage. They are well suited to live-cell microscopy and quantitative analysis, and they can be used in multiple imaging modes, including one- or two-photon fluorescence intensity or lifetime microscopy. Although not nearly complete, there now exists a substantial set of genetically encoded reporters that can be used to monitor many aspects of neuronal and glial biology, and these biosensors can be used to visualize synaptic transmission and activity-dependent signaling *in vitro* and *in vivo*. In this review, we present an overview of design strategies for engineering biosensors, including sensor designs using circularly permuted fluorescent proteins and using fluorescence resonance energy transfer between fluorescent proteins. We also provide examples of indicators that sense small ions (e.g., pH, chloride, zinc), metabolites (e.g., glutamate, glucose, ATP, cAMP, lipid metabolites), signaling pathways (e.g., G protein-coupled receptors, Rho GTPases), enzyme activities (e.g., protein kinase A, caspases), and reactive species. We focus on examples where these genetically encoded indicators have been applied to brain-related studies and used with live-cell fluorescence microscopy.

Keywords: genetically encoded; biosensor; fluorescent protein; circularly permuted; resonance energy transfer; FRET; live-cell microscopy.

Introduction

Fluorescent proteins (FPs) have proven to be an incredibly versatile platform for engineering sensors of enzyme activities, membrane voltage, ions, molecules, and proteins (Frommer et al., 2009;

*Corresponding author.
Tel.: +1-617-432-0137; Fax: +1-617-432-0121
E-mail: gary_yellen@hms.harvard.edu

DOI: 10.1016/B978-0-444-59426-6.00012-4

Knopfel et al., 2010; Newman et al., 2011). Since the first Ca^{2+} sensors were reported (Miyawaki et al., 1997; Romoser et al., 1997), the number and variety of FP-based sensors have grown immensely, and many extensive reviews on their design and application in cell biology have been written (Chudakov et al., 2010; Frommer et al., 2009; Newman et al., 2011). In this chapter, we provide examples of FP-based sensors that have been applied specifically to brain-related studies. Advances in FP engineering, calcium sensors, and voltage sensors are covered in other chapters of this volume, and therefore, we do not focus on these topics. We first discuss major strategies for engineering genetically targeted FP-based sensors, and then we provide several examples organized by sensor target.

Genetically encoded indicators

Several design strategies have been used to engineer genetically encoded sensors that provide a fluorescence readout for the level of an analyte (small molecule or ion) or the activity of a signaling system (Fig. 1). We refer to this broad class of sensors as genetically encoded indicators (GEIs), borrowing terminology from genetically encoded calcium indicators (GECIs). A GEI has two functional units, the *sensor domain* and the *reporter domain*. The reporter domain is a single FP or a pair of FPs that provides the fluorescence readout. The sensing domain can be a peptide motif, a full protein, or a combination of the two that senses the target of interest. In practice, the *coupling* between the sensor and the reporter domains influences most of the GEI characteristics, including fluorescence spectra, type of response, and the dynamic range. We discuss different GEI design and sensing parameters in this section.

Once a target of interest has been identified, the choice and design of the sensor domain is the first consideration for GEI construction. In order to elicit a fluorescence response from the GEI, ligand binding or enzymatic modification must cause a change in molecular conformation. Although some artificial sensing scaffolds have been engineered, these often lack specificity for the intended target (Vinkenborg et al., 2010). Rather, a highly successful strategy relies on naturally occurring sensor domains that can be adapted for GEI construction. For example, the Cameleon family of Ca^{2+} sensors exploits Ca^{2+}-dependent binding of calmodulin (CaM) and the M13 peptide (Miyawaki et al., 1997; Romoser et al., 1997). The physical change induced by Ca^{2+} is coupled to a perturbation of the reporting domain; that is, the Ca^{2+}·CaM-M13 binding energy is used to do work to alter the GEI fluorescence.

The choice of the color and the number of FPs for the reporter domain is the next design consideration. We consider reporter domains that consist of either one or two FPs. In single FP GEIs, the physical change in the sensor domain causes a chemical or structural change in the local environment of the FP chromophore, altering its intrinsic fluorescence characteristics. The sensitivity to this structural change can be natural, engineered through point mutations or engineered by a circular permutation strategy. In dual FP GEIs, the physical change in the sensor domain causes a change in resonance energy transfer between the two FPs. We discuss each of these strategies in greater detail in later sections.

After choosing sensor and reporter domains, engineering the coupling between the domains is the most critical process of GEI development, and coupling must be optimized to obtain desirable GEI characteristics. Structure-guided design can significantly aid the rational development of GEIs (Akerboom et al., 2009; Wang et al., 2008a), but empirical discovery remains a substantial component of the process. For example, optimizing the length and composition of the peptide linkers connecting the sensor and reporter domains is often a critical process, but it is also a major hurdle that largely relies on random trial and error. The process of optimizing the coupling between domains can entail screening many libraries of mutants and iteratively improving the sensor's characteristics. In this

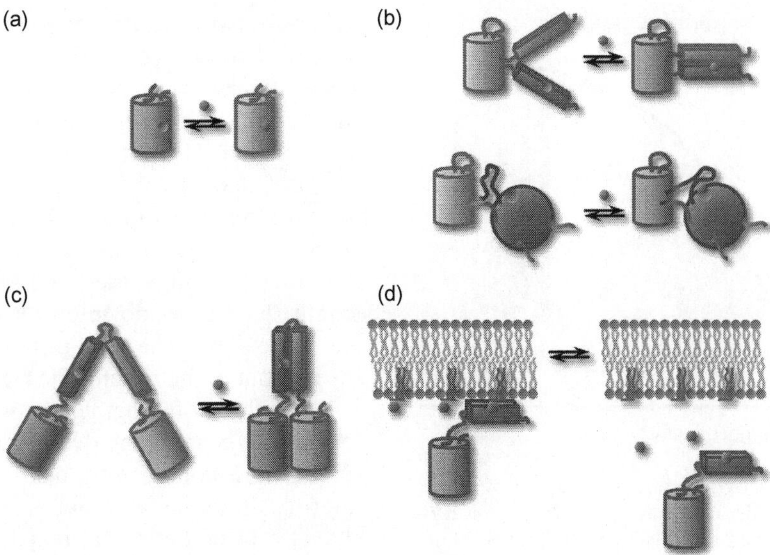

Fig. 1. Design strategies for engineering genetically encoded indicators. Fluorescent proteins are shown as cylinders. Sensor domains are boxes or spheres in orange or purple. Target ligands are shown in red. Examples are shown for indicators of small-molecule analytes, but the designs can also be applied to engineer indicators of enzyme activity or protein activation. (a) A single environmentally sensitive fluorescent protein acts as the sensor and reporter. (b) Circular-permutation of a fluorescent protein renders its fluorescence sensitive to a conformational change in a sensor domain. The sensor domain may consist of two proteins or peptides attached to the new termini of the circularly permuted fluorescent protein (top). Alternatively, the circularly permuted fluorescent protein may be inserted into a sensor protein in a region that experiences a conformation change (bottom). (c) Fluorescent proteins are attached to two parts of a sensor domain that experience a relative motion, and the conformational change can alter the distance or orientation between the fluorescent proteins, changing the FRET efficiency. (d) A fluorescent protein is attached to a sensor domain that localizes differently depending on ligand binding or activation. In the example shown, the indicator localizes to the plasma membrane or cytosol depending on the ligand status. (For interpretation of the references to color in this figure legend, the reader is referred to the Web version of this chapter.)

development stage, several parameters are considered, including the GEI's affinity for the target, the sensing range, the sensing kinetics, the type of fluorescence response, the dynamic range of the fluorescence response, the specificity of the response, and the perturbations expression of the sensor may cause.

The levels and dynamics of ions, metabolites, and enzymatic substrates are tightly regulated in biological systems, and thus the GEI's target affinity, sensing range, and sensing kinetics must be well tuned to respond to physiological changes. The GEI's apparent affinity for its target (K_{app}) describes the point of half-maximal activation or saturation by a target ligand. The sensing range describes the range of target concentrations or activities over which distinct signals can be detected. Both are described by the GEI's dose–response curve (Fig. 2). The sensing range is dictated by the steepness of the dose–response. For example, a sensor domain with multiple ligand binding sites may exhibit large positive cooperativity and a steep response curve, resulting in a narrow sensing range. This type of sensing may be useful for a binary readout or an "on–off" sensor, but it is not ideal for monitoring graded responses. Related to the affinity, the sensing kinetics are determined by the on and off rates of the target, the rate of the conformational change of the sensor domain, and the rate of coupling

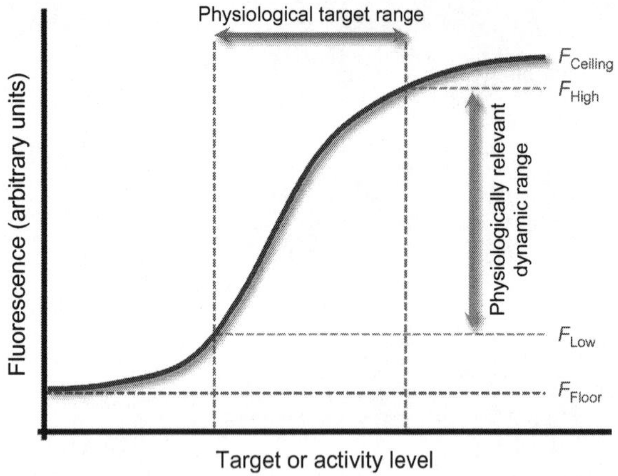

Fig. 2. A hypothetical GEI dose–response curve. The physiologically relevant dynamic range (F_{High}/F_{Low}) is usually smaller than the GEI's maximum fluorescence dynamic range ($F_{Ceiling}/F_{Floor}$) because only a portion of the total sensing range is sampled in the physiological scenario. (For color version of this figure, the reader is referred to the Web version of this chapter.)

between the sensor and reporter domains. Ideally, the K_{app} and sensing range will match the physiological midpoint and target level range, and the response time will be faster than the kinetics of endogenous signaling.

When optically monitoring GEIs, the fluorescence response can be either a simple intensity change or a ratiometric change that affects the relative intensity at different wavelengths. Intensity-based measurements use a single excitation wavelength and monitor fluorescence changes at a single emission wavelength. However, the fluorescence intensity also depends on the concentration of the fluorophore, and comparison of absolute intensity changes is usually not possible because of cell-to-cell and experiment-to-experiment variability in GEI expression levels. Instead, intensity measurements generally are used to monitor relative changes during an experimental manipulation. Ratiometric measurements eliminate the dependence on fluorophore concentration, enabling direct comparisons between experiments that facilitate quantitative analysis. Ratiometric measurements can either use two excitation wavelengths and monitor a single emission (excitation ratiometric) or use a single excitation and monitor two emission wavelengths (emission ratiometric). For example, an excitation ratiometric GEI exhibits two peaks in its basal fluorescence excitation spectrum. When the GEI senses its target, the excitation spectrum changes in a specific manner: one peak increases, the other peak decreases, and there exists a wavelength that shows no intensity change called the isosbestic point. The ratio of the peak intensities is the readout of the ratiometric GEI.

Regardless of whether it is an intensity or ratiometric change, the dynamic range of the fluorescence response must also be maximized to improve the signal-to-noise and detection of changes. The fluorescence dynamic range describes the maximal detectable change in the fluorescence response. Although the two are functionally related, the dynamic range of the fluorescence response and the sensing range are distinct parameters (Fig. 2). It is important to note that the dynamic range of the fluorescence response may differ substantially between purified protein in solution and protein expressed intracellularly because of environmental factors.

In the process of optimizing the GEI's characteristics, it is critical to verify the specificity of the response to the target. The sensor domain may have substantial affinity for structurally related analytes or closely related enzyme activities, or there may be naturally occurring allosteric modulators. Rational mutagenesis and screening in some cases can reduce off-target interference, especially if structure–function studies of the endogenous sensor domain are available to guide mutagenesis. At the very least, a reasonable effort should be made to identify interfering factors so that the proper experimental controls can be conducted when using the GEI. Interference can also result from environmental sensitivity of the FPs, for example, to pH, as will be discussed in a later section.

Finally, care must be taken when using naturally occurring sensing domains because they can interact with and alter endogenous processes, especially if the sensing domain has a natural enzymatic activity. To diminish interference by Cameleon, the CaM and M13 were mutated in parallel to maintain recognition for one another but abrogate recognition by endogenous CaM (Palmer et al., 2006). Another approach is to choose a sensing domain from a species different from that under study. For example, bacterial periplasmic proteins have been exploited for sensing analytes in mammalian systems to reduce the possibility of biological cross talk (Okumoto et al., 2008). A related concern is that the GEI itself can cause pathological buffering of the analyte, interfering with endogenous signaling by changing the free concentration. FPs and GEIs may reach micromolar concentrations when expressed in cells, and analytes whose intracellular concentration is near or less than this may be affected. Some analytes occur at low free concentrations but have a much higher total concentrations due to endogenous buffering, and these analytes may be less affected by sensor expression. However, pathological buffering by the GEI can be a large problem for analytes that occur at low total concentrations.

Clearly, engineering an optimal GEI with respect to all of these parameters is not trivial, but many useful GEIs have been developed that have provided important insight into brain function, despite deficiencies in one or more of their sensing properties. We next describe in greater detail design strategies based on single FP environmental sensitivity, circular permutation of a single FP, resonance energy transfer between two FPs, or translocation of FPs.

GEIs using environment-sensitive fluorescent proteins

Although the FP β-barrel structure substantially shields the interior chromophore from the environment, it is important to realize that FPs still can respond to external changes, and in particular, all FPs exhibit some degree of pH sensitivity (Chudakov et al., 2010; Day and Davidson, 2009). Environmentally sensitive intrinsic fluorescence can be a result of the acid–base chemistry of the chromophore (Bizzarri et al., 2009). Channels for solvent access or proton relay networks can couple the interior environment near the chromophore to changes in the external solution (Jayaraman et al., 2000; Wachter et al., 1998). It is well known that pH often changes with neuronal activity, metabolism, and intracellular signaling (Bizzarri et al., 2009; Casey et al., 2010; Chesler and Kaila, 1992; Srivastava et al., 2007). This is a critical issue for GEIs because the fluorescence change due to a pH transient can be of the same magnitude as the maximum fluorescence response to its target, and thus the GEI fluorescence response can be aliased with pH transients. In this section, we briefly describe how chromophore acid–base chemistry affects the intrinsic fluorescence of FPs. We also describe the related phenomenon of excited-state proton transfer. Finally, we discuss how natural or engineered environmental sensitivity can be exploited in GEIs.

The wild-type green fluorescent protein (GFP) chromophore can exist in different ionization states, and the protonated neutral form has distinct photophysical properties from the deprotonated anionic form (Bizzarri et al., 2009; Tsien, 1998). The GFP chromophore is a *p*-hydroxybenzylidene-5-imidazolidinone (*p*-HBI), containing a phenol group derived from Tyr66 that can be deprotonated to form an anionic chromophore (Fig. 3). In solution, the pK_a of a phenol hydroxyl proton is ~10; however, the pK_a of the *p*-HBI phenol is lowered by the conjugated π-system of the chromophore and the protein environment, shifting the pK_a to a value of ~8 or less depending on the specific FP. In GFP, a bound water, Ser205, and Glu222 form a proton relay network that can accept a proton from the chromophore, and Thr203 can rotate to stabilize the phenolate anion. The protonated neutral state ("A" state) and the deprotonated anionic state ("B" state) are important to consider because protonation of the chromophore changes its absorbance and fluorescence properties. The A state of GFP has an absorbance peak at 395nm with a

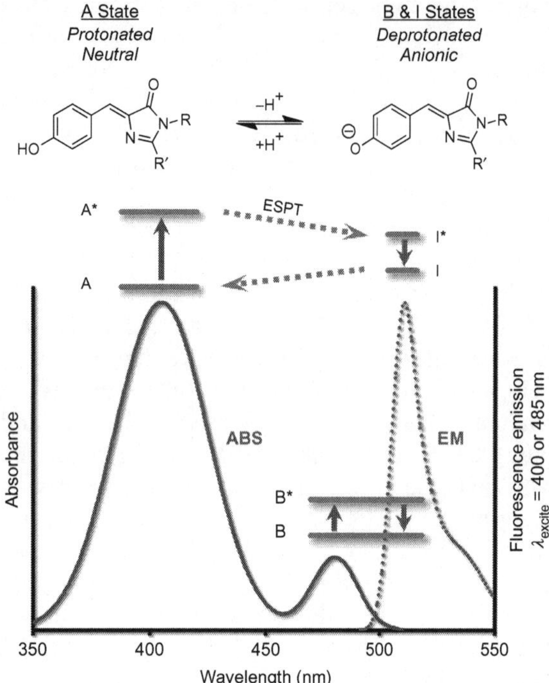

Fig. 3. Protonation of the fluorescent protein chromophore and its effects on the absorbance and fluorescence spectra. The example is theoretical but based on the wild-type green fluorescent protein chromophore and spectra. In the physiological pH range, the chromophore can be protonated in a neutral "A" state or ionized in an anionic "B" state (top). Two peaks in the absorbance spectrum (solid blue curve) reflect the two states. A single peak is observed in the fluorescence emission spectrum (dotted green curve) when exciting the B state, and a similar emission spectrum is observed when exciting the A state because excited-state proton transfer occurs, creating an anionic excited state I* very similar to the B* anionic excited state. Simplified Jablonski diagrams are superimposed to represent energy transitions following absorption (blue arrow) and resulting in fluorescence emission (green arrow). The B and B* energy levels are shown lower in the diagram for visual clarity and are not indicative of actual energy differences with the A or I states. (For interpretation of the references to color in this figure legend, the reader is referred to the Web version of this chapter.)

fluorescence emission peak at 508nm, but the B state has an absorbance peak at 475nm with a fluorescence emission peak at 503nm. Peak absorbance of the B state occurs at a longer wavelength because deprotonation of the chromophore decreases the transition energy between the ground and excited state. The internal conformation of wild-type GFP favors the A state, and there is only a minor population of the B state in neutral solution. In contrast, the S65T mutation of the enhanced GFP (EGFP) mutant favors the B state, and as a result, EGFP peak excitation and fluorescence emission occur at 488 and 508nm, respectively. Many engineered GFP variants show significant coupling between the external pH and chromophore protonation, with apparent pK_a values of 5–6.5 (Chudakov et al., 2010). Therefore, the chromophore may exist in an equilibrium of ionization states that is sensitive to changes in solution pH, resulting in pH-dependent fluorescence.

The protonated neutral state of wild-type GFP has a large Stokes shift (the difference between the excitation and emission wavelengths) because of excited-state proton transfer (ESPT) (Bizzarri et al., 2009; Tsien, 1998). Excitation at 395nm generates an excited state of the protonated neutral chromophore (A*), and in this excited state, the *p*-HBI phenol is more acidic. The phenol deprotonates through ESPT to the aforementioned proton relay network, creating an excited-state anion (I*). The transition energy is decreased for relaxation from the I* anionic excited state, and peak fluorescence emission after ESPT is red-shifted to 508nm. In the absence of ESPT, relaxation from the A* protonated excited state is a higher energy transition with peak fluorescence emission at 460nm; however, ESPT is efficient and occurs on the timescale of ~10ps in GFP, resulting in effectively a single emission peak at 508nm. Of note, the I* excited-state anion is distinct from the B* excited-state anion. As previously mentioned, the B state is stabilized by conformational rearrangements including rotation of Thr203, and the A and B states represent not only distinct ionization states but also distinct internal protein conformations. After ESPT from A* to create the I* excited-state anion, fluorescence emission occurs before a conformational change can transpire. Thus, peak fluorescence emission from I* and B* excited states is very similar but not exactly equal in energy.

For yellow fluorescent proteins (YFPs), chloride and pH sensitivity are interrelated (Jayaraman et al., 2000). In YFPs, Thr203 is mutated to an aromatic amino acid such as Tyr. The aromatic residue can π-stack with the chromophore, decreasing the energy gap between the anionic B and B* states. The mutation also creates an accessible halide binding site close to the chromophore. When a halide such as Cl^- binds, the additional negative charge stabilizes the protonated A state of the chromophore. Thus, Cl^- binding can shift the apparent pK_a of the YFP, making its fluorescence sensitive to changes in Cl^- concentration (Griesbeck et al., 2001; Wachter et al., 2000).

Sensitivity to pH and Cl^- can clearly affect the intrinsic fluorescence of FPs, but this can be exploited to create GEIs precisely for H^+ and Cl^-. This type of GEI uses only a single environmentally sensitive FP that acts as both the sensor and the reporter domains (Fig. 1a). Mutagenesis can be used to tune its sensing characteristics, to alter the chromophore acid–base chemistry, or to alter the ESPT reaction to create excitation or emission ratiometric GEIs (Hanson et al., 2002; Miesenbock et al., 1998). Further, mutagenesis can also introduce additional, nonnatural environmental sensitivity. For example, by introducing cysteines in the exterior β-barrel near the chromophore, the roGFPs and rxYFPs were engineered to be sensitive to reduction and oxidation (Hanson et al., 2004; Ostergaard et al., 2001). We discuss specific examples of pH, Cl^-, and reduction–oxidation (redox)-sensitive FPs of this type in the later sections. More commonly, the sensing domain is a distinct unit, and the remaining designs co-opt natural peptides or proteins for this purpose.

GEIs using circularly permuted fluorescent proteins

Circular permutation of a single FP is one strategy used to couple a conformational change in the sensor domain to a perturbation of the chromophore. The native FP termini are spatially distant from the chromophore, and typically movements of the termini do not alter the chromophore or the intrinsic fluorescence. In a circularly permuted FP (cpFP), the original N- and C-termini are ligated by a short peptide linker, and new termini are created (Baird et al., 1999). If the new termini are located spatially close to the chromophore, physical movements of the termini may alter the local environment around the chromophore. The altered cpFP conformation often increases solvent accessibility to the chromophore or shifts the acid–base equilibrium, causing a change in the intrinsic fluorescence.

For example, in the calcium sensor GCaMP, Ca^{2+}-dependent binding between CaM and the M13 peptide alters the fluorescence intensity of cpEGFP (Nakai et al., 2001). In this sensor, EGFP is circularly permuted at residue 149, and the new termini are attached to CaM and the M13 peptide (Fig. 1b). In the absence of Ca^{2+}, apo-GCaMP2 has a peak absorbance at 399nm characteristic of the GFP protonated A state; however, it has very low fluorescence, suggesting that nonradiative decay is favorable. Bound to Ca^{2+}, the peak absorbance shifts to 488nm, and a conformational change in the CaM–M13 binding pair stabilizes a fluorescent deprotonated B state (Akerboom et al., 2009; Wang et al., 2008a). Circular permutation was exploited here to couple a conformational change in the sensor domain to a shift in the cpEGFP acid–base equilibrium, creating a GEI that responds with simple fluorescence intensity changes.

Circular permutation can also be used to create GEIs that respond with a ratiometric fluorescence change, and the cpEYFP-based Ca^{2+}-sensor Pericam is one example (Nagai et al., 2001). Ratiometric Pericam exhibits Ca^{2+}-dependent changes in its absorbance spectrum similar to GCaMP2, but excitation of the A state produces fluorescence emission. Mutagenesis of cpEYFP at His148 and Thr203 created a variant in which nonradiative decay after excitation of the A state is decreased and in which ESPT is efficient. Excitation of both the protonated A (418nm) and deprotonated B (494nm) states produces a peak fluorescence emission at 511 nm. Thus, binding of Ca^{2+} produces a shift in the fluorescence excitation spectrum of Ratiometric Pericam.

While circular permutation is a versatile and important design strategy for engineering GEIs, one major disadvantage of using cpFPs is that circular permutation can increase pH sensitivity. Analyte binding to a cpFP-based GEI typically affects the protonation of the chromophore, producing a shift in the A and B state populations. When measured at constant pH, the fluorescence response reflects analyte concentration, but when measured as a function of pH, the binding of analyte is seen simply to shift the pK_a. This means that a simple change in pH can mimic the effect of a change in analyte concentration when in fact there is none. Artifacts due to pH changes can plague both intensity and ratiometric cpFP-based GEIs (Nagai et al., 2001). Proper use of these sensors thus requires either very tight control of pH or simultaneous measurement of pH using an additional sensor (see the section below on pH sensors).

GEIs using Förster resonance energy transfer

Förster-type resonance energy transfer (FRET) between two FPs is another strategy used to couple a conformational change in the sensor domain to a fluorescence response. Unlike the circular permutation strategy, FRET-based sensing does not involve a structural or chemical change in the chromophore of either FP. Instead, the efficiency of energy transfer between the two chromophores is altered by a conformational change in the sensor domain. After light absorption, an excited-state fluorophore typically relaxes back to the ground state by fluorescence emission or by a nonradiative decay mechanism (such as collision with a solvent molecule). When an excited-state donor fluorophore is in close proximity to an acceptor chromophore, the donor can additionally relax by transferring energy to the acceptor via a dipole–dipole interaction, if the donor's emission spectrum overlaps with the acceptor's absorbance spectrum. When the acceptor itself is a fluorophore, the FRET-induced acceptor excited state can subsequently relax by acceptor fluorescence emission. Note that the acceptor chromophore does not itself need to be fluorescent, and FRET sensors with dark acceptors are useful for fluorescence lifetime applications (Murakoshi et al., 2008). The efficiency of FRET is very sensitive to the distance and the relative orientation between the donor and acceptor. GEIs can exploit FRET by attaching compatible FPs (a FRET pair) to different sensor domains that interact (intermolecular FRET) or by attaching the FRET pair to a single sensor domain (intramolecular FRET).

Intermolecular FRET has been used to detect protein–protein interactions because of its strong distance dependence. A key parameter is the Förster radius, R_0, which is the distance between the acceptor and donor at which the FRET efficiency is 50%. For FP pairs, R_0 is approximately 4–5 nm (Patterson et al., 2000). When two proteins are tagged each with one FP of a FRET pair, substantial FRET suggests close proximity and a potential interaction between the proteins. It is generally assumed that a decrease in FRET indicates dissociation of the proteins. Interpretation of a FRET decrease must be treated with care, however, because FRET is so sensitive to the distance and relative orientation between the FRET pair. That is, a conformational change could also decrease FRET efficiency without disrupting a stable protein–protein interaction. For example, interpretation of FRET data has led to some controversy over G protein-coupled receptor (GPCR) signaling and is discussed in a later section (Ciruela et al., 2010).

In GEIs that use intramolecular FRET, a conformational change in the sensor domain changes the efficiency of energy transfer between an attached FRET pair. For example, Cameleon is a FRET-based calcium sensor that exploits Ca^{2+}-dependent CaM–M13 binding (Miyawaki et al., 1997; Romoser et al., 1997). The sensor domain consists of CaM and the M13 peptide expressed as a fusion that is flanked on its ends by the FRET pair of enhanced blue FP (EBFP) and EGFP (Fig. 1c). In the absence of Ca^{2+}, the sensor domain is extended, but when Ca^{2+} binds to Cameleon, Ca^{2+}·CaM binds the M13 peptide. This conformational change brings the EBFP and EGFP closer together, and the FRET efficiency increases. For intramolecular FRET-based GEIs, the FRET efficiency can be measured by exciting the donor and comparing the donor and acceptor fluorescence emission intensities. Therefore, intramolecular FRET sensors are also ratiometric sensors.

Compared to cpFP-based GEIs, FRET sensors can show better pH insensitivity if the right FRET pair is chosen. There are several FPs that have low pK_a values and whose fluorescence is pH insensitive in the physiological pH range (Chudakov et al., 2010). However, it is still important to clearly test pH sensitivity of the final construct because the sensor domain also can confer pH sensitivity to the GEI.

One disadvantage of intramolecular FRET-based GEIs is that there is a tendency for the fluorescence response to change only subtly. For intramolecular FRET sensors, there can be substantial basal FRET because the physical linkage of the FPs requires that they stay within a certain proximity to one another. Additionally, the size of the FPs limits the maximum absolute FRET efficiency. The FP β-barrel is ~4nm in height and ~2.5nm in diameter (Day and Davidson, 2009; Tsien, 1998), and the Förster radius for FP pairs is typically 4–5nm. Therefore, the axial diameter of the β-barrel limits the closest possible distance between the acceptor and donor chromophores. It has been estimated that the maximum absolute FRET efficiency for FP pairs is ~50% (Patterson et al., 2000). Interestingly, cpFPs have been used to improve the dynamic range of the FRET response (Nagai et al., 2004). Here, unlike cpFPs discussed previously, circular permutation is used to alter the apparent relative orientation of the FRET pair and not to induce an environmentally responsive chromophore (Baird et al., 1999).

GEIs using translocation responses

Changes in the spatial distribution of FP-labeled proteins also can be used as a readout of analyte concentration or protein activity (Fig. 1d). Translocation GEIs have been used extensively to monitor activation of GPCRs and downstream signaling-related lipid metabolism (Ciruela et al., 2010; Newman et al., 2011). For example, EGFP can be fused to a pleckstrin homology (PH) domain that binds phosphatidylinositol-(4,5)-bisphosphate (PIP_2) in the plasma membrane and also binds inositol-(1,4,5)-trisphosphate (IP_3) in the cytosol. When membrane PIP_2 concentration is high and cytosolic IP_3 concentration is low, PH-EGFP enriches in the plasma membrane. When a phospholipase is activated, PIP_2 is hydrolyzed to IP_3 and diacylglycerol (DAG), and PH-EGFP translocates to the cytosol. Thus, changes in the spatial distribution of PH-EGFP indicate changes in PIP_2 and IP_3 levels or activity of a phospholipase. The use of translocation GEIs requires microscopy with sufficient spatial resolution.

In the next sections, we provide examples in which GEIs have been used in brain-related studies. The sections are organized by the target analyte or protein activity being monitored. When available, we focus on examples of live-cell imaging of GEIs.

pH Sensing

Although several pH-sensitive FPs have been developed (Bizzarri et al., 2009), the pHluorin pH-sensors have been predominantly used to monitor synaptic transmission (Miesenbock et al., 1998). The pHluorins are single GFPs that contain mutations conferring significant pH sensitivity to their fluorescence, and they were engineered using structure-guided design (Miesenbock et al., 1998). Ratiometric pHluorin exhibits two peaks in its fluorescence excitation spectrum that respond ratiometrically to pH. Ecliptic pHluorin is an intensity-based GEI that is dark at $pH<6$ and becomes fluorescent upon alkalinization.

Ecliptic pHluorin was engineered to decrease background fluorescence, favoring a drastic intensity change and sacrificing a ratiometric response. The folding and fluorescence properties of ecliptic pHluorin were improved with two additional mutations, producing the now extensively used variant, superecliptic pHluorin (SEP) (Sankaranarayanan et al., 2000). At chemical synapses, neurotransmitters are loaded into acidic secretory vesicles. When a vesicle fuses with the plasma membrane to release its cargo, the vesicle lumen equilibrates with the extracellular environment causing a sharp alkalinization. When SEP is targeted to the lumen of synaptic vesicles, its fluorescence reports this pH transition, and hence SEP can be used to visualize synaptic vesicle cycling and synaptic transmission.

SEP has been targeted to the lumen of synaptic vesicles by fusing it to vesicle proteins such as VAMP2 (named synaptopHluorin or spH) (Sankaranarayanan and Ryan, 2001) or synaptophysin (named sypHy) (Granseth et al., 2006). New insights into the recycling modes of distinct presynaptic vesicle pools have been gained using these GEIs, but disputes over the kiss-and-run versus full fusion mechanisms persist, often due to controversy over the methodology (Fernandez-Alfonso and Ryan, 2004; Gandhi and Stevens, 2003; Granseth et al., 2006; Zhang et al., 2009). Presynaptic vesicle cycling has been visualized using FM styryl dyes, pHluorin, and recently quantum dots, each with advantages and disadvantages. pHluorin overcame many problems associated with the FM styryl dyes. The FM dyes require complicated loading protocols, they can lack specificity, and they can result in complicated response kinetics that confound interpretation (Ryan, 2001). Unlike FM dyes, pHluorin can be used to discriminate between the kinetics of endocytosis and exocytosis using an alkaline trapping method (Sankaranarayanan and Ryan, 2001). Quantum dots have improved signal over noise compared to both FM dyes and pHluorin; however, quantum dots, like FM dyes, still require loading with stimulation protocols and are not easily loaded deep into tissue. pHluorins, on the other hand, have the powerful advantage of being amenable to genetic targeting and *in vivo* monitoring (Li et al., 2005).

Despite the controversy surrounding synaptic vesicle cycling, pHluorin fusions have been broadly used to monitor both pre- and postsynaptic function during synaptic transmission. Vesicle-targeted pHluorins provide an optical assay for investigating regulators of synaptic transmission, including small molecules (Thorsen et al., 2010) like isoflurane (Hemmings et al., 2005) and specific proteins. For example, a VGLUT1-pHluorin fusion was used to demonstrate that increased α-synuclein inhibits neurotransmitter release and decreases the size of the presynaptic vesicle pool that recycles (Nemani et al., 2010). On the postsynaptic side, a GluR1-pHluorin fusion was used to demonstrate that the AAA-ATPase Thorase is important for AMPA receptor endocytosis following NMDA stimulation and Thorase may inhibit recycling of glutamate receptors by receptor disassembly (Zhang et al., 2011). In these examples, visualization of pHluorin was used in conjunction with genetic models to assess synaptic protein function with resolution of single neurons and synapses.

Populations of cells have also been imaged with pHluorin to study olfactory circuits and plasticity in live transgenic *Drosophila* and mice (McGann et al., 2005; Shang et al., 2007; Yu et al., 2004). When pHluorin was genetically targeted to the olfactory bulbs in mice or antennal lobes in flies, fluorescence changes correlated with population activity even though single synapses could not be visualized. The fluorescence readout provided a noninvasive optical method to investigate odorant coding. The use of pHluorin in this manner is powerful because it provides a means to monitor the activity of an intact network in a live animal.

In addition to synaptic transmission, many cellular processes involved with neuronal activity cause or rely on pH changes (Casey et al., 2010; Chesler and Kaila, 1992). pHluorin has had a significant impact in brain research, but its primary use has not been to measure pH, perhaps overshadowing other GEIs of pH. Recently, the pH sensor mitoSypHer was used in astrocytes, and a decrease in mitochondrial pH was observed upon glutamate stimulation (Azarias et al., 2011). This GEI was engineered by modifying the hydrogen peroxide sensor HyPer (see section "Reactive oxygen species"). HyPer contains a pH-sensitive YFP, and removing peroxide sensitivity created a ratiometric pH sensor (Poburko et al., 2011). Both cytoplasmic pH and mitochondrial matrix pH could be simultaneously monitored using the small-molecule pH indicator SNARF-1 and mitoSypHer in the same cell.

Chloride Sensing

Chloride is a critical ion that influences membrane potential, cell volume, and pH. The FRET-based GEI for Cl^- called Clomeleon was first developed to measure intracellular chloride in hippocampal neurons and monitor GABAergic signaling (Kuner and Augustine, 2000). It is a tandem construct consisting of a cyan FP (CFP) and a chloride-sensitive YFP (Topaz), resulting in chloride-sensitive FRET. Clomeleon has several known performance issues: it has a low affinity (K_{app}~90mM) compared to the intracellular Cl^- concentration (~10mM); Topaz is pH sensitive; and CFP and Topaz photobleach at different rates causing the FRET signal to change during strong or repeated illumination.

Despite these problems, Clomeleon has been useful for visualizing Cl^- in both cultured neurons and brain slices from transgenic mice. For example, developmental changes in neuronal chloride concentration as well as transient GABA-evoked changes in intracellular chloride have been imaged with Clomeleon (Berglund et al., 2006; Berglund et al., 2008). In particular, Clomeleon was used to provide direct evidence for spatial Cl^- gradients in retinal bipolar ON cells (Duebel et al., 2006). Perforated patch recordings had previously suggested that a Cl^- gradient existed, but the electrophysiology only provided indirect evidence. Using two-photon imaging of retinal preparations, an increased Cl^- concentration in dendrites compared to the soma was observed. The spatial gradient imaged with Clomeleon correlated with the functional consequence that GABA stimulation caused Cl^- influx at the soma but efflux at the dendrite.

Several studies using Clomeleon exemplify how GEIs and two-photon microscopy can be a very effective combination for imaging thick samples or deep in tissues. Preparations of brain slices or intact tissues are desirable because, compared to dissociated cultures, they better maintain physiologically relevant cellular environments, network connections, and anatomical relationships. Unfortunately, single-cell electrophysiology in thick preparations can be technically challenging and limit the number of cells that can be interrogated. Loading small-molecule indicators deep within tissues can also be inefficient. In contrast, GEIs such as Clomeleon can be genetically targeted, overcoming loading issues, and two-photon microscopy facilitates optically monitoring deep in tissues with the potential to interrogate several cells simultaneously. For example, imaging cerebellar slices from Clomeleon transgenic mice was used to visualize tonic inhibition of cerebellar granule cells by glial GABA release (Lee et al., 2010). Whole hippocampi and hippocampal slices from Clomeleon transgenic mice also have been used to observe pathological chloride accumulation in response to ischemia and epileptiform activity (Dzhala et al., 2010; Glykys et al., 2009; Pond et al., 2006).

Clomeleon has been used most often for Cl^- visualization, but it is not the only GEI for Cl^-. An improved Clomeleon called Cl-sensor has been created using a similar FRET strategy. In Cl-sensor, Topaz has been replaced by a mutated YFP that increases the chloride affinity (K_{app}~30 mM) and decreases the pH sensitivity (Markova et al., 2008). Intracellular chloride dynamics have been imaged in cultured neurons and retinal slices using Cl-sensor (Waseem et al., 2010). Additionally, a fusion of Cl-sensor with the glycine receptor

GlyR has been used to monitor chloride conduction and ion channel activity (Mukhtarov et al., 2008). Finally, transgenic mice that express a chloride-sensitive EYFP under control of the $K_V3.1$ promoter have also been generated (Metzger et al., 2002). Glutamate-evoked increases in Cl^- were imaged using hippocampal neuron cultures from these transgenic mice (Slemmer et al., 2004).

Zinc Sensing

Zn^{2+} is another biologically important ion that has been proposed to play a role in normal and pathological brain processes such as synaptic plasticity and excitotoxicity. Zn^{2+} has been found in glutamate containing synaptic vesicles in the cerebral cortex, accumulates in neurons after excitotoxic injury, and is an allosteric modulator of glutamate and GABA receptors (Frederickson et al., 2005). Chelation or exogenous application of Zn^{2+} can induce seizures and activate cell death programs. Physiological Zn^{2+} concentrations are tightly buffered, and free extracellular concentrations range from 1 to 10nM while free intracellular concentrations may be in the nanomolar to picomolar range. Because it is highly buffered, Zn^{2+} is a challenging analyte to monitor (Vinkenborg et al., 2010).

A handful of both nonratiometric intensity and ratiometric FRET GEIs of Zn^{2+} have been engineered (Vinkenborg et al., 2010). Zinc finger motifs containing His_4 and Cys_2His_2 binding sites have been used for sensor domains, but the GEIs using these binding motifs have affinities of ~2–200μM, quite low compared to the expected free Zn^{2+} concentrations (Dittmer et al., 2009). Despite the low affinity, imaging the Cys_2His_2 sensor targeted to the mitochondrial matrix provided evidence that there is a high resting Zn^{2+} concentration in the mitochondria of cultured hippocampal neurons. Further, simultaneous imaging of mitochondrial Cys_2His_2-based sensor and the small-molecule Zn^{2+} indicator FluoZin3 revealed that the neuronal response to glutamate depends on extracellular Zn^{2+}. In the absence of extracellular Zn^{2+}, glutamate stimulation causes mitochondrial Zn^{2+} release, but in the presence of extracellular Zn^{2+}, glutamate stimulation causes uptake and mitochondrial accumulation of Zn^{2+}. Improved GEIs with subnanomolar affinity for Zn^{2+} have been engineered, and these sensors should prove useful for future imaging of Zn^{2+} neurobiology (Qin et al., 2011; Vinkenborg et al., 2010).

Glutamate Sensing

Glutamate is the major excitatory neurotransmitter, and the dynamics of glutamate release and clearance affect synapse and network level signaling as well as neuron–glia metabolic coupling. Two similar FRET-based GEIs, FLIPE (Deuschle et al., 2005) and SuperGluSnFr (Hires et al., 2008), have been constructed using the same bacterial periplasmic binding protein ybeJ/GltI. This bacterial protein undergoes a hinge-like motion when glutamate binds to it. When ECFP and EYFP are fused to the termini of ybeJ/GltI so that they flank the binding protein, the glutamate-induced conformational change results in a change in FRET efficiency.

FLIPE and SuperGluSnFr have been used to visualize glutamate release in response to electrical stimulation and potassium chloride depolarization. Extracellular glutamate was monitored by expressing these GEIs on the surface of cultured hippocampal neurons (Hires et al., 2008; Okumoto et al., 2005). The imaging data suggested that reuptake by glial glutamate transporters mediates fast clearance following release, but at least in culture, significant spillover of several hundred nanomolar glutamate can still persist. Previous methods lacked the spatial and temporal resolution to perform this type of measurement.

These GEIs may prove valuable for evaluating glutamatergic synaptic transmission *in vivo*, but as of yet no genetic model or transient expression model has been published. There are published examples in which purified sensor was applied directly onto brain slices (Dulla et al., 2008). Despite the low resolution of these experiments, fluorescence transients that were imaged during stimulated activity correlated with electrical field potential recordings (Dulla et al., 2008). This method was used with an acute epileptiform brain slice model to show correlations among the interstimulus interval, the magnitude of the integrated field potential, and the size of the area for which GEI fluorescence reported glutamate release (Tani et al., 2010).

Glucose Sensing

Glucose is the primary energy source for the brain, and glucose-dependent metabolic rate is the basis for technologies such as FDG-PET scanning. Glucose metabolism classically has been observed through biochemical techniques that lack single-cell resolution such as isotope tracing or microdialysis. An engineered glucose-sensitive GEI provides much greater resolution of single-cell metabolism (Fehr et al., 2003). Like the GEIs for glutamate, the glucose sensor is one of a series of other stereotyped sensors based on bacterial periplasmic binding proteins used as the sensor domains (Okumoto et al., 2008). The bacterial MglB protein is flanked by ECFP and the YFP Citrine, and a glucose-induced conformational change alters the FRET efficiency. This glucose sensor has been significantly optimized (Takanaga et al., 2008), and a recent version FLII^{12}Pglu-600μδ6 (also called FLII^{12}-Pglu-700μδ6) has been used to measure glucose metabolism in many cell types including cultured astrocytes and neurons (Bittner et al., 2010).

Direct measurements of glucose transport and metabolism have begun to emerge using this optimized glucose sensor. Independent measurements suggest that cultured astrocytes maintain their intracellular glucose concentration at half of the extracellular concentration in the steady state. Upon inhibition of glucose transport, intracellular glucose rapidly decreases, indicative of hexokinase activity (Bittner et al., 2010). Interestingly, depolarization with potassium causes a rapid decline in intracellular glucose within seconds, and there is a rapid reversal upon removal of excess potassium (Bittner et al., 2011). In contrast, glutamate does not cause an acute decrease in intracellular glucose (Bittner et al., 2011; Prebil et al., 2011a; Prebil et al., 2011b), but it does cause a slow increase in glucose concentration and glycolytic rate after several minutes (Bittner et al., 2011). Noradrenaline stimulation causes a rapid increase in intracellular glucose which can be partially attenuated by inhibition of glycogen phosphorylase (Prebil et al., 2011b). Future imaging of FLII^{12}Pglu-600μδ6 *in vivo* should help clarify many details of energy metabolism in neurons versus glia.

ATP Sensing

The dynamics of ATP production and hydrolysis in neurons and astrocytes is a critical aspect of brain energy metabolism and signaling. Recently, GEIs have been developed to visualize ATP. The sensor Perceval was engineered by inserting the circularly permuted YFP cpVenus into the bacterial protein GlnK (Berg et al., 2009). When Mg·ATP binds to GlnK, there is protein loop movement at the binding site. By inserting the cpVenus into the loop region, the conformational change induced by Mg·ATP binding is coupled to a perturbation of the cpVenus chromophore. This perturbation causes a ratiometric shift in the Perceval's fluorescence excitation spectrum. The binding affinity for Mg·ATP is very high (K_{app}~0.1 μM), and ADP can competitively bind to GlnK with lower affinity (K_{app}~0.2 μM); however, ADP binding does not induce a loop movement. Further, intracellular ADP and ATP concentrations are near 0.5 and 5 mM. As a result, Perceval is always bound by either ADP or Mg·ATP, but the two nucleotides have distinct effects on the fluorescence. The combined result of these properties is that Perceval faithfully responds to changes in the ATP/ADP ratio independent of the total nucleotide level. The ATP/

ADP ratio is indicative of the cellular energy status, and several ATP-dependent processes are sensitive to the ATP/ADP ratio. Perceval has proven useful for imaging energy depletion during metabolic inhibition of cultured cells.

Another GEI exploits the conformational change that occurs upon ATP binding to the ε subunit of the F_0F_1-ATP synthase (Imamura et al., 2009). By attaching the CFP mseCFP and cpVenus to the bacterial ε subunit, a series of GEIs called the ATeam sensors were engineered with varying affinities for ATP. These GEIs are insensitive to ADP and report absolute ATP concentrations. Similar to Perceval, ATeams were used to image ATP depletion during metabolic inhibition of cultured cells. Perceval and the ATeams are complementary GEIs, and in the future, they should provide a great deal of information on neuronal and glial energy metabolism, especially if they are applied in conjunction with other GEIs such as the glucose sensor.

GPCR and heterotrimeric G protein signaling

GPCRs are critical components of the membrane-bound signal transduction machinery that affect synaptic transmission and plasticity. Ligand binding to the GPCR causes a conformational change to an active form. The activated GPCR acts as a guanine nucleotide exchange factor, activating heterotrimeric G proteins ($G_{\alpha\beta\gamma}$) by promoting the replacement of bound GDP with GTP in the G_α protein. The activated G_α can subsequently activate downstream signaling before its intrinsic GTPase activity hydrolyzes the bound GTP, causing auto-inactivation. Two major downstream pathways involve activation of phospholipase C (PLC) or activation/inhibition of adenylate cyclase. Activation of GPCRs has been monitored with several imaging probes (Lohse et al., 2008), including both intermolecular and intramolecular FRET-based GEIs. An important caveat of FP-based GEIs is that the bulky FP can interfere with protein–protein interactions, and the expressed GEI can interfere with endogenous signaling downstream of GPCR activation. Despite this caveat, FRET-based GEIs have been instrumental in delineating previously inaccessible GPCR and G protein activation kinetics.

To monitor GPCR activation by intermolecular FRET, the separate components of the signaling complex, including the GPCR itself and the $G_{\alpha\beta\gamma}$ proteins, are fused to FRET compatible FPs such as CFP and YFP. GPCR activation causes a conformational change or dissociation of the GPCR, G_α, and $G_{\beta\gamma}$ proteins. When an appropriate pair of the signaling components is fused to FPs of a FRET pair, activation causes a decrease in FRET (Janetopoulos et al., 2001), but detailed interpretations of intermolecular FRET studies can be complicated. There are controversies concerning whether G proteins dissociate from the GPCR following activation and whether a GPCR/G protein complex preassembles (Bunemann et al., 2003; Hein et al., 2005; Janetopoulos et al., 2001; Nobles et al., 2005). This intermolecular FRET strategy can be extended to monitor GPCR interactions with other important regulatory proteins such as β-arrestins, revealing differential kinetics and timing between GPCR activation and inactivation (Vilardaga et al., 2003). It has also been used to monitor downstream targets such as G protein-gated inwardly rectifying potassium channels (GIRKs), combining FRET imaging and electrophysiology to provide evidence for a G protein-mediated GIRK dose response (Hommers et al., 2010).

Intermolecular FRET has been used extensively to monitor the activation dynamics of metabotropic glutamate receptors (mGluRs) and $GABA_B$ receptors. These GPCRs have large extracellular ligand binding domains and function as receptor dimers (Kunishima et al., 2000). Ligand binding causes conformational changes in the extracellular domains that are transduced to the intracellular domains. By inserting CFP and YFP into intracellular loops of separate receptors of a receptor dimer, a ligand-induced conformational change can cause a change in FRET efficiency. Hence, mGluR and $GABA_B$ receptor activation can be optically monitored (Matsushita et al., 2010; Tateyama et al., 2004). Using this imaging strategy, the activation kinetics of mGluR1β were measured. Despite the slow signaling downstream from mGluR activation as compared to an ionotropic receptor, mGluR1β activation itself is fast, suggesting that downstream events are rate limiting (Marcaggi et al., 2009). Importantly, these kinetic parameters were used to build a model in order to evaluate the contribution of transient mGluR activation to shaping synaptic transmission. FRET imaging also showed that different ligands can induce distinct structural changes, resulting in distinct downstream G protein signaling. mGluR1α activation by glutamate or the exogenous ligand Gd^{3+} was imaged. Activation by the two ligands shows different FRET responses, and the downstream responses are different. Glutamate increases both intracellular calcium and cAMP levels, but Gd^{3+} increases only the calcium levels. The imaging data suggest that glutamate activates both $G_{\alpha s}$ and $G_{\alpha q}$, but Gd^{3+} only activates $G_{\alpha q}$ (Tateyama et al., 2004).

An alternative FRET imaging strategy exploits a relative movement between the third intracellular loop and C-terminus of GPCRs (Lohse et al., 2007). With CFP inserted into Loop III and a YFP into the C-terminus, activation of the GPCR causes a decrease in FRET efficiency (Rochais et al., 2007; Vilardaga et al., 2003). This strategy no longer makes assumptions about the preassembly of the signaling complex, a controversial issue. Both intermolecular and intramolecular FRET strategies have been used in conjunction with electrophysiology to conduct a comprehensive study of the kinetic steps in M1 muscarinic acetylcholine receptor (M1 mAChR) activation, downstream PLC activation, and subsequent inhibition of Kv7.2/7.3 M currents (Jensen et al., 2009). By systematically tagging different components of the M1 mAChR–$G_{\alpha\beta\gamma}$–PLC pathway, imaging data suggest that PIP_2 hydrolysis is rate limiting due to the expression levels of PLC rather than the intrinsic kinetics of PLC. Further, kinetic parameters measured in the imaging experiments were used to build a kinetic model, facilitating a quantitative comparison of the empirical observations (Falkenburger et al., 2010; Jensen et al., 2009).

Lipid metabolism linked signaling: PIPs, IPs, DAG

Downstream from GPCR activation, metabolism of phosphoinositide lipids is intimately linked to intracellular signaling. Phosphoinositides can be hydrolyzed to inositol phosphates and DAG in a receptor-mediated manner, and these phospholipid metabolites can subsequently modulate ion channel activity, synaptic transmission, and plasticity. To better dissect this GPCR-related signaling branch, several translocation GEIs have been developed.

PH domains have been extensively used as sensing motifs to construct translocation GEIs of PIP_2 (Varnai and Balla, 2006). For example, the PH domain from PLC δ1 has been fused to EGFP ($PH_{PLC\delta 1}$-EGFP) to create a translocation probe. $PH_{PLC\delta 1}$-EGFP binds PIP_2 in the membrane, but upon PIP_2 hydrolysis to DAG and inositol-(1,4,5)-trisphosphate (IP_3), $PH_{PLC\delta 1}$-EGFP translocates to the cytoplasm (Stauffer et al., 1998; Varnai and Balla, 1998). The $PH_{PLC\delta 1}$-EGFP construct has been used in neuroblastoma cells and cultured hippocampal, striatal, and cerebellar granule neurons to study PLC regulation of muscarinic receptors and mGluRs (Billups et al., 2006; Butcher et al., 2009; Kumar et al., 2008; van der Wal et al., 2001; Willets et al., 2007; Xu et al., 2003). It has also recently been applied in organotypic brain slice to verify that carbachol can induce PIP_2 hydrolysis in cortical pyramidal neurons (Yan et al., 2009). In an early study, activity-dependent PIP_2 metabolism was visualized at presynaptic terminals in cultured hippocampal neurons (Micheva et al., 2001). A combination of $PH_{PLC\delta 1}$-EGFP and FM4-64 live-cell imaging with *post hoc* immunochemistry suggested that postsynaptic NMDAR activation is involved in activity-dependent PIP_2 metabolism at the presynaptic terminal.

It has been observed, however, that $PH_{PLC\delta1}$-EGFP has a 20-fold higher selectivity for IP_3, and therefore, translocation following PIP_2 hydrolysis is primarily due to increased cytosolic IP_3 (Hirose et al., 1999). Taken into account, this GEI can then be used to monitor IP_3 metabolism (Okubo et al., 2001), or an alternative sensing domain can be sought. To this end, the Tubby domain, a PIP_2-binding motif of a mammalian transcription factor, was found to be selective for PIP_2 and insensitive to IP_3 (Quinn et al., 2008; Santagata et al., 2001; Szentpetery et al., 2009). The Tubby-EGFP probe translocates to the cytoplasm due to PIP_2 hydrolysis and not to increased cytosolic IP_3.

FRET-based GEIs selective for IP_3 have also been developed using the ligand binding domains of IP_3 receptors in order to overcome the sensitivity of PH domains to both PIP_2 and IP_3 (Matsuura et al., 2006; Remus et al., 2006; Sato et al., 2005; Tanimura et al., 2004). These probes have been useful for specifically investigating the kinetics of IP_3 production at the plasma membrane, in the cytosol, and even in the dendrites of cultured hippocampal neurons following mAChR or mGluR activation (Sato et al., 2005; Tanimura et al., 2004).

Finally, translocation and FRET-based DAG-selective GEIs have been engineered. These GEIs exploit the DAG-dependent membrane association of the C1 domains of protein kinase C (PKC) γ and β (Nishioka et al., 2008; Oancea et al., 1998; Violin et al., 2003). The DAGR sensor combines both DAG-dependent translocation and environmentally sensitive FRET for a quantitative fluorescence response. DAGR is an intramolecular FRET construct in which the C1 domains of PKCβII are sandwiched between CFP and Citrine. In contrast to the PH and Tubby probes, increased DAG in the membrane causes translocation of DAGR from the cytosol to the membrane. The membrane environment adds additional conformational constraints on the relative orientation and distance between the FPs, causing an increase in FRET associated with translocation (Violin et al., 2003). DAGR has been used in single astrocytes, for example, to demonstrate that PLCε is necessary for thrombin-induced DAG production and that carbachol-induced DAG production occurs independently because separate PLCs are activated (Citro et al., 2007).

These GEIs form a useful family of sensors that can be used to monitor the timing of GPCR modulation of synaptic activity. For example, the entire set of $PH_{PLC\delta1}$-EGFP, Tubby-EGFP, and $C1_{PKC\gamma}$-EGFP GEIs was used to dissect the kinetics of Ca^{2+}-dependent changes in PIP_2/IP_3, PIP_2, and DAG concentrations in SH-SY5Y neuroblastoma cells and cultured hippocampal neurons after stimulation of mAChRs (Nelson et al., 2008). By combining imaging and electrophysiology, these GEIs have also been extensively used to study GPCR-mediated PLC modulation of Kv7.2/7.3 M-currents, K_{ATP} currents, and TRPV2 currents (Jensen et al., 2009; Mercado et al., 2010; Quinn et al., 2008; Suh et al., 2004). In one comprehensive study of PLC regulation of M-currents, PIP_2/IP_3/DAG metabolism was imaged with $PH_{PLC\delta1}$-EGFP and $C1A_{PKC}$-EGFP, Ca^{2+} was also imaged, and electrophysiology was conducted in a series of parallel experiments. The combined data demonstrated that Ca^{2+} is required and that there is a feedback loop for PLC regulation of Kv7.2/7.3 currents (Horowitz et al., 2005).

Cyclic nucleotides and downstream kinases

Cyclic nucleotides such as cyclic adenosine monophosphate (cAMP) and cyclic guanosine monophosphate (cGMP) are well-known second messengers in intracellular signaling cascades. Their temporal and spatial dynamics are highly regulated to ensure normal physiology, and many

GEIs have been engineered for live-cell imaging of cAMP, cGMP, and their downstream signaling cascades (Vincent et al., 2008; Willoughby and Cooper, 2008). Most of these GEIs are FRET based. Initial sensors used intermolecular FRET strategies (Zaccolo et al., 2000), but recent sensors have used an intramolecular FRET strategy to provide a more quantitative and robust fluorescence readout.

There are several intramolecular FRET-based GEIs for imaging cAMP. For instance, a cAMP-binding protein Epac has been used as the scaffold between a FRET pair, and changes in the FRET response report changes in cAMP levels (DiPilato et al., 2004; Nikolaev et al., 2004). Further, transgenic animals have been made for *in vivo* imaging of cAMP dynamics in intact networks. For example, Epac1-camps has been stably expressed in *Drosophila* neurons to study how neuropeptides modulate the clock network via cAMP-dependent signaling (Shafer et al., 2008).

Protein kinase activities downstream from cAMP can also be monitored with GEIs. To report the activity of the downstream cAMP-dependent protein kinase A (PKA), a cognate peptide substrate and a binding domain have been used as the scaffold between a FRET pair. Upon PKA phosphorylation of the cognate peptide, its subsequent interaction with the adjacent binding domain changes the FRET response (Nagai et al., 2000; Zhang et al., 2001). Imaging two GEIs, ICUE2 for cAMP and AKAR2 for PKA activity, has been used to study oscillations in second-messenger signaling cascades in dissociated cultures of retinal neurons (Dunn et al., 2006). Further, using viral delivery of the AKAR2 reporter, nuclear and cytosolic PKA dynamics have been studied in individual neurons in mouse brain slice preparations (Gervasi et al., 2007). Similar to GEIs for PKA, by choosing different cognate peptide and binding domains, sensors of other protein kinases, such as Akt/protein kinase B, PKC, and AMP-dependent protein kinase, have been generated (Sasaki et al., 2003; Tsou et al., 2011; Violin et al., 2003). Their signals often reflect a balance between the phosphorylation reaction by the kinase and the dephosphorylation reaction by cellular phosphatases. When using these GEIs, it is important to validate their sensitivity and specificity (Vincent et al., 2008), and a GEI variant with spoiled phosphorylation site is often a useful negative control.

To report cellular cGMP dynamics, FRET-based and cpFP-based GEIs have been designed (Honda et al., 2001; Nausch et al., 2008). Further, to detect the upstream signaling molecule nitric oxide (NO), a FRET-based cGMP reporter can be used in combination with soluble guanylate cyclase, an NO-dependent enzyme; exposure to minute NO leads to formation of several cGMP molecules and thus amplifies the fluorescent signal. This amplification scheme has been used to detect picomolar NO released from cultured hippocampal neurons (Sato et al., 2006).

Small GTPases

Small GTPases regulate cytoskeletal dynamics at the neurite growth cone (Dickson, 2001). Small GTPases are G proteins that function similar to the α-subunits of heterotrimeric G proteins. Small GTPases can be activated by exchange of GTP for GDP, promoted by guanine nucleotide exchange factors, and they are inactivated when GTP is hydrolyzed to GDP, promoted by GTPase activating proteins. To image the activated conformation of a small GTPase, one successful strategy uses the target GTPase and a binding partner labeled with the respective partners of a FRET pair. The binding partner selectively recognizes the GTP-bound, activated conformation of the GTPase, leading to a dynamic FRET signal that provides a readout for activation (Kraynov et al., 2000). This strategy has been adapted to engineer GEIs based on intramolecular FRET, and sensors have been developed to detect the activation of Ras and Rho family members—Ras, Rap1, RhoA, Rac, Cdc42—as well as for other small GTPases (Aoki et al., 2008; Itoh et al., 2002; Mochizuki et al., 2001; Pertz et al., 2006; Sabouri-Ghomi et al., 2008; Yoshizaki et al., 2003).

Neurite outgrowth in the PC12 cell line has been examined extensively using GEIs for Ras and Rho small GTPases (Nakamura et al., 2008). The Raichu series of GEIs allow neurite outgrowth to be monitored in parallel with G protein activity, revealing signaling requirements for the successful action of outgrowth signals such as nerve growth factor or dibutyryl-cAMP (Goto et al., 2011). Quantitative live-cell imaging has also provided spatial and temporal parameters characterizing the Rac1 and cdc42 signaling pathway for neurite outgrowth (Aoki et al., 2004, 2005, 2007). These parameters were used to create a signaling model, predicting that a PI-5phophatase negatively regulates outgrowth (Aoki et al., 2007). Experimental verification of this prediction shows that this type of quantitative analysis and modeling may provide valuable insights into the convergence and divergence of signaling pathways governing neurite outgrowth and guidance.

These GEIs have also been used to show that Rac and RhoA regulate dendritic spine morphogenesis in cultured hippocampal neurons. Both overexpression of a constitutively active RhoA mutant and knockdown of the polarity protein PAR-6 inhibit dendritic spine formation. Imaging the Raichu GEI for RhoA demonstrated that PAR-6 knockdown caused increased RhoA activity, suggesting that PAR-6 promotes spine morphogenesis by negative upstream regulation of RhoA (Zhang and Macara, 2008). Similarly, imaging Rac activity suggested a delineation of several steps of NMDAR regulation of spine morphogenesis. The guanine nucleotide exchange factor PIX activates Rac, and PIX interaction with Rac is promoted by CaMKK/CaMKI. Ras imaging showed that Ras activity in dendritic spines can be increased by depolarization in a CaMK-dependent manner, and overexpression of a phosphorylation-deficient PIX dominant negative can attenuate activation (Saneyoshi et al., 2008). These imaging data may suggest that NMDAR activation causes calcium influx and CaMK activation, and these signaling events could lead to local phosphorylation of PIX and subsequently local activation of Rac to promote spine morphogenesis.

Imaging RhoA activity in cultured cortical neurons and in brain slices has also shown that the subcellular localization of RhoA activation is important for the regulation of cortical neuron migration during development (Pacary et al., 2011). The regulatory proteins Rnd2 and Rnd3 promote neuronal migration by inhibiting RhoA activity, but knockdown has different pathological phenotypes. In this study, Raichu was genetically targeted to different subcellular compartments, and imaging RhoA activity demonstrated that only Rnd3 knockdown increases RhoA activity at the plasma membrane. The spatial resolution resulting from this strategy was instrumental for showing that spatial organization of RhoA activity existed and that compartmentation of signaling components may be a key regulatory feature in neuronal migration.

Proteases

Protease activity regulates numerous signaling events including neuronal cell death, and FRET-based GEIs have long been used to monitor proteolytic cleavage of substrates. Sensors of proteolysis can be created by fusing a FRET pair using a peptide linker that contains recognition and cleavage sites. For example, FRET-based GEIs containing the Asp-Glu-Val-Asp (DEVD) caspase cleavage sequence have been used to monitor apoptosis (Xu et al., 1998). In the absence of caspase activity, the FPs remain covalently attached, and their close proximity results in a high FRET state. Upon proteolytic cleavage, the FPs can diffuse away from one another resulting in a loss of FRET, indicating caspase activation and apoptosis.

These FRET-based GEIs of protease activities have greatly complemented biochemical and genetic techniques in the study of neuronal cell death. GEIs containing the DEVD cleavage site have been used to investigate the role of Puma, a BH3-only proapoptotic protein, in caspase-dependent and -independent neuronal cell death (Tuffy et al., 2010). The DEVD cleavage site can also be replaced by other cleavage motifs, and the cleavage domain can even be an entire protein such as the full-length BID protein, another BH3-only proapoptotic protein (Ward et al., 2006). Glutamate excitotoxicity can induce caspase cleavage of full-length BID, and the truncated BID product can activate mitochondrial apoptosis pathways. Imaging a CFP-BID-YFP construct revealed the kinetics of caspase proteolysis of BID as well as the translocation of both full-length and truncated BID to mitochondria in cerebellar granule neurons (Ward et al., 2006).

Glutamate receptor activation can also lead to activation of other proteases such as calpains. Calpains are Ca^{2+}-dependent proteases, and two types have been identified that have micromolar (μ-calpain) and millimolar (m-calpain) affinity for Ca^{2+}. Calcium influx following synaptic activity has been proposed to activate calpains, causing proteolytic cleavage of cytoskeletal, receptor, and cytosolic substrates. The identity of these substrates suggests that calpain activity may modulate morphology, synaptic plasticity, and cell death (Zadran et al., 2010). A GEI of calpain activity was designed using a fusion between ECFP and EYFP linked by the calpain recognition site from α-spectrin (Vanderklish et al., 2000). The sensor was also targeted to postsynaptic dendritic spines using the Shaker PDZ domain. Imaging this sensor in cultured hippocampal neurons and organotypic hippocampal slices revealed that activation of glutamate receptors decreases FRET within minutes, demonstrating activity-dependent calpain activation (Vanderklish et al., 2000). Imaging experiments also showed that chronic glutamate exposure causes a large sustained increase in cytosolic Ca^{2+} that precedes calpain activation. These imaging data suggest that calpain cleavage of the sodium–calcium exchanger is not responsible for the loss of calcium homeostasis and eventually cell death, as had been previously hypothesized (Gerencser et al., 2009).

Reactive oxygen species and redox potential

Reactive oxygen species (ROS) and free radicals are generated in diverse metabolic processes, and they can often act as messengers in signal transduction by altering the redox status of various signaling proteins (Winterbourn, 2008). Considering the importance of ROS and redox signaling in cellular physiology, numerous chemical-based probes (Wardman, 2007) and genetically encoded GFP-based biosensors (Meyer and Dick, 2010) have been devised. These reporters can assess either the cellular redox status, in general, or a particular ROS or redox metabolite.

As reporters of general cellular redox status, redox-sensitive yellow FPs (rxYFPs) (Ostergaard et al., 2001) and redox-sensitive green FPs (roGFPs) (Hanson et al., 2004) have been created by engineering two surface-exposed cysteine residues near the chromophore. The thiol oxidation status of these cysteines is influenced by cellular oxidants and antioxidants, and redox status affects the FP conformation and the fluorescence response. Compared with the rxYFP reporter, roGFP is more commonly used for two reasons: first, unlike rxYFP, roGFP is ratiometric and thus provides a more quantitative measurement; second, some roGFP variants are notably pH resistant (Meyer and Dick, 2010), while rxYFP is pH sensitive in the physiological pH range (Ostergaard et al., 2001). By using targeting sequences to various cellular compartments, these redox-sensitive reporters can report the general redox states of mitochondria, nuclei, and endoplasmic reticulum (Dooley et al., 2004; Merksamer et al., 2008), and the GEIs have been expressed in dissociated neuronal cultures for bioenergetics studies (Vesce et al., 2005). In addition, transgenic mice have been generated to express the mito-roGFP biosensor under the control of the tyrosine hydroxylase promoter. By imaging acute brain slices with two-photon microscopy, mitochondrial oxidant stress levels were monitored in substantia nigra dopaminergic neurons during activity (Guzman et al., 2010).

Although roGFP variants have undefined redox specificity, other fluorescent sensors have been engineered to address this issue by linking GFP variants to specific ROS binding protein domains. For instance, a fluorescent reporter of hydrogen peroxide, named HyPer, has been created by linking a cpYFP to a bacterial H_2O_2 binding domain OxyR (Belousov et al., 2006). Unfortunately, HyPer is very pH sensitive like many cpFP-based GEIs, and a pH fluctuation of 0.3units can mimic the entire excursion of the sensor response. It is thus important to monitor pH to correct for artifacts in the sensor measurements. Fortuitously, it has been found that cpYFP by itself, even with no linkage to any ligand binding domains, could report superoxide. While cpYFP has been exploited to monitor superoxide burst in single mitochondria (Wang et al., 2008b), it is unclear how superoxide interacts with the cpYFP chromophore, and whether this interaction is specific for superoxide.

In addition to ligand binding domains, redox enzymes can be linked to GFP variants to create fluorescent reporters with defined redox specificity. For instance, while roGFP2 by itself reports the general cellular redox status, by linking roGFP2 to a human thioltransferase Grx1, a fluorescent reporter specific for glutathione redox has been created (Gutscher et al., 2008). Not only is Grx1-roGFP2 pH resistant, but it is also fast in its response to changes in glutathione redox. Using a similar strategy, fluorescent sensors of other redox equilibria can be created as well (Gutscher et al., 2009).

Challenges for using GEIs

We believe that GEIs provide a unique and powerful method for monitoring neuronal and glial activity in live cells, but there are many important challenges for using GEIs that should be reiterated, including specificity- and expression-induced pathophysiology.

Engineering and validating the specificity of a GEI is both a critical and nontrivial task in sensor development. As mentioned throughout this review, pH sensitivity is a pervasive problem for GEI specificity. Many GEIs are significantly pH sensitive and require correction in order to discriminate real changes of interest from signal artifacts caused by pH changes (Berg et al., 2009). pH is not the only culprit, however, and many other environmental factors can cause signal artifacts or changes in a GEI's sensing characteristics (Takanaga et al., 2008; Willemse et al., 2007). For example, interpretation of cpYFP responses in mitochondria as an indication of superoxide flashes (Wang et al., 2008b) has been questioned because of alternative interpretations of pharmacological data (Muller, 2009) and the inherent pH sensitivity of the probe (Schwarzlander et al., 2011). This case illustrates how, even with significant sensor characterization (Wang et al., 2008b), questions of specificity can arise as the GEI is used, and equivocal sensing behavior may merit deeper investigation. Therefore, due diligence in the characterization of interferences *in vitro* and in the validation of specificity *in vivo* is required in order to establish GEIs as reliable tools in neurobiology; otherwise uncertainty over the specificity of a GEI can cause confusion and controversy over the interpretation of data.

Pathophysiology induced by expression of GEIs is also an undesired outcome that should be considered. As previously mentioned, alteration or interaction with endogenous signaling pathways by expressed GEIs can seriously compromise the utility of a reporter (e.g., GEIs that buffer low-level analytes or bind signaling proteins with greater occupancy than endogenous binding partners). It is always possible that the expression of a foreign protein can have unexpected and unwanted effects on the anatomy and physiology of the system. While GFPs have generally been shown and assumed to be inert reporters *in vitro* and *in vivo* (Feng et al., 2000; Zuo et al., 2004), there are several examples in the literature demonstrating that the innocuousness of GFP or GEI expression should be explicitly tested. Early reports indicated that overexpression of GFP in cell lines can be cytotoxic with increased apoptosis (Hanazono et al., 1997; Liu et al., 1999). Targeted expression of GFP can cause fatal cardiomyopathy in mice (Huang et al., 2000), and GFP has been shown to interfere with excitation–contraction coupling by interacting with myosin (Agbulut et al., 2006). It has also been reported that GFP can unexpectedly attenuate NFκB signaling and polyubiquitination (Baens et al., 2006). Importantly, GFP expression in the brain can increase the expression of stress response-associated genes (Comley et al., 2011) and may be neurotoxic causing significant neuronal death (Klein et al., 2006; Krestel et al., 2004). These examples serve as a reminder that all observation methods, including optical monitoring of GEIs, perturb the system to some extent. While the literature suggests that GFP and GEIs can be used with minimal perturbation of cellular function, explicit experimental validation of this supposition should not be neglected.

Finally, and most trivially, low fluorescence or small fluorescence changes can be a practical limitation on the use of many GEIs in cells: the real-world signal-to-noise in the presence of instrumentation noise and background fluorescence may just be too small for detecting subtle changes in the physiological targets of the indicators.

Future directions

Many GEIs have been engineered for monitoring metabolic and signaling dynamics in the brain. While these provide a significant tool set, it is far from complete. Obviously, continued engineering of new GEIs for different targets is a major challenge. Compared to the tens of thousands of unique metabolites and proteins, currently only a tiny fraction can be directly imaged with GEIs. While some general design principles have emerged, numerous challenges remain in the engineering and application of GEIs. These challenges extend to all levels of the methodology from basic FP chemistry up to systems level modeling.

There is no doubt that GEI development will benefit from an increased understanding of FP chemistry and photophysics. Biological factors such as changing pH and ROS levels are unavoidable, and a mechanistic understanding of how these factors affect FP fluorescence will aid the optimization of GEIs. Likewise, a mechanistic understanding of photophysical properties such as photoconversion, photobleaching, and photoisomerization will hopefully lead to the continued engineering of new FPs with improved characteristics. The development of new FP color variants will also greatly facilitate the diversification of GEI color schemes, a key requirement for multiplex imaging of several GEIs simultaneously.

It will also be important to promote the development of multiphoton microscopy techniques for imaging of GEIs in thick samples and *in vivo*. Combining GEIs with multiphoton imaging can be a powerful, noninvasive method for visualizing signaling dynamics in intact networks (Svoboda and Yasuda, 2006). Alternative methods such as two-photon fluorescence lifetime microscopy also show promise for quantitative imaging of FPs and GEIs *in vivo* (Drobizhev et al., 2011). Hopefully, the techniques and instrumentation will become more accessible so that imaging experiments can be carried out more routinely in live animals, providing better physiological parameters.

Finally, quantitative analysis of data from parallel imaging studies will be important for building systems level models. Models can provide a great deal of insight into the control and behavior of intricate metabolic and signaling networks, but models are limited by the empirical parameters they are built from. Imaging GEIs *in vivo* will be a critical step in building physiologically relevant models of brain function.

Conclusion

FP-based sensors provide a powerful tool for quantitative live-cell imaging, and consequently, the field of protein engineering and GEIs has grown immensely. Here, we have outlined several design strategies for the development of genetically encoded indicators, and we have provided several examples of their use in brain-related studies. These sensors provide a high level of spatial and temporal resolution, and the ability to monitor signaling dynamics on either the single-cell or population level has revealed many new details about biological networks. While the development and optimization of new sensors can require several years of engineering work, we believe the payoff is well worth the effort.

Acknowledgments

M. T. is supported by a Postdoctoral Fellowship from the NIH (F32NS066613), and this work was supported by a research grant from the NIH (R01NS055031) to G. Y.

References

Agbulut, O., Coirault, C., Niederlander, N., Huet, A., Vicart, P., Hagege, A., et al. (2006). GFP expression in muscle cells impairs actin-myosin interactions: Implications for cell therapy. *Nature Methods*, *3*, 331.

Akerboom, J., Rivera, J. D., Guilbe, M. M., Malave, E. C., Hernandez, H. H., Tian, L., et al. (2009). Crystal structures of the GCaMP calcium sensor reveal the mechanism of fluorescence signal change and aid rational design. *The Journal of Biological Chemistry*, *284*, 6455–6464.

Aoki, K., Kiyokawa, E., Nakamura, T., & Matsuda, M. (2008). Visualization of growth signal transduction cascades in living cells with genetically encoded probes based on Forster resonance energy transfer. *Philosophical Transactions of the Royal Society of London B: Biological Sciences*, *363*, 2143–2151.

Aoki, K., Nakamura, T., Fujikawa, K., & Matsuda, M. (2005). Local phosphatidylinositol 3,4,5-trisphosphate accumulation recruits Vav2 and Vav3 to activate Rac1/Cdc42 and initiate neurite outgrowth in nerve growth factor-stimulated PC12 cells. *Molecular Biology of the Cell*, *16*, 2207–2217.

Aoki, K., Nakamura, T., Inoue, T., Meyer, T., & Matsuda, M. (2007). An essential role for the SHIP2-dependent negative feedback loop in neuritogenesis of nerve growth factor-stimulated PC12 cells. *The Journal of Cell Biology*, *177*, 817–827.

Aoki, K., Nakamura, T., & Matsuda, M. (2004). Spatio-temporal regulation of Rac1 and Cdc42 activity during nerve growth factor-induced neurite outgrowth in PC12 cells. *The Journal of Biological Chemistry*, *279*, 713–719.

Azarias, G., Perreten, H., Lengacher, S., Poburko, D., Demaurex, N., Magistretti, P. J., et al. (2011). Glutamate transport decreases mitochondrial pH and modulates oxidative metabolism in astrocytes. *The Journal of Neuroscience*, *31*, 3550–3559.

Baens, M., Noels, H., Broeckx, V., Hagens, S., Fevery, S., Billiau, A. D., et al. (2006). The dark side of EGFP: Defective polyubiquitination. *PLoS One*, *1*, e54.

Baird, G. S., Zacharias, D. A., & Tsien, R. Y. (1999). Circular permutation and receptor insertion within green fluorescent proteins. *Proceedings of the National Academy of Sciences of the United States of America*, *96*, 11241–11246.

Belousov, V. V., Fradkov, A. F., Lukyanov, K. A., Staroverov, D. B., Shakhbazov, K. S., Terskikh, A. V., et al. (2006). Genetically encoded fluorescent indicator for intracellular hydrogen peroxide. *Nature Methods*, *3*, 281–286.

Berg, J., Hung, Y. P., & Yellen, G. (2009). A genetically encoded fluorescent reporter of ATP:ADP ratio. *Nature Methods*, *6*, 161–166.

Berglund, K., Schleich, W., Krieger, P., Loo, L. S., Wang, D., Cant, N. B., et al. (2006). Imaging synaptic inhibition in transgenic mice expressing the chloride indicator, Clomeleon. *Brain Cell Biology*, *35*, 207–228.

Berglund, K., Schleich, W., Wang, H., Feng, G., Hall, W. C., Kuner, T., et al. (2008). Imaging synaptic inhibition throughout the brain via genetically targeted Clomeleon. *Brain Cell Biology*, *36*, 101–118.

Billups, D., Billups, B., Challiss, R. A., & Nahorski, S. R. (2006). Modulation of Gq-protein-coupled inositol trisphosphate and Ca2+ signaling by the membrane potential. *The Journal of Neuroscience*, *26*, 9983–9995.

Bittner, C. X., Loaiza, A., Ruminot, I., Larenas, V., Sotelo-Hitschfeld, T., Gutierrez, R., et al. (2010). High resolution measurement of the glycolytic rate. *Frontiers in Neuroenergetics*, *2*, 26.

Bittner, C. X., Valdebenito, R., Ruminot, I., Loaiza, A., Larenas, V., Sotelo-Hitschfeld, T., et al. (2011). Fast and reversible stimulation of astrocytic glycolysis by k+ and a delayed and persistent effect of glutamate. *The Journal of Neuroscience*, *31*, 4709–4713.

Bizzarri, R., Serresi, M., Luin, S., & Beltram, F. (2009). Green fluorescent protein based pH indicators for *in vivo* use: A review. *Analytical and Bioanalytical Chemistry*, *393*, 1107–1122.

Bunemann, M., Frank, M., & Lohse, M. J. (2003). Gi protein activation in intact cells involves subunit rearrangement rather than dissociation. *Proceedings of the National Academy of Sciences of the United States of America, 100*, 16077–16082.

Butcher, A. J., Torrecilla, I., Young, K. W., Kong, K. C., Mistry, S. C., Bottrill, A. R., et al. (2009). N-methyl-D-aspartate receptors mediate the phosphorylation and desensitization of muscarinic receptors in cerebellar granule neurons. *The Journal of Biological Chemistry, 284*, 17147–17156.

Casey, J. R., Grinstein, S., & Orlowski, J. (2010). Sensors and regulators of intracellular pH. *Nature Reviews Molecular Cell Biology, 11*, 50–61.

Chesler, M., & Kaila, K. (1992). Modulation of pH by neuronal activity. *Trends in Neurosciences, 15*, 396–402.

Chudakov, D. M., Matz, M. V., Lukyanov, S., & Lukyanov, K. A. (2010). Fluorescent proteins and their applications in imaging living cells and tissues. *Physiological Reviews, 90*, 1103–1163.

Ciruela, F., Vilardaga, J. P., & Fernandez-Duenas, V. (2010). Lighting up multiprotein complexes: Lessons from GPCR oligomerization. *Trends in Biotechnology, 28*, 407–415.

Citro, S., Malik, S., Oestreich, E. A., Radeff-Huang, J., Kelley, G. G., Smrcka, A. V., et al. (2007). Phospholipase Cepsilon is a nexus for Rho and Rap-mediated G protein-coupled receptor-induced astrocyte proliferation. *Proceedings of the National Academy of Sciences of the United States of America, 104*, 15543–15548.

Comley, L. H., Wishart, T. M., Baxter, B., Murray, L. M., Nimmo, A., Thomson, D., et al. (2011). Induction of cell stress in neurons from transgenic mice expressing yellow fluorescent protein: Implications for neurodegeneration research. *PLoS One, 6*, e17639.

Day, R. N., & Davidson, M. W. (2009). The fluorescent protein palette: Tools for cellular imaging. *Chemical Society Reviews, 38*, 2887–2921.

Deuschle, K., Okumoto, S., Fehr, M., Looger, L. L., Kozhukh, L., & Frommer, W. B. (2005). Construction and optimization of a family of genetically encoded metabolite sensors by semirational protein engineering. *Protein Science, 14*, 2304–2314.

Dickson, B. J. (2001). Rho GTPases in growth cone guidance. *Current Opinion in Neurobiology, 11*, 103–110.

DiPilato, L. M., Cheng, X., & Zhang, J. (2004). Fluorescent indicators of cAMP and Epac activation reveal differential dynamics of cAMP signaling within discrete subcellular compartments. *Proceedings of the National Academy of Sciences of the United States of America, 101*, 16513–16518.

Dittmer, P. J., Miranda, J. G., Gorski, J. A., & Palmer, A. E. (2009). Genetically encoded sensors to elucidate spatial distribution of cellular zinc. *The Journal of Biological Chemistry, 284*, 16289–16297.

Dooley, C. T., Dore, T. M., Hanson, G. T., Jackson, W. C., Remington, S. J., & Tsien, R. Y. (2004). Imaging dynamic redox changes in mammalian cells with green fluorescent protein indicators. *The Journal of Biological Chemistry, 279*, 22284–22293.

Drobizhev, M., Makarov, N. S., Tillo, S. E., Hughes, T. E., & Rebane, A. (2011). Two-photon absorption properties of fluorescent proteins. *Nature Methods, 8*, 393–399.

Duebel, J., Haverkamp, S., Schleich, W., Feng, G., Augustine, G. J., Kuner, T., et al. (2006). Two-photon imaging reveals somatodendritic chloride gradient in retinal ON-type bipolar cells expressing the biosensor Clomeleon. *Neuron, 49*, 81–94.

Dulla, C., Tani, H., Okumoto, S., Frommer, W. B., Reimer, R. J., & Huguenard, J. R. (2008). Imaging of glutamate in brain slices using FRET sensors. *Journal of Neuroscience Methods, 168*, 306–319.

Dunn, T. A., Wang, C. T., Colicos, M. A., Zaccolo, M., DiPilato, L. M., Zhang, J., et al. (2006). Imaging of cAMP levels and protein kinase A activity reveals that retinal waves drive oscillations in second-messenger cascades. *The Journal of Neuroscience, 26*, 12807–12815.

Dzhala, V. I., Kuchibhotla, K. V., Glykys, J. C., Kahle, K. T., Swiercz, W. B., Feng, G., et al. (2010). Progressive NKCC1-dependent neuronal chloride accumulation during neonatal seizures. *The Journal of Neuroscience, 30*, 11745–11761.

Falkenburger, B. H., Jensen, J. B., & Hille, B. (2010). Kinetics of M1 muscarinic receptor and G protein signaling to phospholipase C in living cells. *The Journal of General Physiology, 135*, 81–97.

Fehr, M., Lalonde, S., Lager, I., Wolff, M. W., & Frommer, W. B. (2003). *In vivo* imaging of the dynamics of glucose uptake in the cytosol of COS-7 cells by fluorescent nanosensors. *The Journal of Biological Chemistry, 278*, 19127–19133.

Feng, G., Mellor, R. H., Bernstein, M., Keller-Peck, C., Nguyen, Q. T., Wallace, M., et al. (2000). Imaging neuronal subsets in transgenic mice expressing multiple spectral variants of GFP. *Neuron, 28*, 41–51.

Fernandez-Alfonso, T., & Ryan, T. A. (2004). The kinetics of synaptic vesicle pool depletion at CNS synaptic terminals. *Neuron, 41*, 943–953.

Frederickson, C. J., Koh, J. Y., & Bush, A. I. (2005). The neurobiology of zinc in health and disease. *Nature Reviews Neuroscience, 6*, 449–462.

Frommer, W. B., Davidson, M. W., & Campbell, R. E. (2009). Genetically encoded biosensors based on engineered fluorescent proteins. *Chemical Society Reviews, 38*, 2833–2841.

Gandhi, S. P., & Stevens, C. F. (2003). Three modes of synaptic vesicular recycling revealed by single-vesicle imaging. *Nature, 423*, 607–613.

Gerencser, A. A., Mark, K. A., Hubbard, A. E., Divakaruni, A. S., Mehrabian, Z., Nicholls, D. G., et al. (2009). Real-time visualization of cytoplasmic calpain activation and calcium deregulation in acute glutamate excitotoxicity. *Journal of Neurochemistry*, *110*, 990–1004.

Gervasi, N., Hepp, R., Tricoire, L., Zhang, J., Lambolez, B., Paupardin-Tritsch, D., et al. (2007). Dynamics of protein kinase A signaling at the membrane, in the cytosol, and in the nucleus of neurons in mouse brain slices. *The Journal of Neuroscience*, *27*, 2744–2750.

Glykys, J., Dzhala, V. I., Kuchibhotla, K. V., Feng, G., Kuner, T., Augustine, G., et al. (2009). Differences in cortical versus subcortical GABAergic signaling: A candidate mechanism of electroclinical uncoupling of neonatal seizures. *Neuron*, *63*, 657–672.

Goto, A., Hoshino, M., Matsuda, M., & Nakamura, T. (2011). Phosphorylation of STEF/Tiam2 by protein kinase a is critical for Rac1 activation and neurite outgrowth in dibutyryl cAMP-treated PC12D cells. *Molecular Biology of the Cell*, *22*, 1780–1790.

Granseth, B., Odermatt, B., Royle, S. J., & Lagnado, L. (2006). Clathrin-mediated endocytosis is the dominant mechanism of vesicle retrieval at hippocampal synapses. *Neuron*, *51*, 773–786.

Griesbeck, O., Baird, G. S., Campbell, R. E., Zacharias, D. A., & Tsien, R. Y. (2001). Reducing the environmental sensitivity of yellow fluorescent protein. Mechanism and applications. *The Journal of Biological Chemistry*, *276*, 29188–29194.

Gutscher, M., Pauleau, A. L., Marty, L., Brach, T., Wabnitz, G. H., Samstag, Y., et al. (2008). Real-time imaging of the intracellular glutathione redox potential. *Nature Methods*, *5*, 553–559.

Gutscher, M., Sobotta, M. C., Wabnitz, G. H., Ballikaya, S., Meyer, A. J., Samstag, Y., et al. (2009). Proximity-based protein thiol oxidation by H2O2-scavenging peroxidases. *The Journal of Biological Chemistry*, *284*, 31532–31540.

Guzman, J. N., Sanchez-Padilla, J., Wokosin, D., Kondapalli, J., Ilijic, E., Schumacker, P. T., et al. (2010). Oxidant stress evoked by pacemaking in dopaminergic neurons is attenuated by DJ-1. *Nature*, *468*, 696–700.

Hanazono, Y., Yu, J. M., Dunbar, C. E., & Emmons, R. V. (1997). Green fluorescent protein retroviral vectors: Low titer and high recombination frequency suggest a selective disadvantage. *Human Gene Therapy*, *8*, 1313–1319.

Hanson, G. T., Aggeler, R., Oglesbee, D., Cannon, M., Capaldi, R. A., Tsien, R. Y., et al. (2004). Investigating mitochondrial redox potential with redox-sensitive green fluorescent protein indicators. *The Journal of Biological Chemistry*, *279*, 13044–13053.

Hanson, G. T., McAnaney, T. B., Park, E. S., Rendell, M. E., Yarbrough, D. K., Chu, S., et al. (2002). Green fluorescent protein variants as ratiometric dual emission pH sensors. 1. Structural characterization and preliminary application. *Biochemistry*, *41*, 15477–15488.

Hein, P., Frank, M., Hoffmann, C., Lohse, M. J., & Bunemann, M. (2005). Dynamics of receptor/G protein coupling in living cells. *The EMBO Journal*, *24*, 4106–4114.

Hemmings, H. C., Jr., Yan, W., Westphalen, R. I., & Ryan, T. A. (2005). The general anesthetic isoflurane depresses synaptic vesicle exocytosis. *Molecular Pharmacology*, *67*, 1591–1599.

Hires, S. A., Zhu, Y., & Tsien, R. Y. (2008). Optical measurement of synaptic glutamate spillover and reuptake by linker optimized glutamate-sensitive fluorescent reporters. *Proceedings of the National Academy of Sciences of the United States of America*, *105*, 4411–4416.

Hirose, K., Kadowaki, S., Tanabe, M., Takeshima, H., & Iino, M. (1999). Spatiotemporal dynamics of inositol 1,4,5-trisphosphate that underlies complex Ca2+ mobilization patterns. *Science*, *284*, 1527–1530.

Hommers, L. G., Klenk, C., Dees, C., & Bunemann, M. (2010). G proteins in reverse mode: Receptor-mediated GTP release inhibits G protein and effector function. *The Journal of Biological Chemistry*, *285*, 8227–8233.

Honda, A., Adams, S. R., Sawyer, C. L., Lev-Ram, V., Tsien, R. Y., & Dostmann, W. R. (2001). Spatiotemporal dynamics of guanosine 3',5'-cyclic monophosphate revealed by a genetically encoded, fluorescent indicator. *Proceedings of the National Academy of Sciences of the United States of America*, *98*, 2437–2442.

Horowitz, L. F., Hirdes, W., Suh, B. C., Hilgemann, D. W., Mackie, K., & Hille, B. (2005). Phospholipase C in living cells: Activation, inhibition, Ca2+ requirement, and regulation of M current. *The Journal of General Physiology*, *126*, 243–262.

Huang, W. Y., Aramburu, J., Douglas, P. S., & Izumo, S. (2000). Transgenic expression of green fluorescence protein can cause dilated cardiomyopathy. *Nature Medicine*, *6*, 482–483.

Imamura, H., Nhat, K. P., Togawa, H., Saito, K., Iino, R., Kato-Yamada, Y., et al. (2009). Visualization of ATP levels inside single living cells with fluorescence resonance energy transfer-based genetically encoded indicators. *Proceedings of the National Academy of Sciences of the United States of America*, *106*, 15651–15656.

Itoh, R. E., Kurokawa, K., Ohba, Y., Yoshizaki, H., Mochizuki, N., & Matsuda, M. (2002). Activation of rac and cdc42 video imaged by fluorescent resonance energy transfer-based single-molecule probes in the membrane of living cells. *Molecular and Cellular Biology*, *22*, 6582–6591.

Janetopoulos, C., Jin, T., & Devreotes, P. (2001). Receptor-mediated activation of heterotrimeric G-proteins in living cells. *Science*, *291*, 2408–2411.

Jayaraman, S., Haggie, P., Wachter, R. M., Remington, S. J., & Verkman, A. S. (2000). Mechanism and cellular applications of a green fluorescent protein-based halide sensor. *The Journal of Biological Chemistry*, *275*, 6047–6050.

Jensen, J. B., Lyssand, J. S., Hague, C., & Hille, B. (2009). Fluorescence changes reveal kinetic steps of muscarinic receptor-mediated modulation of phosphoinositides and Kv7.2/7.3 K+ channels. *The Journal of General Physiology*, *133*, 347–359.

Klein, R. L., Dayton, R. D., Leidenheimer, N. J., Jansen, K., Golde, T. E., & Zweig, R. M. (2006). Efficient neuronal gene transfer with AAV8 leads to neurotoxic levels of tau or green fluorescent proteins. *Molecular Therapy*, *13*, 517–527.

Knopfel, T., Lin, M. Z., Levskaya, A., Tian, L., Lin, J. Y., & Boyden, E. S. (2010). Toward the second generation of optogenetic tools. *The Journal of Neuroscience*, *30*, 14998–15004.

Kraynov, V. S., Chamberlain, C., Bokoch, G. M., Schwartz, M. A., Slabaugh, S., & Hahn, K. M. (2000). Localized Rac activation dynamics visualized in living cells. *Science*, *290*, 333–337.

Krestel, H. E., Mihaljevic, A. L., Hoffman, D. A., & Schneider, A. (2004). Neuronal co-expression of EGFP and beta-galactosidase in mice causes neuropathology and premature death. *Neurobiology of Disease*, *17*, 310–318.

Kumar, V., Jong, Y. J., & O'Malley, K. L. (2008). Activated nuclear metabotropic glutamate receptor mGlu5 couples to nuclear Gq/11 proteins to generate inositol 1,4,5-trisphosphate-mediated nuclear Ca2+ release. *The Journal of Biological Chemistry*, *283*, 14072–14083.

Kuner, T., & Augustine, G. J. (2000). A genetically encoded ratiometric indicator for chloride: Capturing chloride transients in cultured hippocampal neurons. *Neuron*, *27*, 447–459.

Kunishima, N., Shimada, Y., Tsuji, Y., Sato, T., Yamamoto, M., Kumasaka, T., et al. (2000). Structural basis of glutamate recognition by a dimeric metabotropic glutamate receptor. *Nature*, *407*, 971–977.

Lee, S., Yoon, B. E., Berglund, K., Oh, S. J., Park, H., Shin, H. S., et al. (2010). Channel-mediated tonic GABA release from glia. *Science*, *330*, 790–796.

Li, Z., Burrone, J., Tyler, W. J., Hartman, K. N., Albeanu, D. F., & Murthy, V. N. (2005). Synaptic vesicle recycling studied in transgenic mice expressing synaptopHluorin. *Proceedings of the National Academy of Sciences of the United States of America*, *102*, 6131–6136.

Liu, H. S., Jan, M. S., Chou, C. K., Chen, P. H., & Ke, N. J. (1999). Is green fluorescent protein toxic to the living cells? *Biochemical and Biophysical Research Communications*, *260*, 712–717.

Lohse, M. J., Bunemann, M., Hoffmann, C., Vilardaga, J. P., & Nikolaev, V. O. (2007). Monitoring receptor signaling by intramolecular FRET. *Current Opinion in Pharmacology*, *7*, 547–553.

Lohse, M. J., Nikolaev, V. O., Hein, P., Hoffmann, C., Vilardaga, J. P., & Bunemann, M. (2008). Optical techniques to analyze real-time activation and signaling of G-protein-coupled receptors. *Trends in Pharmacological Sciences*, *29*, 159–165.

Marcaggi, P., Mutoh, H., Dimitrov, D., Beato, M., & Knopfel, T. (2009). Optical measurement of mGluR1 conformational changes reveals fast activation, slow deactivation, and sensitization. *Proceedings of the National Academy of Sciences of the United States of America*, *106*, 11388–11393.

Markova, O., Mukhtarov, M., Real, E., Jacob, Y., & Bregestovski, P. (2008). Genetically encoded chloride indicator with improved sensitivity. *Journal of Neuroscience Methods*, *170*, 67–76.

Matsushita, S., Nakata, H., Kubo, Y., & Tateyama, M. (2010). Ligand-induced rearrangements of the GABA(B) receptor revealed by fluorescence resonance energy transfer. *The Journal of Biological Chemistry*, *285*, 10291–10299.

Matsu-ura, T., Michikawa, T., Inoue, T., Miyawaki, A., Yoshida, M., & Mikoshiba, K. (2006). Cytosolic inositol 1,4,5-trisphosphate dynamics during intracellular calcium oscillations in living cells. *The Journal of Cell Biology*, *173*, 755–765.

McGann, J. P., Pirez, N., Gainey, M. A., Muratore, C., Elias, A. S., & Wachowiak, M. (2005). Odorant representations are modulated by intra- but not interglomerular presynaptic inhibition of olfactory sensory neurons. *Neuron*, *48*, 1039–1053.

Mercado, J., Gordon-Shaag, A., Zagotta, W. N., & Gordon, S. E. (2010). Ca2+-dependent desensitization of TRPV2 channels is mediated by hydrolysis of phosphatidylinositol 4,5-bisphosphate. *The Journal of Neuroscience*, *30*, 13338–13347.

Merksamer, P. I., Trusina, A., & Papa, F. R. (2008). Real-time redox measurements during endoplasmic reticulum stress reveal interlinked protein folding functions. *Cell*, *135*, 933–947.

Metzger, F., Repunte-Canonigo, V., Matsushita, S., Akemann, W., Diez-Garcia, J., Ho, C. S., et al. (2002). Transgenic mice expressing a pH and Cl-sensing yellow-fluorescent protein under the control of a potassium channel promoter. *The European Journal of Neuroscience*, *15*, 40–50.

Meyer, A. J., & Dick, T. P. (2010). Fluorescent protein-based redox probes. *Antioxidants & Redox Signaling*, *13*, 621–650.

Micheva, K. D., Holz, R. W., & Smith, S. J. (2001). Regulation of presynaptic phosphatidylinositol 4,5-biphosphate by neuronal activity. *The Journal of Cell Biology*, *154*, 355–368.

Miesenbock, G., De Angelis, D. A., & Rothman, J. E. (1998). Visualizing secretion and synaptic transmission with pH-sensitive green fluorescent proteins. *Nature*, *394*, 192–195.

Miyawaki, A., Llopis, J., Heim, R., McCaffery, J. M., Adams, J. A., Ikura, M., et al. (1997). Fluorescent indicators for Ca2+ based on green fluorescent proteins and calmodulin. *Nature*, *388*, 882–887.

Mochizuki, N., Yamashita, S., Kurokawa, K., Ohba, Y., Nagai, T., Miyawaki, A., et al. (2001). Spatio-temporal images of growth-factor-induced activation of Ras and Rap1. *Nature*, *411*, 1065–1068.

Mukhtarov, M., Markova, O., Real, E., Jacob, Y., Buldakova, S., & Bregestovski, P. (2008). Monitoring of chloride and activity of glycine receptor channels using genetically encoded fluorescent sensors. *Philosophical Transactions. Series A, Mathematical, Physical, and Engineering Sciences*, *366*, 3445–3462.

Muller, F. L. (2009). A critical evaluation of cpYFP as a probe for superoxide. *Free Radical Biology & Medicine*, *47*, 1779–1780.

Murakoshi, H., Lee, S. J., & Yasuda, R. (2008). Highly sensitive and quantitative FRET-FLIM imaging in single dendritic spines using improved non-radiative YFP. *Brain Cell Biology*, *36*, 31–42.

Nagai, Y., Miyazaki, M., Aoki, R., Zama, T., Inouye, S., Hirose, K., et al. (2000). A fluorescent indicator for visualizing cAMP-induced phosphorylation *in vivo*. *Nature Biotechnology*, *18*, 313–316.

Nagai, T., Sawano, A., Park, E. S., & Miyawaki, A. (2001). Circularly permuted green fluorescent proteins engineered to sense Ca2+. *Proceedings of the National Academy of Sciences of the United States of America*, *98*, 3197–3202.

Nagai, T., Yamada, S., Tominaga, T., Ichikawa, M., & Miyawaki, A. (2004). Expanded dynamic range of fluorescent indicators for Ca(2+) by circularly permuted yellow fluorescent proteins. *Proceedings of the National Academy of Sciences of the United States of America*, *101*, 10554–10559.

Nakai, J., Ohkura, M., & Imoto, K. (2001). A high signal-to-noise Ca(2+) probe composed of a single green fluorescent protein. *Nature Biotechnology*, *19*, 137–141.

Nakamura, T., Aoki, K., & Matsuda, M. (2008). FRET imaging and in silico simulation: Analysis of the signaling network of nerve growth factor-induced neuritogenesis. *Brain Cell Biology*, *36*, 19–30.

Nausch, L. W., Ledoux, J., Bonev, A. D., Nelson, M. T., & Dostmann, W. R. (2008). Differential patterning of cGMP in vascular smooth muscle cells revealed by single GFP-linked biosensors. *Proceedings of the National Academy of Sciences of the United States of America*, *105*, 365–370.

Nelson, C. P., Nahorski, S. R., & Challiss, R. A. (2008). Temporal profiling of changes in phosphatidylinositol 4,5-bisphosphate, inositol 1,4,5-trisphosphate and diacylglycerol allows comprehensive analysis of phospholipase C-initiated signalling in single neurons. *Journal of Neurochemistry*, *107*, 602–615.

Nemani, V. M., Lu, W., Berge, V., Nakamura, K., Onoa, B., Lee, M. K., et al. (2010). Increased expression of alpha-synuclein reduces neurotransmitter release by inhibiting synaptic vesicle reclustering after endocytosis. *Neuron*, *65*, 66–79.

Newman, R. H., Fosbrink, M. D., & Zhang, J. (2011). Genetically encodable fluorescent biosensors for tracking signaling dynamics in living cells. *Chemistry Review*, *111*, 3614–3666.

Nikolaev, V. O., Bunemann, M., Hein, L., Hannawacker, A., & Lohse, M. J. (2004). Novel single chain cAMP sensors for receptor-induced signal propagation. *The Journal of Biological Chemistry*, *279*, 37215–37218.

Nishioka, T., Aoki, K., Hikake, K., Yoshizaki, H., Kiyokawa, E., & Matsuda, M. (2008). Rapid turnover rate of phosphoinositides at the front of migrating MDCK cells. *Molecular Biology of the Cell*, *19*, 4213–4223.

Nobles, M., Benians, A., & Tinker, A. (2005). Heterotrimeric G proteins precouple with G protein-coupled receptors in living cells. *Proceedings of the National Academy of Sciences of the United States of America*, *102*, 18706–18711.

Oancea, E., Teruel, M. N., Quest, A. F., & Meyer, T. (1998). Green fluorescent protein (GFP)-tagged cysteine-rich domains from protein kinase C as fluorescent indicators for diacylglycerol signaling in living cells. *The Journal of Cell Biology*, *140*, 485–498.

Okubo, Y., Kakizawa, S., Hirose, K., & Iino, M. (2001). Visualization of IP(3) dynamics reveals a novel AMPA receptor-triggered IP(3) production pathway mediated by voltage-dependent Ca(2+) influx in Purkinje cells. *Neuron*, *32*, 113–122.

Okumoto, S., Looger, L. L., Micheva, K. D., Reimer, R. J., Smith, S. J., & Frommer, W. B. (2005). Detection of glutamate release from neurons by genetically encoded surface-displayed FRET nanosensors. *Proceedings of the National Academy of Sciences of the United States of America*, *102*, 8740–8745.

Okumoto, S., Takanaga, H., & Frommer, W. B. (2008). Quantitative imaging for discovery and assembly of the metabo-regulome. *New Phytologist*, *180*, 271–295.

Ostergaard, H., Henriksen, A., Hansen, F. G., & Winther, J. R. (2001). Shedding light on disulfide bond formation: Engineering a redox switch in green fluorescent protein. *The EMBO Journal*, *20*, 5853–5862.

Pacary, E., Heng, J., Azzarelli, R., Riou, P., Castro, D., Lebel-Potter, M., et al. (2011). Proneural transcription factors regulate different steps of cortical neuron migration through Rnd-mediated inhibition of RhoA signaling. *Neuron*, *69*, 1069–1084.

Palmer, A. E., Giacomello, M., Kortemme, T., Hires, S. A., Lev-Ram, V., Baker, D., et al. (2006). Ca2+ indicators based on computationally redesigned calmodulin-peptide pairs. *Chemical Biology*, *13*, 521–530.

Patterson, G. H., Piston, D. W., & Barisas, B. G. (2000). Forster distances between green fluorescent protein pairs. *Analytical Biochemistry*, *284*, 438–440.

Pertz, O., Hodgson, L., Klemke, R. L., & Hahn, K. M. (2006). Spatiotemporal dynamics of RhoA activity in migrating cells. *Nature*, *440*, 1069–1072.

Poburko, D., Santo-Domingo, J., & Demaurex, N. (2011). Dynamic regulation of the mitochondrial proton gradient during cytosolic calcium elevations. *The Journal of Biological Chemistry*, *286*, 11672–11684.

Pond, B. B., Berglund, K., Kuner, T., Feng, G., Augustine, G. J., & Schwartz-Bloom, R. D. (2006). The chloride transporter Na(+)-K(+)-Cl- cotransporter isoform-1 contributes to intracellular chloride increases after *in vitro* ischemia. *The Journal of Neuroscience*, *26*, 1396–1406.

Prebil, M., Chowdhury, H. H., Zorec, R., & Kreft, M. (2011). Changes in cytosolic glucose level in ATP stimulated live astrocytes. *Biochemical and Biophysical Research Communications*, *405*, 308–313.

Prebil, M., Vardjan, N., Jensen, J., Zorec, R., & Kreft, M. (2011). Dynamic monitoring of cytosolic glucose in single astrocytes. *Glia*, *59*, 903–913.

Qin, Y., Dittmer, P. J., Park, J. G., Jansen, K. B., & Palmer, A. E. (2011). Measuring steady-state and dynamic endoplasmic reticulum and Golgi Zn2+ with genetically encoded sensors. *Proceedings of the National Academy of Sciences of the United States of America*, *108*, 7351–7356.

Quinn, K. V., Behe, P., & Tinker, A. (2008). Monitoring changes in membrane phosphatidylinositol 4,5-bisphosphate in living cells using a domain from the transcription factor tubby. *The Journal of Physiology*, *586*, 2855–2871.

Remus, T. P., Zima, A. V., Bossuyt, J., Bare, D. J., Martin, J. L., Blatter, L. A., et al. (2006). Biosensors to measure inositol 1,4,5-trisphosphate concentration in living cells with spatiotemporal resolution. *The Journal of Biological Chemistry*, *281*, 608–616.

Rochais, F., Vilardaga, J. P., Nikolaev, V. O., Bunemann, M., Lohse, M. J., & Engelhardt, S. (2007). Real-time optical recording of beta1-adrenergic receptor activation reveals supersensitivity of the Arg389 variant to carvedilol. *The Journal of Clinical Investigation*, *117*, 229–235.

Romoser, V. A., Hinkle, P. M., & Persechini, A. (1997). Detection in living cells of Ca2+-dependent changes in the fluorescence emission of an indicator composed of two green fluorescent protein variants linked by a calmodulin-binding sequence. A new class of fluorescent indicators. *The Journal of Biological Chemistry*, *272*, 13270–13274.

Ryan, T. A. (2001). Presynaptic imaging techniques. *Current Opinion in Neurobiology*, *11*, 544–549.

Sabouri-Ghomi, M., Wu, Y., Hahn, K., & Danuser, G. (2008). Visualizing and quantifying adhesive signals. *Current Opinion in Cell Biology*, *20*, 541–550.

Saneyoshi, T., Wayman, G., Fortin, D., Davare, M., Hoshi, N., Nozaki, N., et al. (2008). Activity-dependent synaptogenesis: Regulation by a CaM-kinase kinase/CaM-kinase I/betaPIX signaling complex. *Neuron*, *57*, 94–107.

Sankaranarayanan, S., De Angelis, D., Rothman, J. E., & Ryan, T. A. (2000). The use of pHluorins for optical measurements of presynaptic activity. *Biophysical Journal*, *79*, 2199–2208.

Sankaranarayanan, S., & Ryan, T. A. (2001). Calcium accelerates endocytosis of vSNAREs at hippocampal synapses. *Nature Neuroscience*, *4*, 129–136.

Santagata, S., Boggon, T. J., Baird, C. L., Gomez, C. A., Zhao, J., Shan, W. S., et al. (2001). G-protein signaling through tubby proteins. *Science*, *292*, 2041–2050.

Sasaki, K., Sato, M., & Umezawa, Y. (2003). Fluorescent indicators for Akt/protein kinase B and dynamics of Akt activity visualized in living cells. *The Journal of Biological Chemistry*, *278*, 30945–30951.

Sato, M., Nakajima, T., Goto, M., & Umezawa, Y. (2006). Cell-based indicator to visualize picomolar dynamics of nitric oxide release from living cells. *Analytical Chemistry*, *78*, 8175–8182.

Sato, M., Ueda, Y., Shibuya, M., & Umezawa, Y. (2005). Locating inositol 1,4,5-trisphosphate in the nucleus and neuronal dendrites with genetically encoded fluorescent indicators. *Analytical Chemistry*, *77*, 4751–4758.

Schwarzlander, M., Logan, D. C., Fricker, M. D., & Sweetlove, L. (2011). The circularly permuted yellow fluorescent protein cpYFP that has been used as a superoxide probe is highly responsive to pH but not superoxide in mitochondria: Implications for the existence of superoxide 'flashes'. *The Biochemical Journal*, *437*, 381–387.

Shafer, O. T., Kim, D. J., Dunbar-Yaffe, R., Nikolaev, V. O., Lohse, M. J., & Taghert, P. H. (2008). Widespread receptivity to neuropeptide PDF throughout the neuronal circadian clock network of Drosophila revealed by real-time cyclic AMP imaging. *Neuron*, *58*, 223–237.

Shang, Y., Claridge-Chang, A., Sjulson, L., Pypaert, M., & Miesenbock, G. (2007). Excitatory local circuits and their implications for olfactory processing in the fly antennal lobe. *Cell*, *128*, 601–612.

Slemmer, J. E., Matsushita, S., De Zeeuw, C. I., Weber, J. T., & Knopfel, T. (2004). Glutamate-induced elevations in intracellular chloride concentration in hippocampal cell cultures derived from EYFP-expressing mice. *The European Journal of Neuroscience*, *19*, 2915–2922.

Srivastava, J., Barber, D. L., & Jacobson, M. P. (2007). Intracellular pH sensors: Design principles and functional significance. *Physiology (Bethesda, MD)*, *22*, 30–39.

Stauffer, T. P., Ahn, S., & Meyer, T. (1998). Receptor-induced transient reduction in plasma membrane PtdIns(4,5)P2 concentration monitored in living cells. *Current Biology*, *8*, 343–346.

Suh, B. C., Horowitz, L. F., Hirdes, W., Mackie, K., & Hille, B. (2004). Regulation of KCNQ2/KCNQ3 current by G protein cycling: The kinetics of receptor-mediated signaling by Gq. *The Journal of General Physiology*, *123*, 663–683.

Svoboda, K., & Yasuda, R. (2006). Principles of two-photon excitation microscopy and its applications to neuroscience. *Neuron*, *50*, 823–839.

Szentpetery, Z., Balla, A., Kim, Y. J., Lemmon, M. A., & Balla, T. (2009). Live cell imaging with protein domains capable of recognizing phosphatidylinositol 4,5-bisphosphate; a comparative study. *BMC Cell Biology*, *10*, 67.

Takanaga, H., Chaudhuri, B., & Frommer, W. B. (2008). GLUT1 and GLUT9 as major contributors to glucose influx in HepG2 cells identified by a high sensitivity intramolecular FRET glucose sensor. *Biochimica et Biophysica Acta*, *1778*, 1091–1099.

Tani, H., Dulla, C. G., Huguenard, J. R., & Reimer, R. J. (2010). Glutamine is required for persistent epileptiform activity in the disinhibited neocortical brain slice. *The Journal of Neuroscience*, *30*, 1288–1300.

Tanimura, A., Nezu, A., Morita, T., Turner, R. J., & Tojyo, Y. (2004). Fluorescent biosensor for quantitative real-time measurements of inositol 1,4,5-trisphosphate in single living cells. *The Journal of Biological Chemistry*, *279*, 38095–38098.

Tateyama, M., Abe, H., Nakata, H., Saito, O., & Kubo, Y. (2004). Ligand-induced rearrangement of the dimeric metabotropic glutamate receptor 1alpha. *Nature Structural & Molecular Biology*, *11*, 637–642.

Thorsen, T. S., Madsen, K. L., Rebola, N., Rathje, M., Anggono, V., Bach, A., et al. (2010). Identification of a small-molecule inhibitor of the PICK1 PDZ domain that inhibits hippocampal LTP and LTD. *Proceedings of the National Academy of Sciences of the United States of America*, *107*, 413–418.

Tsien, R. Y. (1998). The green fluorescent protein. *Annual Review of Biochemistry*, *67*, 509–544.

Tsou, P., Zheng, B., Hsu, C. H., Sasaki, A. T., & Cantley, L. C. (2011). A fluorescent reporter of AMPK activity and cellular energy stress. *Cell Metabolism*, *13*, 476–486.

Tuffy, L. P., Concannon, C. G., D'Orsi, B., King, M. A., Woods, I., Huber, H. J., et al. (2010). Characterization of Puma-dependent and Puma-independent neuronal cell death pathways following prolonged proteasomal inhibition. *Molecular and Cellular Biology*, *30*, 5484–5501.

van der Wal, J., Habets, R., Varnai, P., Balla, T., & Jalink, K. (2001). Monitoring agonist-induced phospholipase C activation in live cells by fluorescence resonance energy transfer. *The Journal of Biological Chemistry*, *276*, 15337–15344.

Vanderklish, P. W., Krushel, L. A., Holst, B. H., Gally, J. A., Crossin, K. L., & Edelman, G. M. (2000). Marking synaptic activity in dendritic spines with a calpain substrate exhibiting fluorescence resonance energy transfer. *Proceedings of the National Academy of Sciences of the United States of America*, *97*, 2253–2258.

Varnai, P., & Balla, T. (1998). Visualization of phosphoinositides that bind pleckstrin homology domains: Calcium- and agonist-induced dynamic changes and relationship to myo-[3H]inositol-labeled phosphoinositide pools. *The Journal of Cell Biology*, *143*, 501–510.

Varnai, P., & Balla, T. (2006). Live cell imaging of phosphoinositide dynamics with fluorescent protein domains. *Biochimica et Biophysica Acta*, *1761*, 957–967.

Vesce, S., Jekabsons, M. B., Johnson-Cadwell, L. I., & Nicholls, D. G. (2005). Acute glutathione depletion restricts mitochondrial ATP export in cerebellar granule neurons. *The Journal of Biological Chemistry*, *280*, 38720–38728.

Vilardaga, J. P., Bunemann, M., Krasel, C., Castro, M., & Lohse, M. J. (2003). Measurement of the millisecond activation switch of G protein-coupled receptors in living cells. *Nature Biotechnology*, *21*, 807–812.

Vincent, P., Gervasi, N., & Zhang, J. (2008). Real-time monitoring of cyclic nucleotide signaling in neurons using genetically encoded FRET probes. *Brain Cell Biology*, *36*, 3–17.

Vinkenborg, J. L., Koay, M. S., & Merkx, M. (2010). Fluorescent imaging of transition metal homeostasis using genetically encoded sensors. *Current Opinion in Chemical Biology*, *14*, 231–237.

Violin, J. D., Zhang, J., Tsien, R. Y., & Newton, A. C. (2003). A genetically encoded fluorescent reporter reveals oscillatory phosphorylation by protein kinase C. *The Journal of Cell Biology*, *161*, 899–909.

Wachter, R. M., Elsliger, M. A., Kallio, K., Hanson, G. T., & Remington, S. J. (1998). Structural basis of spectral shifts in the yellow-emission variants of green fluorescent protein. *Structure*, *6*, 1267–1277.

Wachter, R. M., Yarbrough, D., Kallio, K., & Remington, S. J. (2000). Crystallographic and energetic analysis of binding of selected anions to the yellow variants of green fluorescent protein. *Journal of Molecular Biology*, *301*, 157–171.

Wang, W., Fang, H., Groom, L., Cheng, A., Zhang, W., Liu, J., et al. (2008). Superoxide flashes in single mitochondria. *Cell*, *134*, 279–290.

Wang, Q., Shui, B., Kotlikoff, M. I., & Sondermann, H. (2008). Structural basis for calcium sensing by GCaMP2. *Structure*, *16*, 1817–1827.

Ward, M. W., Rehm, M., Duessmann, H., Kacmar, S., Concannon, C. G., & Prehn, J. H. (2006). Real time single cell analysis of Bid cleavage and Bid translocation during caspase-dependent and neuronal caspase-independent apoptosis. *The Journal of Biological Chemistry*, *281*, 5837–5844.

Wardman, P. (2007). Fluorescent and luminescent probes for measurement of oxidative and nitrosative species in cells and tissues: Progress, pitfalls, and prospects. *Free Radical Biology & Medicine*, *43*, 995–1022.

Waseem, T., Mukhtarov, M., Buldakova, S., Medina, I., & Bregestovski, P. (2010). Genetically encoded Cl-sensor as a tool for monitoring of Cl-dependent processes in small neuronal compartments. *Journal of Neuroscience Methods*, *193*, 14–23.

Willemse, M., Janssen, E., de Lange, F., Wieringa, B., & Fransen, J. (2007). ATP and FRET—A cautionary note. *Nature Biotechnology*, *25*, 170–172.

Willets, J. M., Nelson, C. P., Nahorski, S. R., & Challiss, R. A. (2007). The regulation of M1 muscarinic acetylcholine receptor desensitization by synaptic activity in cultured hippocampal neurons. *Journal of Neurochemistry*, *103*, 2268–2280.

Willoughby, D., & Cooper, D. M. (2008). Live-cell imaging of cAMP dynamics. *Nature Methods*, *5*, 29–36.

Winterbourn, C. C. (2008). Reconciling the chemistry and biology of reactive oxygen species. *Nature Chemical Biology*, *4*, 278–286.

Xu, X., Gerard, A. L., Huang, B. C., Anderson, D. C., Payan, D. G., & Luo, Y. (1998). Detection of programmed cell death using fluorescence energy transfer. *Nucleic Acids Research*, *26*, 2034–2035.

Xu, C., Watras, J., & Loew, L. M. (2003). Kinetic analysis of receptor-activated phosphoinositide turnover. *The Journal of Cell Biology*, *161*, 779–791.

Yan, H. D., Villalobos, C., & Andrade, R. (2009). TRPC channels mediate a muscarinic receptor-induced afterdepolarization in cerebral cortex. *The Journal of Neuroscience*, *29*, 10038–10046.

Yoshizaki, H., Ohba, Y., Kurokawa, K., Itoh, R. E., Nakamura, T., Mochizuki, N., et al. (2003). Activity of Rho-family GTPases during cell division as visualized with FRET-based probes. *The Journal of Cell Biology*, *162*, 223–232.

Yu, D., Ponomarev, A., & Davis, R. L. (2004). Altered representation of the spatial code for odors after olfactory classical conditioning; memory trace formation by synaptic recruitment. *Neuron*, *42*, 437–449.

Zaccolo, M., De Giorgi, F., Cho, C. Y., Feng, L., Knapp, T., Negulescu, P. A., et al. (2000). A genetically encoded, fluorescent indicator for cyclic AMP in living cells. *Nature Cell Biology*, *2*, 25–29.

Zadran, S., Jourdi, H., Rostamiani, K., Qin, Q., Bi, X., & Baudry, M. (2010). Brain-derived neurotrophic factor and epidermal growth factor activate neuronal m-calpain via mitogen-activated protein kinase-dependent phosphorylation. *The Journal of Neuroscience*, *30*, 1086–1095.

Zhang, Q., Li, Y., & Tsien, R. W. (2009). The dynamic control of kiss-and-run and vesicular reuse probed with single nanoparticles. *Science*, *323*, 1448–1453.

Zhang, J., Ma, Y., Taylor, S. S., & Tsien, R. Y. (2001). Genetically encoded reporters of protein kinase A activity reveal impact of substrate tethering. *Proceedings of the National Academy of Sciences of the United States of America*, *98*, 14997–15002.

Zhang, H., & Macara, I. G. (2008). The PAR-6 polarity protein regulates dendritic spine morphogenesis through p190 RhoGAP and the Rho GTPase. *Developmental Cell*, *14*, 216–226.

Zhang, J., Wang, Y., Chi, Z., Keuss, M. J., Pai, Y. M., Kang, H. C., et al. (2011). The AAA(+) ATPase thorase regulates AMPA receptor-dependent synaptic plasticity and behavior. *Cell*, *145*, 284–299.

Zuo, Y., Lubischer, J. L., Kang, H., Tian, L., Mikesh, M., Marks, A., et al. (2004). Fluorescent proteins expressed in mouse transgenic lines mark subsets of glia, neurons, macrophages, and dendritic cells for vital examination. *The Journal of Neuroscience*, *24*, 10999–11009.

Subject Index

Note: Page numbers followed by "*f*" indicate figures, and "*t*" indicate tables.

Other volumes in PROGRESS IN BRAIN RESEARCH

Volume 149: Cortical Function: A View from the Thalamus, by V.A. Casagrande, R.W. Guillery and S.M. Sherman (Eds.) – 2005 ISBN 0-444-51679-4.
Volume 150: The Boundaries of Consciousness: Neurobiology and Neuropathology, by Steven Laureys (Ed.) – 2005, ISBN 0-444-51851-7.
Volume 151: Neuroanatomy of the Oculomotor System, by J.A. Büttner-Ennever (Ed.) – 2006, ISBN 0-444-51696-4.
Volume 152: Autonomic Dysfunction after Spinal Cord Injury, by L.C. Weaver and C. Polosa (Eds.) – 2006, ISBN 0-444-51925-4.
Volume 153: Hypothalamic Integration of Energy Metabolism, by A. Kalsbeek, E. Fliers, M.A. Hofman, D.F. Swaab, E.J.W. Van Someren and R.M. Buijs (Eds.) – 2006, ISBN 978-0-444-52261-0.
Volume 154: Visual Perception, Part 1, Fundamentals of Vision: Low and Mid-Level Processes in Perception, by S. Martinez-Conde, S.L. Macknik, L.M. Martinez, J.M. Alonso and P.U. Tse (Eds.) – 2006, ISBN 978-0-444-52966-4.
Volume 155: Visual Perception, Part 2, Fundamentals of Awareness, Multi-Sensory Integration and High-Order Perception, by S. Martinez-Conde, S.L. Macknik, L.M. Martinez, J.M. Alonso and P.U. Tse (Eds.) – 2006, ISBN 978-0-444-51927-6.
Volume 156: Understanding Emotions, by S. Anders, G. Ende, M. Junghofer, J. Kissler and D. Wildgruber (Eds.) – 2006, ISBN 978-0-444-52182-8.
Volume 157: Reprogramming of the Brain, by A.R. Møller (Ed.) – 2006, ISBN 978-0-444-51602-2.
Volume 158: Functional Genomics and Proteomics in the Clinical Neurosciences, by S.E. Hemby and S. Bahn (Eds.) – 2006, ISBN 978-0-444-51853-8.
Volume 159: Event-Related Dynamics of Brain Oscillations, by C. Neuper and W. Klimesch (Eds.) – 2006, ISBN 978-0-444-52183-5.
Volume 160: GABA and the Basal Ganglia: From Molecules to Systems, by J.M. Tepper, E.D. Abercrombie and J.P. Bolam (Eds.) – 2007, ISBN 978-0-444-52184-2.
Volume 161: Neurotrauma: New Insights into Pathology and Treatment, by J.T. Weber and A.I.R. Maas (Eds.) – 2007, ISBN 978-0-444-53017-2.
Volume 162: Neurobiology of Hyperthermia, by H.S. Sharma (Ed.) – 2007, ISBN 978-0-444-51926-9.
Volume 163: The Dentate Gyrus: A Comprehensive Guide to Structure, Function, and Clinical Implications, by H.E. Scharfman (Ed.) – 2007, ISBN 978-0-444-53015-8.
Volume 164: From Action to Cognition, by C. von Hofsten and K. Rosander (Eds.) – 2007, ISBN 978-0-444-53016-5.
Volume 165: Computational Neuroscience: Theoretical Insights into Brain Function, by P. Cisek, T. Drew and J.F. Kalaska (Eds.) – 2007, ISBN 978-0-444-52823-0.
Volume 166: Tinnitus: Pathophysiology and Treatment, by B. Langguth, G. Hajak, T. Kleinjung, A. Cacace and A.R. Møller (Eds.) – 2007, ISBN 978-0-444-53167-4.
Volume 167: Stress Hormones and Post Traumatic Stress Disorder: Basic Studies and Clinical Perspectives, by E.R. de Kloet, M.S. Oitzl and E. Vermetten (Eds.) – 2008, ISBN 978-0-444-53140-7.
Volume 168: Models of Brain and Mind: Physical, Computational and Psychological Approaches, by R. Banerjee and B.K. Chakrabarti (Eds.) – 2008, ISBN 978-0-444-53050-9.
Volume 169: Essence of Memory, by W.S. Sossin, J.-C. Lacaille, V.F. Castellucci and S. Belleville (Eds.) – 2008, ISBN 978-0-444-53164-3.
Volume 170: Advances in Vasopressin and Oxytocin – From Genes to Behaviour to Disease, by I.D. Neumann and R. Landgraf (Eds.) – 2008, ISBN 978-0-444-53201-5.
Volume 171: Using Eye Movements as an Experimental Probe of Brain Function—A Symposium in Honor of Jean Büttner-Ennever, by Christopher Kennard and R. John Leigh (Eds.) – 2008, ISBN 978-0-444-53163-6.
Volume 172: Serotonin–Dopamine Interaction: Experimental Evidence and Therapeutic Relevance, by Giuseppe Di Giovanni, Vincenzo Di Matteo and Ennio Esposito (Eds.) – 2008, ISBN 978-0-444-53235-0.
Volume 173: Glaucoma: An Open Window to Neurodegeneration and Neuroprotection, by Carlo Nucci, Neville N. Osborne, Giacinto Bagetta and Luciano Cerulli (Eds.) – 2008, ISBN 978-0-444-53256-5.
Volume 174: Mind and Motion: The Bidirectional Link Between Thought and Action, by Markus Raab, Joseph G. Johnson and Hauke R. Heekeren (Eds.) – 2009, 978-0-444-53356-2.
Volume 175: Neurotherapy: Progress in Restorative Neuroscience and Neurology — Proceedings of the 25th International Summer School of Brain Research, held at the Royal Netherlands Academy of Arts and Sciences, Amsterdam, The Netherlands, August 25–28, 2008, by J. Verhaagen, E.M. Hol, I. Huitinga, J. Wijnholds, A.A. Bergen, G.J. Boer and D.F. Swaab (Eds.) –2009, ISBN 978-0-12-374511-8.
Volume 176: Attention, by Narayanan Srinivasan (Ed.) – 2009, ISBN 978-0-444-53426-2.
Volume 177: Coma Science: Clinical and Ethical Implications, by Steven Laureys, Nicholas D. Schiff and Adrian M. Owen (Eds.) – 2009, 978-0-444-53432-3.
Volume 178: Cultural Neuroscience: Cultural Influences On Brain Function, by Joan Y. Chiao (Ed.) – 2009, 978-0-444-53361-6.
Volume 179: Genetic models of schizophrenia, by Akira Sawa (Ed.) – 2009, 978-0-444-53430-9.
Volume 180: Nanoneuroscience and Nanoneuropharmacology, by Hari Shanker Sharma (Ed.) – 2009, 978-0-444-53431-6.

Volume 181: Neuroendocrinology: The Normal Neuroendocrine System, by Luciano Martini, George P. Chrousos, Fernand Labrie, Karel Pacak and Donald W. Pfaff (Eds.) – 2010, 978-0-444-53617-4.

Volume 182: Neuroendocrinology: Pathological Situations and Diseases, by Luciano Martini, George P. Chrousos, Fernand Labrie, Karel Pacak and Donald W. Pfaff (Eds.) – 2010, 978-0-444-53616-7.

Volume 183: Recent Advances in Parkinson's Disease: Basic Research, by Anders Björklund and M. Angela Cenci (Eds.) – 2010, 978-0-444-53614-3.

Volume 184: Recent Advances in Parkinson's Disease: Translational and Clinical Research, by Anders Björklund and M. Angela Cenci (Eds.) – 2010, 978-0-444-53750-8.

Volume 185: Human Sleep and Cognition Part I: Basic Research, by Gerard A. Kerkhof and Hans P.A. Van Dongen (Eds.) – 2010, 978-0-444-53702-7.

Volume 186: Sex Differences in the Human Brain, their Underpinnings and Implications, by Ivanka Savic (Ed.) – 2010, 978-0-444-53630-3.

Volume 187: Breathe, Walk and Chew: The Neural Challenge: Part I, by Jean-Pierre Gossard, Réjean Dubuc and Arlette Kolta (Eds.) – 2010, 978-0-444-53613-6.

Volume 188: Breathe, Walk and Chew; The Neural Challenge: Part II, by Jean-Pierre Gossard, Réjean Dubuc and Arlette Kolta (Eds.) – 2011, 978-0-444-53825-3.

Volume 189: Gene Expression to Neurobiology and Behaviour: Human Brain Development and Developmental Disorders, by Oliver Braddick, Janette Atkinson and Giorgio M. Innocenti (Eds.) – 2011, 978-0-444-53884-0.

Volume 190: Human Sleep and Cognition Part II: Clinical and Applied Research, by Hans P.A. Van Dongen and Gerard A. Kerkhof (Eds.) – 2011, 978-0-444-53817-8.

Volume 191: Enhancing Performance for Action and perception: Multisensory Integration, Neuroplasticity and Neuroprosthetics: Part I, by Andrea M. Green, C. Elaine Chapman, John F. Kalaska and Franco Lepore (Eds.) – 2011, 978-0-444-53752-2.

Volume 192: Enhancing Performance for Action and Perception: Multisensory Integration, Neuroplasticity and Neuroprosthetics: Part II, by Andrea M. Green, C. Elaine Chapman, John F. Kalaska and Franco Lepore (Eds.) – 2011, 978-0-444-53355-5.

Volume 193: Slow Brain Oscillations of Sleep, Resting State and Vigilance, by Eus J.W. Van Someren, Ysbrand D. Van Der Werf, Pieter R. Roelfsema, Huibert D. Mansvelder and Fernando H. Lopes da Silva (Eds.) – 2011, 978-0-444-53839-0.

Volume 194: Brain Machine Interfaces: Implications For Science, Clinical Practice And Society, by Jens Schouenborg, Martin Garwicz and Nils Danielsen (Eds.) – 2011, 978-0-444-53815-4.

Volume 195: Evolution of the Primate Brain: From Neuron to Behavior, by Michel A. Hofman and Dean Falk (Eds.) – 2012, 978-0-444-53860-4.